SEEING
FURTHER

THE ROYAL
SOCIETY

EDITED & INTRODUCED BY
BILL BRYSON
CONTRIBUTING EDITOR JON TURNEY

SEEING FURTHER

THE STORY OF SCIENCE
& THE ROYAL SOCIETY

Harper
Press

The endpapers show pages from the Charter Book, probably the Society's most important single historical document, created in 1663 after the second Royal Charter, establishing the structure of the Royal Society, was granted. Since the earliest days of the Society its vellum pages have recorded the signatures of each new Fellow and Foreign Member, as well as those of each Royal patron, as they were elected year by year.

Front: The founding signatures.
Back: Recent signatures from 2001 and 2002, including some of the contributors to this book.

HarperPress
An imprint of HarperCollins*Publishers*
77–85 Fulham Palace Road
Hammersmith, London W6 8JB
www.harpercollins.co.uk

Visit our authors' blog: www.fifthestate.co.uk
Love this book? www.bookarmy.com

First published in Great Britain by HarperPress in 2010

1

A catalogue record for this book
is available from the British Library

ISBN 978-0-00-730256-7

Design by 'OME Design

Printed and bound in Spain by
Gráficas Estella

Mixed Sources
Product group from well-managed
forests and other controlled sources
www.fsc.org Cert no. SW-COC-1806
© 1996 Forest Stewardship Council
FSC

FSC is a non-profit international organisation established to promote the responsible management of the world's forests. Products carrying the FSC label are independently certified to assure consumers that they come from forests that are managed to meet the social, economic and ecological needs of present or future generations.

Find out more about HarperCollins and the environment at www.harpercollins.co.uk/green

---

The following fonts have been used in the design of this book:

TRAJAN IS AN OLD STYLE SERIF TYPEFACE DESIGNED IN 1989 BY CAROL TWOMBLY FOR ADOBE. THE DESIGN IS BASED ON THE LETTERFORMS OF CAPITALIS MONUMENTALIS OR ROMAN SQUARE CAPITALS, AS USED FOR THE INSCRIPTION AT THE BASE OF TRAJAN'S COLUMN FROM WHICH THE TYPEFACE TAKES ITS NAME.

Garamond is the name given to a group of old style serif typefaces named for the punch-cutter Claude Garamond (circa 1480–1561). A majority of the typefaces named Garamond are more closely related to the work of a later punch-cutter Jean Jannon, who was key to it's development in the early to mid 1600s – the same time the Royal Society was formed.

CONTENTS

If the following observations do not seem to you to be too minute
should esteem it as a favor if you wou'd please to communicate th
the royal society

It has been asserted by some eminent Mathematicians, the su
that
it² logarithms of the numbers 1.2.3.4.&c to z is equal
½ log, c + z+½ × Log, z lessened by the series

$$\frac{1}{12z} - \frac{1}{360z^3} + \frac{1}{1260z^5} - \frac{1}{1680z^7} + \frac{1}{1188z^9} + \&c \quad if$$

circumference of a circle whose radius is unity. And
that this expression will very nearly approach to the v
um when z is large, & you take in only a proper num
terms of the foregoing series: but the whole series
press any quantity at all; because after the 5th ter
begin to increase, & they afterwards increase at a
what can be compensated by the increase of the
represent as numbers ever so large, it will
the following manner in which the coefficient
formed. Ta = ½
13 f = 21a   8 g = 2 fa
B = 2   C = 2 × 3 × 4 c   D   5 × 6
&c on, & A, B, C,   will
of the foregoing series: from   asily
in the series after the
from the 1st term n,
be greater than n ×
terms of this
increase in infinitu
can that series
y taking n = 1 & is suppos
square root of the periphery
that is said containing the foregoi
much in the same manner, concerning the

$$p(\theta \mid y) = \frac{p(\theta)\,p(y \mid \theta)}{\int p(\eta)\,p(y \mid \eta)\,d\eta}$$

BILL
BRYSON

INTRODUCTION

Bill Bryson is the internationally bestselling author of *The Lost Continent, Mother Tongue, Neither Here Nor There, Made in America, Notes from a Small Island, A Walk in the Woods, Notes from a Big Country, Down Under, The Life and Times of the Thunderbolt Kid* and *A Short History of Nearly Everything*, which was shortlisted for the Samuel Johnson Prize, won the Aventis Prize for Science Books in 2004, and was awarded the Descartes Science Communication Prize in 2005.

I CAN TELL YOU AT ONCE THAT MY FAVOURITE FELLOW OF THE ROYAL SOCIETY WAS THE REVEREND THOMAS BAYES, FROM TUNBRIDGE WELLS IN KENT, WHO LIVED FROM ABOUT 1701 TO 1761. HE WAS BY ALL ACCOUNTS A HOPELESS PREACHER, BUT A BRILLIANT MATHEMATICIAN. AT SOME POINT – IT IS NOT CERTAIN WHEN – HE DEVISED THE COMPLEX MATHEMATICAL EQUATION THAT HAS COME TO BE KNOWN AS THE BAYES THEOREM, WHICH LOOKS LIKE THIS:

$$ p\left(\theta\,|\,y\right) = \frac{p\left(\theta\right)p\left(y\,|\,\theta\right)}{\int p\left(\eta\right)p\left(y\,|\,\eta\right)d\eta} $$

People who understand the formula can use it to work out various probability distributions – or inverse probabilities, as they are sometimes called. It is a way of arriving at statistical likelihoods based on partial information. The remarkable feature of Bayes' theorem is that it had no practical applications in his own lifetime. Although simple cases yield simple sums, most uses demand serious computational power to do the volume of calculations. So in Bayes' day it was simply an interesting but largely pointless exercise.

Bayes evidently thought so little of his theorem that he didn't bother to publish it. It was a friend who sent it to the Royal Society in London in 1763, two years after Bayes' death, where it was published in the Society's *Philosophical Transactions* with the modest title of 'An Essay Towards Solving a Problem in the Doctrine of Chances'. In fact, it was a milestone in the history of mathematics. Today, with the aid of supercomputers, Bayes' theorem is used routinely in the modelling of climate change and weather forecasting generally, in interpreting radiocarbon dates, in social policy, astrophysics, stock market analysis, and wherever else probability is a problem. And its discoverer is remembered today simply because nearly 250 years ago someone at the Royal Society decided it was worth preserving his work, just in case.

The Royal Society has been doing interesting and heroic things like this since 1660 when it was founded, one damp weeknight in late November, by a dozen men who had gathered in rooms at Gresham College in London to hear Christopher Wren, twenty-eight years old and not yet generally famous, give a lecture on astronomy. It seemed to them a good idea to form a Society – that is all they called it at first – to assist and promote the accumulation of useful knowledge.

Nobody had ever done anything quite like this before, or would ever do it half as well again. The Royal Society (it became royal with the granting of a charter by Charles II in 1662) invented scientific publishing and peer review. It made English the primary language of scientific discourse, in place of Latin. It systematised experimentation. It promoted – indeed, insisted upon – clarity of expression in place of high-flown rhetoric. It brought together the best thinking from all over the world. It created modern science.

Nothing, it seems, was beneath its attention. Society members took an early interest in microscopy, woodland management, architectural load bearing,

Read at R.S.
24 Novemb. 1763.

Sr

If the following observations do not seem to you to be too minute, I shou'd esteem it as a favor if you wou'd please to communicate them to the royal society

It has been asserted by some eminent Mathematicians, that the sum of ye logarithms of the numbers 1. 2. 3. 4. &c to z is equal to

$\frac{1}{2}$ Log, c + $\overline{z+\frac{1}{2}}$ × Log, z lessened by the series

$$z - \frac{1}{12\,z} + \frac{1}{360\,z^3} - \frac{1}{1260\,z^5} + \frac{1}{1680\,z^7} - \frac{1}{1188\,z^9} + \&c$$ if c denote

the circumference of a circle whose radius is unity. And it is true that this expression will very nearly approach to the value of that sum when z is large, & you take in only a proper number of the first terms of the foregoing series: but the whole series can never properly express any quantity at all; because after the 5th term the coefficients begin to increase, & they afterwards increase at a greater rate than what can be compensated by the increase of the powers of z: tho' z represent a number ever so large, as will be evident by considering the following manner in which the coefficients of that series may be formed. Take $a = \frac{1}{12}$ $5b = a^2$ $7c = 2ba$ $9d = 2ca + b^2$ $11e = 2da + 2cb$ $13f = 2ea + 2db + c^2$ $15g = 2fa + 2eb + 2dc$. & so on, then take $A = a$ $B = 2b$ $C = 2 \times 3 \times 4\,c$ $D = 2 \times 3 \times 4 \times 5 \times 6\,d$ $E = 2 \times 3 \times 4 \times 5 \times 6 \times 7 \times 8\,e$ & so on, & A, B, C, D, E, F, &c will be the coefficients of the foregoing series: from whence it easily follows that if any term in the series after the 3 first be called y & its distance from the 1st term n, the next term immediately following will be greater than $n \times \dfrac{\overline{2n-1}}{6n+9} \times \dfrac{y}{z^2}$. Wherefore at length the subsequent terms of this series are greater than the preceeding ones & increase in infinitu & therefore the whole series can have no ultimate value whatsoever.

Much less can that series have any ultimate value, which is deduced from it by taking z = 1 & is supposed to be equal to the logarithm of the square root of the periphery of a circle whose radius is unity, & what is said concerning the foregoing series is true & appears to be so, much in the same manner concerning the series for finding out the sum of of the logarithms of the odd numbers 3. 5. 7. &c. z & those that are given for finding out the sum of the infinite progressions in which the several terms have the same numerator whilst their denominators are any certain power of numbers increasing in arithmetical proportion. But

Opposite:
Letter from
Thomas Bayes to
John Canton
concerning
logarithms, 24
November 1763.

the behaviour of gases, the development of the pocket watch, the thermal expansion of glass. Before most people had ever tasted a potato, the Royal Society debated the practicality of making it a staple crop in Ireland (ironically, as a hedge against famine). Two years after its formation, Christopher Merret, one of the founding Fellows, demonstrated a method for fermenting wine twice over, endowing it with a pleasing effervescence. He had, in short, invented champagne. The next year John Aubrey contributed a paper on the ancient stone monuments at Avebury, and so effectively created archaeology. John Locke contributed a paper on the poisonous fish of the Bahamas. And so it went on, decade after productive decade. When Benjamin Franklin flew his kite in a thunderstorm it was for the Royal Society that he very nearly killed himself. When a gas holder in Woolwich exploded with devastating consequences or gunpowder repeatedly failed to ignite or the navy needed a cure for scurvy, the Royal Society was called in to advise.

At least three things have always set the Society apart. First, from the outset, it was truly international. In 1665, Henry Oldenburg, himself German born, became editor of the Society's first journal (now one of seven), which was given the full and satisfying name *Philosophical Transactions: Giving some Accompt of the Present Undertakings, Studies and Labours of the Ingenious in many Considerable Parts of the World*. No words from the Society's early annals have more significance than that phrase 'many Considerable Parts of the World'.

'The international aspect was clearly a central part of what made it a success so early,' says Stephen Cox, the Society's genial chief executive. 'Right from the start we were getting papers from people like Marcello Malpighi and Christiaan Huygens, so very early on it had become a place where ideas from all over could be exchanged – a kind of early version of the Internet really.' As Cox likes to note, the Royal Society had a foreign secretary a hundred years before the British government did.

In an age when sabres hardly ever ceased rattling, the Society became the least nationalistic of national institutions. The name itself is telling.

Royal Society of London describes a location, not an allegiance. Had it been the Royal Society of *Great Britain* it would have been a very different organisation whether it wished it or not. So throughout its history it has been the most admirably neutral and cosmopolitan of entities. When Benjamin Franklin was a voice of revolution against Great Britain, he was still an esteemed and welcome member of the Society; and when Captain James Cook circumnavigated the globe in British ships in the name of knowledge he did so with perfect assurances that he would not be molested by any American vessels he encountered. During the Napoleonic wars, Humphry Davy was able to travel on scientific business across Europe thanks to a letter of dispensation from Napoleon that he carried in his pocket. The Société Philomathique gave him a dinner in Paris and drank the health of the Royal Society, if not the king. In like spirit, the Society refused to expel Fellows from enemy nations during either of the world wars, and was one of the first bodies to re-establish links after them.

Engraving of Antoni van Leeuwenhoek by Verkolje.

A letter from
Antoni van
Leeuwenhoek to
the Royal Society
regarding
observations of
duckweed, its roots
and reproduction,
25 December 1702.

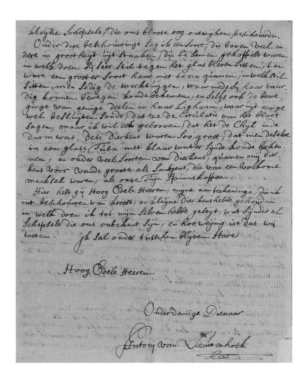

Quite as remarkable as its cosmopolitanism was a second distinctive characteristic of the Royal Society – namely, that it wasn't necessary to be well
born to be part of it. Having wealth and title didn't hurt, of course, but
being scientifically conscientious and experimentally clever were far more
important. No one better illustrated this than a retiring linen draper from
Delft named Antoni van Leeuwenhoek. Over a period of fifty years – a
period that began when he was already past forty – Leeuwenhoek submitted some two hundred papers to the Royal Society, all accompanied by the
most excellent and exacting drawings, of the things he found by looking
through his hand-wrought microscopes. These were tiny wooden paddles
with a little bubble of glass embedded in them. How he managed to work
them is something of a wonder even now, but he achieved magnifications
of up to 275 times and discovered the most incredible things: protozoa,
bacteria and other wriggling life where no life was thought to be. The idea
that there were whole worlds in a drop of fluid was a positive astonishment.

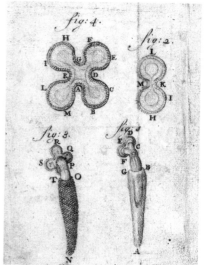

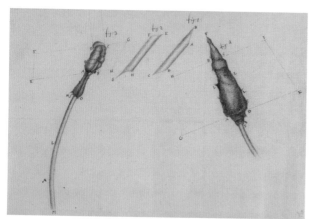

Above left:
A replica of
Leeuwenhoek's
microscope.

Left:
Leeuwenhoek's
observations of his
own facial hair, 22
February 1676.

Above:
Leeuwenhoek's
observations of
rotifers and their
parasitic worms,
4 November 1704.

Leeuwenhoek had practically no education. He filed his reports in Low Dutch because he had no English and no Latin. He didn't even have High Dutch, it appears. But none of that mattered. What mattered was that he had a genius for microscopy and a profound respect for knowledge.

In 350 years, the Royal Society has had a mere 8,200 members, but what a roll call of names. In no very particular order they include Isaac Newton, Christopher Wren, Edmond Halley, Robert Boyle, Robert Hooke, Benjamin Franklin, John Locke, Humphry Davy, Charles Darwin, Ernest Rutherford, Isambard Kingdom Brunel, Joseph Banks, T.H. Huxley, James Watt, Joseph Lister, Henry Cavendish, Michael Faraday, James Clerk Maxwell, Lawrence

Bragg, Paul Dirac, Peter Medawar, Alexander Fleming, James Chadwick, Lord Rayleigh, William Ramsey, Lord Kelvin, Kathleen Lonsdale, Dorothy Hodgkin, Miriam Rothschild, Anne McLaren and literally hundreds more who changed the world by changing our understanding of it. To be part of such an establishment is an extraordinary achievement. This isn't just the most venerable learned society in the world, it is the finest club.

Throughout its busy history, the Society has demonstrated an almost uncanny knack for selecting people before they gave any particular hint of the greatness that would make them immortal. Edmond Halley was made a Fellow *before* he received his degree from Oxford. Charles Darwin, elected in 1839 only three years after his youthful *Beagle* voyage, was not even known for his work on barnacles, much less on evolution. William Henry Fox Talbot became an FRS a good two years before the first vague notion of photography flitted through his head. And of course there was Thomas Bayes, scribbling a theorem that the world would have to wait nearly 250 years to use.

The Society has also demonstrated a heroic, and indeed endearing, tendency to recognise the unsung. The example that leaps to mind for me here is that of Hermann Sprengel, the forgotten father of electric lighting. Everyone thanks Joseph Swan and Thomas Edison for giving us the homely glow of incandescent lighting, but in fact Sir William Grove (who, it more or less goes without saying, was himself a Fellow) had demonstrated a working incandescent bulb well over thirty years before them – seven years before Edison was even born. It's just that Grove's bulb didn't last very long. What was needed was a vacuum that would allow a filament to burn for long periods. Sprengel, a German chemist working in London, invented a pump that could drain the air from a glass chamber down to one-millionth of its normal volume, allowing filaments to burn for hours and making electric lighting a commercial possibility at last. Edison and Swan found the filaments and got the glory. Sprengel was forgotten almost at once by everyone except the Royal Society, which made him a Fellow in 1878, nearly fifteen years before he was recognised by any institution in his native Germany.

The best place I know to get some sense of what the Royal Society is and has achieved is a modest, crowded storeroom in the basement of its headquarters in Carlton House Terrace in London. Here, neatly shelved or tucked into drawers and cabinets, are three and a half centuries of accumulated treasures – Newton's manuscript copy of the *Principia*, the Shelton Regulator clock used by Captain Cook to time the transit of Venus on the *Endeavour* voyage, Joseph Priestley's folding spectacles, Leeuwenhoek's precious drawings, the papers of Robert Hooke and Robert Boyle – representing the moments of birth of some of the most enormous ideas human minds have ever had.

Keith Moore, the Society's librarian, reaches into an anonymous-looking metal cupboard and, with an air of gentleness and care, brings out a white box. Inside it, resting delicately, is an object that automatically provokes an awed hush: the death mask of Isaac Newton. Only by a remarkable chance did the mask come into the Society's possession. It had been lost for many years when, in 1839, a Mr Christie, a Fellow of the Society, developed a sudden desire to have a bust of Newton on his shelves and called in at a curio shop on Tichborne Street in London, near his place of work, to ask if they had anything. The shopkeeper replied that he had no statues, but they had a curious mask, which his father had bought many years before. After some rooting around, he found it and brought it to Christie to examine. It was Newton's death mask. It had sat unregarded on a shelf for at least half a century, and in all likelihood would eventually have

Isaac Newton's death mask.

been lost altogether had Christie not made his lucky enquiry.

The mask is a transfixing object, not surprisingly, but what is more unexpectedly moving is a small, exquisite piece of apparatus that sits on the shelf alongside it: a reflecting telescope made by Newton himself in 1669. It is only six inches long but beautifully fashioned. Newton ground the glass himself, designed the swivelling socket, turned the wood with his own hand. In its time this was an absolute technological marvel, but it is also a thing of lustrous beauty. Nowhere could you find an item that more vividly demonstrates the beauty as well as the wonder of science.

Keith shows me some papers he has just been cataloguing. They are letters from Thomas Thorpe, an English chemist, written to his wife, Emma, during an 1878 Royal Society expedition to the American west. The purpose of the expedition was to view a solar eclipse, which, among other things, would allow them to confirm or disprove the existence of the planet Vulcan. The papers are irresistibly absorbing, partly because Thorpe brings a scientist's curiosity to everything he sees – the quality of US trout, the character of the town of Cheyenne (home of '6,000 of the biggest scoundrels the world contains'), the climate, geology, everything – but also because they so vividly and charmingly catalogue the difficulties and discomforts necessary to do science in the field in the nineteenth century (or possibly any time).

The reflecting telescope made by Newton in 1669.

When you look along the stacks or peek into the drawers, it is impossible not to be struck with wonder at how much aggregated human effort – how much thought and toil and nights under canvas – is embedded in what we know about the world and universe and how they are put together.

'This is only a small part of it,' Keith tells me. 'There are eight thousand more boxes in storage in Wiltshire.' He smiles. 'You generate a lot of material in 350 years.'

Which brings me to my third remarkable fact about the Royal Society: it's still there. More than that, it is still there *and* it is still important. How many enterprises can you name that are still doing today what they were formed to do 350 years ago?

It has had its moments of faltering, goodness knows. At times its quench-less curiosity has threatened to give way to mere morbidity. In the early days it was particularly fascinated with monstrous births and that kind of thing, and sometimes it engaged in experiments that were patently imprudent.

One such was in November 1667 when a penurious student named Arthur Coga was induced to let two Fellows transfuse sheep's blood into him in return for the payment of a guinea. No one had any idea what would happen – whether it would kill him or fill him with boundless energy – and this degree of uncertainty left some of the more reflective members feeling distinctly uneasy. In the event, the transfusion didn't do much of anything. Before an audience that included the Bishop of Salisbury, 14 ounces of blood were pumped out of the sheep and into Coga. It seemed to do him no harm. Afterwards, one of those present reported, 'the patient was well and merry, and drank a glass or two of canary, and took a pipe of tobacco'. He went home, slept well and reported no ill effects. Just under two weeks later, the operation was repeated for a new audience. Soon afterwards, however, reports began to trickle in from all over Europe that the experiment had been tried several times elsewhere, often with fatal results. The Society, happily, never tried anything like that again.

If the Royal Society had done nothing after Newton, its fame would be secure. In fact, there were times when it looked as if it might not do much. Twenty years after Newton's reign, it had a president, Martin Folkes, who was famous for slumbering through meetings, and financial difficulties that threatened to become insoluble. By 1740, barely half the Fellows could be counted on to pay their dues, and some were so severely in arrears that the Society's accumulated deficit had risen to over £1,800 – a worrying sum for a private body of modest size. Partly to restore the balance sheet, it began taking in members who were distinguished but not terribly scientific. By the end of the century, Fellows included Edward Gibbon, Warren Hastings and even Lord Byron. Without actually ceasing to be worthy, it could easily have declined into something more peripheral and much less important.

Clearly that didn't happen. At every critical moment throughout its history there has always been an Isaac Newton, a Joseph Banks, a Humphry Davy, a T.H. Huxley, a Lord Rutherford to give the Society clout and lustre, and to keep it firmly attached to scientific endeavour at the highest level.

Today the Royal Society's interests remain an inspiration to recite. It provides 350 research fellowships and its grants support the work of 3,000 scientists all over the world. It bestows great numbers of medals and prizes, maintains an active programme of lectures and debates, and holds a beloved Summer Science Exhibition, which no one who appreciates science and can get to London should miss. It acts as the scientific conscience of the nation. It publishes seven journals, and an endless stream of papers. It remains emphatically international in its outlook, maintaining close links with ninety-one science academies around the world. If we have an Earth worth living on a hundred years from now, the Royal Society will be one of the organisations our grandchildren will wish to thank.

Poke your head through any door in the Royal Society building and what you are likely to find is people in meetings. They meet endlessly at the Royal Society. My own involvement, like that of most outsiders, has been as a member of committees – in my case a committee to select the winners of the annual books prize and another involved with the 350th anniversary celebrations – and on almost every visit to the building I have opened three or four wrong doors to find other people meeting. For a long time I wondered what they could possibly all be meeting about. Then I was given a copy of an extraordinary volume – a sturdy hardback called the *Royal Society Year Book*, which in about 500 pages summarises all that the Royal Society does in a year.

Flick through it at random and you find that it is involved in an impossibly varied range of activities. There is a Dorothy Hodgkin Fellowships Committee, a Hooke Committee, a Trans-Antarctic Association UK Advisory Committee, a Darwin Correspondence Project, a Sir Harold

Hartley Lecture Committee, a Scientific Unions Committee, a South East Asia Rainforest Research Committee, a Newton International Fellowships Committee, a Rosalind Franklin Award Committee, and dozens and dozens more. There is even an Anatomy, Physiology, Endocrinology and Pharmacology (Except Clinical Aspects) of Animal Systems, Neurosciences, Psychology and Reproductive Biology, and Relevant Agricultural Studies Committee (known informally, and perhaps a bit mercifully, as 'Panel 8').

Altogether at the Royal Society there are ninety-six committees, all devoted to promoting important research, honouring an achievement, improving education, badgering governments into behaving intelligently, or otherwise effecting an enhancement to what we know or an improvement to how we proceed.

The most important committees of all are the ten devoted to electing new Fellows. Today there are 1,400 Fellows, including 69 Nobel laureates, and it is they who run the Society. 'It is,' Stephen Cox tells me, smiling, 'like a company with 1,400 non-executive directors. They set policy and identify key areas of concern. It's *their* society.'

Because of all that it has achieved in its time, there is a tendency to equate the Royal Society with things like atoms and gravity and other bits of hard science, but what impresses me is the boundlessness of its range. Consider the contribution of John Lubbock, friend and neighbour of Charles Darwin. Lubbock was a banker by profession, but was in addition a distinguished botanist, astronomer, expert on the social behaviour of insects, politician and antiquarian. Among much else, he coined the terms *palaeolithic, mesolithic* and *neolithic* in 1865. But his real contribution to life was to push through Parliament the first Ancient Monuments Protection Act, which became law in 1882. People forget how much of Britain's historic fabric was nearly destroyed in the past. Before Lubbock's intervention, half of Avebury was nearly cleared away for housing, and at one point it was even threatened that Stonehenge, then still in private hands, might be dismantled and shipped to America. Without Lubbock,

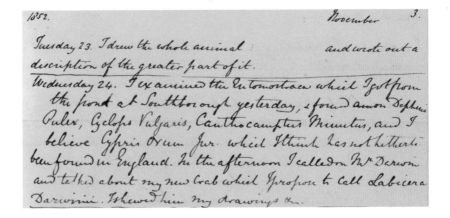

An entry in John Lubbock's diary describing a crab which he intends to name after Charles Darwin, 24 November 1852.

many stone circles, tumuli and other historical features of the landscape would have vanished long ago. Lubbock also, not incidentally, invented the bank holiday. The Royal Society and its Fellows, you see, have long been at the heart of all kinds of things.

It is impossible to list all the ways that the Royal Society has influenced the world, but you can get some idea by typing in 'Royal Society' as a word search in the electronic version of the *Dictionary of National Biography*. That produces 218 pages of results – 4,355 entries, nearly as many as for the Church of England (at 4,500) and considerably more than for the House of Commons (3,124) or House of Lords (2,503). It is more central to the life and history of Great Britain than most people realise.

And as you are about to see, it not only produces the best science, but also some of the very best science writing.

ACKNOWLEDGMENTS

I would like to thank all the contributors, including the President of the Royal Society, for so generously taking part in the making of this book. I also wish to thank the Council of the Royal Society, Aosaf Afzal, Stephen Cox, Julia Higgins, Julie Hodgkinson, Jo Hopkins, Joanne Madders, Keith Moore, Dominic Reid and Martin Taylor.

1

JAMES GLEICK

AT THE BEGINNING: MORE THINGS IN HEAVEN AND EARTH

James Gleick last visited the Royal Society when researching his recent biography *Isaac Newton*. His first book, *Chaos*, was a National Book Award and Pulitzer Prize finalist and an international bestseller, translated into more than twenty languages. His other books include *Genius: The Life and Science of Richard Feynman*, *Faster: The Acceleration of Just About Everything* and *What Just Happened: A Chronicle from the Information Frontier*.

THE FIRST FORMAL MEETING OF WHAT BECAME THE ROYAL SOCIETY WAS HELD IN LONDON ON 28 NOVEMBER 1660. THE DOZEN MEN PRESENT AGREED TO CONSTITUTE THEMSELVES AS A SOCIETY FOR 'THE PROMOTING OF EXPERIMENTAL PHILOSOPHY'. EXPERIMENTAL PHILOSOPHY? WHAT COULD THAT MEAN? AS JAMES GLEICK SHOWS FROM THEIR OWN RECORDS, IT MEANT, AMONG OTHER THINGS, A BOUNDLESS CURIOSITY ABOUT NATURAL PHENOMENA OF ALL KINDS, AND SOMETHING ELSE – A KIND OF EXUBERANCE OF INQUIRY WHICH HAS LASTED INTO OUR OWN DAY.

To invent science was a heavy responsibility, which these gentlemen took seriously. Having declared their purpose to be 'improving' knowledge, they gathered it and they made it – two different things. From their beginnings in the winter of 1660–61, when they met with the King's approval Wednesday afternoons in Laurence Rooke's room at Gresham College, their way of making knowledge was mainly to talk about it.

For accumulating information in the raw, they were well situated in the place that seemed to them the centre of the universe: 'It has a large Intercourse

Memorandum that Novemb: 28 1660. These
persons following according to the usuall Custome of most
of them, Mett together at Gresham Colledge to heare Mr
Wrens Lecture, viz. The Lord Brouncker, Mr Boyle, Mr
Bruce, Sr Robert Moray, Sr Paul Neile, Dr Wilkins, Dr
Goddard, Dr Petty, Mr Ball, Mr Rooke, Mr Wren, Mr Hill.
And after the Lecture was ended they did according to the usuall
Manner, withdrawe for mutuall converse. Where amongst
other matters that were discoursed of, Something was offered
about a designe of founding a Colledge for the Promoting
of Physico-Mathematicall-Experimentall Learning.
And because they had these frequent occasions of meeting wth
one another, it was proposed that some course might be thought
of to improve this meeting to a more regular way of debating
things, & according to the Manner in other Countries where
there were voluntary associations of men into Academies
for the advancement of various parts of learning, So they
might doe something answerable here for the promoting of
Experimentall Philosophy.
In order to which it was agreed, that this Company would
continue their weekely meeting on wensday at 3 of the clock
in the Tearme time at Mr Rookes Chamber at Gresham Colledge.
In the Vacation at Mr Balls Chamber in the Temple. And
towards the defraying of occasionall expences, every one should
at his first admission, pay downe ten shillings, & besides engage
to pay one shilling weekely, whether present or absent, whilest
he shall please to keep his relation to this company.
At this Meeting Dr Wilkins was apointed to the Chaire, Mr
Ball to be Treasurer, & Mr Croone (though Absent) was named for Register.
And to the end that they might the better be inabled to
make a conjecture of how many the elected number of this Society
should consist; therefore it was desired that a list might be taken

ken of the names of such persons, as were knowne to those present
whom they judged willing & fitt to joyne with them in their designe, who, if they should desire it, might be admitted before
any other.

Uppon which this following Catalogue was offerd.

Lord Hatton	Dr Henshaw
Mr Robt. Boyle	Dr finch
Mr Jones	Dr Baines
Mr Coventry	Dr Wren
Mr Brereton	Mr Smith
Sr Kenelme Digby	Mr Ashmole
Sr Ant: Morgan	Mr Newburg
Mr John Vaughan	Mr Austen
Mr Evelijn	Mr Oldenburg
Mr Rawlins	Mr Pett
Mr Matthew Wren	Mr Croone.
Mr Slingsby	
Mr Henshaw	
Mr Denham	
Mr Povey	
Mr Wilde	
Dr Ward	
Dr Wallis.	
Dr Gliston	
Dr Bates.	
Dr Ent	
Dr Scarburgh	
Dr Phrasier	
Dr Coxes	
Dr Merrét	
Dr Whistler	
Dr Clarke.	
Dr Bathurst	
Dr Cowley	
Dr Willis.	

GRESHAM COLLEGE.

with all the Earth: ... a City, where all the Noises and Business in the World do meet: ... the constant place of Residence for that Knowledge, which is to be made up of the Reports and Intelligence of all Countries.' But we who know everything tend to forget how little was known. They were starting from scratch. To the extent that the slate was not blank, it often needed erasure.

At an initial meeting on 2 January their thoughts turned to the faraway island of Tenerife, where stood the great peak known to mariners on the Atlantic trade routes and sometimes thought to be the tallest in the known world. If questions could be sent there (Ralph Greatorex, a maker of mathematical instruments with a shop in the Strand, proposed to make the voyage), what would the new and experimental philosophers want to ask? The Lord Viscount Brouncker and Robert Boyle, who was performing experiments on that invisible fluid the air, composed a list:

- 'Try the quicksilver experiment.' This involved a glass tube, bent into a U, partly filled with mercury, and closed at one end. Boyle believed that air had weight and 'spring' and that these could be measured. The height of the mercury column fluctuated, which he explained by saying, 'there may be strange Ebbings and Flowings, as it were, in the Atmosphere' – from causes unknown. Christopher Wren ('that excellent Mathematician') wondered whether this might correspond to 'those great Flowings and Ebbs of the Sea, that they call the Spring-Tides', since, after all, Descartes said the tides were caused by pressure made on the air by the Moon and the Intercurrent Ethereal Substance. Boyle, having spent many hours watching the mercury rise and fall unpredictably, somewhat doubted it.

- Find out whether a pendulum clock runs faster or slower at the mountain top. This was a problem, though: pendulum clocks were themselves the best measures of time. So Brouncker and Boyle suggested using an hourglass.

- Hobble birds with weights and find out whether they fly better above or below.

- 'Observe the difference of sounds made by a bell, watch, gun, &c. on the top of the hill, in respect to the same below.'

And many more: candles, vials of smoky liquor, sheep's bladders filled with air, pieces of iron and copper, and various living creatures, to be carried thither.

A stew of good questions, but to no avail. Greatorex apparently did not go, nor anyone else of use to the virtuosi, for the next half-century. Then, when Mr J. Edens made an expedition to the top of the peak in August 1715, he was less interested in the air than in the volcanic activity: 'the Sulphur

Opposite:
Portrait of Robert Boyle by Johann Kerseboom.

NOB. ROBERTVS BOYLE ARMIG.
REGIÆ SOCIETATIS SOC.

discharg[ing] its self like a Squib or Serpent made of Gun-powder, the Fire running downwards in a Stream, and the Smoak ascending upwards'. He did wish he had brought a Barometer – the device having by now been invented and named – but he would have had to send all the way to England, and the expense would have come from his own pocket. Nonetheless he was able to say firmly that there was no truth to the report about 'the Difficulty of breathing upon the top of the place; for we breath'd as well as if we had been below'.

No one knew how tall the mountain was anyway, or how to measure it. Sixteenth-century estimates ranged as high as 15 leagues (more than 80,000 metres) and 70 miles (more than 110,000 metres). One method was to measure from a ship at sea; this required a number for the radius of the Earth, which wasn't known itself, though we know that Eratosthenes had got it right. The authoritative *Geographia* of Bernhardus Varenius, published in Cambridge in 1672 with Isaac Newton's help, computed the height as 8 Italian miles (11,840 metres) – '*quae incredibilis fere est*' – and then guessed 4 to 5 miles instead. (An accurate measurement, 3,718 metres, had to wait till the twentieth century.) But interest in Tenerife did not abate – far from it. Curiosity about remote lands was always honoured in Royal Society discourse. 'It was directed,' according to the minutes for 25 March, 'that inquiry should be made, whether there be such little dwarvish men in the vaults of the Canaries, as was reported.' And at the next meeting, 'It was ordered to inquire, whether the flakes of snow are bigger or less in Teneriffe than in England ...'

Reports did arrive from all over. The inaugural issue of the *Philosophical Transactions* featured a report (written by Boyle, at second hand) of 'a very odd Monstrous Calf' born in Hampshire; another 'of a peculiar Lead-Ore of Germany'; and another of 'an Hungarian Bolus', a sort of clay said to have good effects in physick. From Leyden came news of a man who, by star-gazing nightly in the cold, wet air, obstructed the pores of his skin, 'which appeared hence, because that the shirt, he had worn five or six weeks, was then as white as if he had worn it but one day'. The same correspondent described a young maid, about thirteen years old, who ate salt 'as other children doe Sugar: whence she was so dried up, and grown so stiff, that she could not stirre her limbs, and was thereby starved to death'.

Iceland was the source of especially strange rumours: holes, 'which, if a stone be thrown into them, throw it back again'; fire in the sea, and smoking lakes, and green flames appearing on hillsides; a lake near the middle of the isle 'that kills the birds, that fly over it'; and inhabitants that sell winds and converse with spirits. It was ordered that inquiries be sent regarding all these, as well as 'what is said there concerning raining mice'.

The very existence of these published transactions encouraged witnesses to relay the noteworthy and strange, and who could say what was strange and what was normal? Correspondents were moved to share their 'Observables'. Observables upon a monstrous head. Observables in the body of the Earl of Balcarres (his liver very big; the spleen big also). Observables were as ephemeral as vapour in this camera-less world, and the Society's role was to grant them persistence. Many letters were titled simply, 'An Account of a remarkable [object, event, appearance]': a remarkable meteor, fossil, halo; monument unearthed, marine insect captured, ice shower endured; Aurora Borealis, Imperfection of Sight, Darkness at Detroit; appearance in the Moon, agitation of the sea; and a host of remarkable cures. An Account of a remarkable Fish began, 'I herewith take the liberty of sending you a drawing of a very uncommon kind of fish which was lately caught in King-Road …'

> It fought violently against the fisher-man's boat … and was killed with great difficulty. No body here can tell what fish it is … I took the drawing on the spot, and do wish I had had my Indian Ink and Pencils …

From Scotland came a careful report by Robert Moray of unusual tides in the Western Isles. Moray, a confidant of the King and an earnest early member of the Society, had spent some time in a tract of islands for which he had no name – 'called by the Inhabitants, the Long-Island' (the Outer Hebrides, we would say now). 'I *observed* a very strange Reciprocation of the Flux and Re-flux of the Sea,' he wrote, 'and *heard* of another, no less remarkable.' He described them in painstaking detail: the number of days before the full and quarter moons; the current running sometimes eastward but other times westward; flowing from 9½ of the clock to 3½; ebbing and flowing orderly for some days, but then making 'constantly a great and singular variation'. Tides were a Royal Society favourite, and they were a problem. Humanity had been watching them for uncounted thousands of years, and observing the coincidence of their timing with the phases of the Moon, without developing an understanding of their nature – Descartes notwithstanding. No global sense of the tides could be possible when all recorded information was local. And even now, Moray emphasised the peculiarity of his observations; and quailed at the idea of generalising.

> To penetrate into the *Causes* of these strange Reciprocations of the Tides, would require exact descriptions of the Situation, Shape, and Extent of every piece of the adjacent Coasts of Eust and Herris; the Rocks, Sands, Shelves, Promontorys, Bays, Lakes, Depths, and other Circumstances, which I cannot now set down with any certainty, or accurateness; seeing, they are to be found in no Map.

He had drawn a map himself some years earlier, but it was gone. 'Not having copied [it], I cannot adventure to beat it out again.'

As often as they could be arranged, experiments were performed for the assembled virtuosi. Brouncker prosecuted his experiment of the recoiling of guns, Wren his experiment of the pendulum, William Croone his experiment with bladders and water. When Robert Hooke took charge of experiments, they came with some regularity. Even so, many more experiments were described, or wished for, than were carried out at meetings. The grist of the meetings was discourse – animated and edifying. They loved to talk, these men.

They talked about 'magnetical cures' and 'sympathetical cures' and the possibility of 'tormenting a man with the sympathetic powder'. They talked about spontaneous equivocal generation: 'whether all animals, as well vermin and insects as others, are produced by certain seminal principles, determined to bring forth such and no other kinds. Some of the members conceived, that where the animal itself does not immediately furnish the seed, there may be such seeds, or something analogous to them, dispersed through the air, and conveyed to such matter as is fit and disposed to ferment with it, for the production of this or that kind of animal.' They talked about minerals discovered under ground, in 'veins', wondering whether they grew there or had existed since the creation. Some suggested that metals and stones were produced 'by certain subterraneous juices ... passing through the veins of the earth'.

They talked about why it was hotter in summer than in winter; no one knew, but George Ent had a theory. It was ordered to be registered in a 'book of theories, which was directed to be provided'. George Villiers, the Duke of Buckingham, newly admitted to the Society, produced what he promised was the horn of a unicorn. Legend had it that a circle drawn with such a thing would keep a spider trapped until it died, so they performed the experiment: 'A circle was made with powder of unicorn's horn, and a spider set in the middle of it, but it immediately ran out. The trial being repeated several times, the spider once made some stay on the powder.'

Still, the discourse was liberating. 'Their first purpose,' said Thomas Sprat, writing his 'history' of the Society when it was barely fledged, 'was no more, than onely the satisfaction of breathing a freer air, and of conversing in quiet one with another, without being ingag'd in the passions, and madness of that dismal Age'. The rules were clear: nothing about God; nothing about politics; nothing about 'News (other than what concern'd our business of Philosophy)'. And what news was that? John Wallis specified, 'as Physick, Anatomy, Geometry, Astronomy, Navigation, Statics, Mechanics, and Natural Experiments'.

James Long, newly admitted in April 1663, delivered the news, as the amanuensis reported in his minutes, 'that there were ermines in England'. He promised to produce some. 'He mentioned also, that bay-salt being thrown upon toads would kill them … he made mention likewise of a kind of stones with natural screws, and promised to show some of them.'

At the next meeting, Long talked about the generation of ants: they come out of pods full of eggs. He added that he had seen a maggot under a stag's tongue; that land-newts are more noxious than water newts; and that toads become venomous in hot weather and in hot countries such as Italy. Croone mentioned that he had seen a viper with a young one in its belly, and Long added, 'The female viper hath four teeth, two above and two below; but the male only two and those above.' Hooke showed some new drawings he had made from observations with his microscope, including a spider with six eyes – lately he had been bringing something new to almost every meeting. Moray described a watch with particularly hard steel, which reminded Long that he had once seen a breast-piece so tough that a pistol bullet only dented it.

Long was a military man, having been first a captain and then colonel of horse in a Royalist regiment. John Aubrey describes him as a good swordsman and horseman and a devotee of 'astrology, witchcraft and natural magic'. He does seem to have found him rather voluble – 'an admirable extempore orator for a harangue'. They went hawking together, and what Aubrey recalled was that Long never stopped gabbing. He certainly found his voice at the Royal Society. The minute-taker sometimes sounds weary:

Col. Long having related divers considerable observations of his concerning insects …
… said, that an iron back in a chimney well heated, useth to make a noise like that of bell-metal.

… observed, that a bean cut into two or three pieces produces good beans.

… desired farther time to make his collection of insects for a present to the society.

… mentioned, that a lady had …

… related, that a cornet in Scotland …

… mentioned, that he had known wheat …

Until finally, 'having discoursed of his opinion concerning the smut of corn, *viz.*, that it proceeds from the root, and not the mildew, [Long] was desired to give his discourse in writing'.

In these first years a great many animals were cut up, poisoned, or suffocated. 'It is a most acceptable thing to hear their discourse, and see their experiments,' wrote Samuel Pepys in his diary, and he seemed particularly drawn to experiments involving cats and dogs. '… And so out to Gresham College, and saw a cat killed with the Duke of Florence's poyson, and saw it proved that the oyle of tobacco drawn by one of the Society do the same effect … I saw also an abortive child preserved in spirits of salt.'

… And anon to Gresham College, where, among other discourse, there was tried the great poyson of Maccassa upon a dogg, but it had no effect all the time we sat there.

Then to Gresham College, and there did see a kitling killed almost quite …

Chickens were 'choked' and fish were 'gagged'. The members strangled dogs and dissected living cats. Not all had the stomach for these experiments. Robert Boyle did, and he took pride in this. 'I have been so far from that effeminate squeamishness, that one of the philosophical treatises, for which I have been gathering experiments, is of the nature and use of dungs,' he boasted. 'I have not been so nice, as to decline dissecting dogs, wolves, fishes, and even rats and mice, with my own hands. Nor, when I am in my laboratory, do I scruple with them naked to handle lute and charcoal.' The Society's armoury of mechanical instruments was small in these early years, but one that proved endlessly useful was Boyle's air pump, or 'pneumatical engine'. Among the items placed in glass vessels, from which the air was then exhausted, were birds, mice, ducks, vipers, frogs, oysters and crawfish. Typical experiments would bring the creatures 'to Deaths door', whereupon the Society would observe gasping, vomiting and convulsions. Respiration held many mysteries; so did the circulation of the blood. An experiment could last for many hours or could end in seconds: 'I have this to alledge,' wrote Boyle, 'that, having in the presence of some *Virtuosi* provided for the nonce a very small Receiver, wherein yet a Mouse could live sometime, if the Air were left in it, we were able to evacuate it in one suck, and by that advantage we were enabled, to the wonder of the Beholders, to kill the Animal in less than half a minute.' The experimentation was not, for some time, organised or systematic; sometimes the wonder of the beholders was the chief result. The *Philosophical Transactions* served as a progenitor of *Ripley's Believe It or Not* as well as the *Physical Review*.

'There follow topsy-turvy without any order experiments of all sorts,' wrote Goethe more than a century later, 'news of happenings on earth and in the heavens.' Goethe bore the Royal Society no small resentment, which he nursed by devotedly reading its history, as set down by both Thomas Sprat and Thomas Birch. He translated many pages of extracts, and he complained: 'Everybody communicates what happens to be at hand, phenomena of *Naturlehre*, objects of *Naturgeschichte*, technical operations, everything appearing topsy-turvy without order. Many things quite insignificant, others interesting only in outward appearance, others merely curious, are accepted and given a place.'

It was not until late in 1671 that the members heard about a young Lincolnshire man, Isaac Newton, who had invented a new kind of telescope at least ten times more powerful, inch for inch, than any in existence.

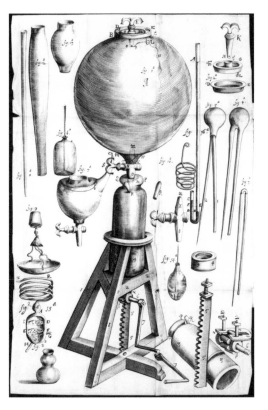

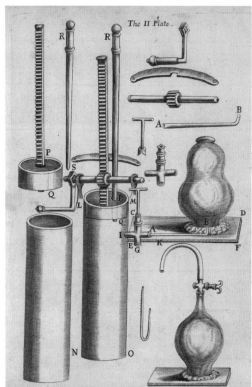

The II Plate.

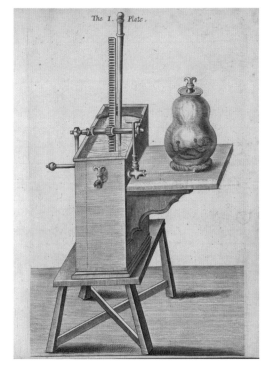

The I. Plate.

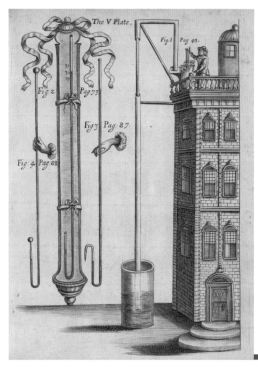

The V Plate.

Fig.1 Pag.43.

Fig.2 Pag.73.

Fig.3 Pag.87.

Fig.4 Pag.88.

He had not sent it to them. He had made it in 1668 or 1669 in Cambridge, where he had just become the new Professor of Mathematics, but kept it mostly to himself. Cambridge being some distance from London, more than two years passed before the news, and then the telescope, reached the Royal Society. As they could see, it was not just a serious scientific advance but a technology with military application. They studied it and showed it to the King. Henry Oldenburg wrote to the twenty-nine-year-old on their behalf. 'Sir,' he began, 'Your Ingenuity is the occasion of this addresse by a hand unknown to you …' In short order they elected him a member, though none had yet met him.

For some time Newton had been reading the Society's reports and taking careful note. News of a fiery mountain: 'Batavia one afternone was covered with a black dust heavyer than gold which is thought came from an hill on Java Major supposed to burne.' Lunar influence: 'Oysters & Crabs are fat at the new moone & leane at the full.' Now he wrote to Oldenburg at the only address he knew – '*Mr Henry Oldenburge at his house about the middle of the old Palmail in St Jamses Fields in Westminster*' – and said he had news of his own. He advertised it enthusiastically: '… in my Judgment the oddest if not the most considerable detection which hath hitherto been made in the operations of Nature.'

The meeting of 8 February 1672 began as usual with the reading out of letters newly arrived. First came a conjecture from John Wallis that the Moon's varying distance to the Earth, its perigee and apogee, might 'much influence the rising and falling of the mercury in the barometer'. He hoped that members of the Society who had barometers would investigate. It was another idea destined for the dustbin.

Next, Tommaso Cornelio wrote from Naples, in Italian, to refute common stories told of the odd effects of the bite of the tarantula. His observations suggested that most such stories were fictitious. (Many, he added soon afterward, come from 'young wanton girles who by some particular indisposition falling into this melancholly madness, perswade themselves according to the vulgar prejudice, to have been stung by a Tarantula'.)

The third letter was more complicated: 'Of Mr Isaac Newton from Cambridge, concerning his discovery of the nature of light, refractions, and colour …' Sunlight, according to this letter, is not homogenous, but consists 'of different rays'. These rays come in pure and indivisible colours. The Society's note-taker wrote this down: 'Some, in their own nature, are disposed to produce red, others green, others blue, others purple, &c.' Newton made a further claim, even more counter-intuitive:

> The most surprising, and wonderful composition was that of *Whiteness*. There is no one sort of Rays which alone can exhibit this. 'Tis ever compounded, and to its composition are requisite all the aforesaid primary Colours, mixed in due proportion. I have often with Admiration beheld, that all the Colours of the Prisme being made to converge, and thereby to be again mixed, … reproduced light, intirely and perfectly white.

This was more interesting, if scarcely more believable, than the Odd Monstrous Calf. It was ordered that the author be solemnly thanked; also that Boyle, Hooke and the Bishop of Salisbury peruse and consider it and report back.

What followed is a story told many times. Newton's experiment of sunlight refracted by two prisms – so ingeniously conceived, carefully performed, and exquisitely narrated – came to be seen as a landmark in the history of science. It established a great truth of nature. It created a template for the art of reasoning from observation to theory. It shines as a beacon from the past so brightly as to cast the rest of the Society's contemporaneous activity into relative shadow.

But this is by definition hindsight. That week in February, thinking nothing of history, Hooke dashed off a critique in a matter of hours. He claimed that he, as Curator of Experiments, had already performed such experiments many hundreds of times. He assured the Society that light is a pulse in the ether and that a prism *adds* colour to whiteness. He infuriated Newton by wielding the word 'hypothesis' as a stiletto. Oldenburg published Newton's entire letter in the *Philosophical Transactions*, and words of admiration began to come from all across Europe, but Newton was peevish and thin-skinned. He had thought the Royal Society would finally be the audience worthy of him: 'For beleive me Sir,' he had told Oldenburg, 'I doe not onely esteem it a duty to concurre with them in the promotion of reall knowledg, but a great privelege that instead of exposing discourses to a prejudic't & censorious multitude (by which means many truths have been bafled and lost) I may with freedom apply my self to so judicious & impartiall an Assembly.' Newton's dispute with Hooke grew into a lifelong enmity. His distaste for wrangling drove him away from the Society for years to come – years spent largely in the secretive study of alchemy and scripture. He did not publish about optics again until he was an old man and Hooke was dead and buried.

It all seemed so innocent at the time. The meeting of 15 February began with a reading of the minutes from the week before. Cornelio's claim about tarantulas needed further discussion: 'some of the members remarking, that it would be hard to accuse of fraud or error Ferdinand Imperato and other good

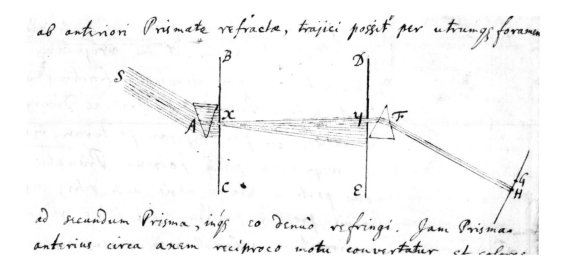

ab anterior Prismate refracta, trajici posset per utrumqs foramen

ad secundum Prisma, inqs co denuo refringi. Jam Prisma anterius circa axem reciproco motu convertatur et cohes.

authors, who had delivered from their own experience, so many mischievous effects of the bite of tarantula's'. They asked Oldenburg to find out what Cornelio had to say in response to those famous men. Then Hooke said that his own observations contradicted Wallis' idea about the closeness of the Moon causing a rise in the mercury of the barometer. Then Hooke presented his comments on Newton. 'Nay,' he said, 'and even those very experiments, which he alledgeth, do seem to me to prove, that *white* is nothing but a pulse or motion, propagated through an homogeneous, uniform, and transparent medium: and that colour is nothing but the disturbance of that light …'

'The same phaenomenon,' Hooke added, 'will be solved by my hypothesis, as well as by his, without any manner of difficulty or straining.' The next week he brought in a candle, to show that, besides the flame and smoke, a continuous stream rose up from it, distinct from the air. Soon after, he showed another phenomenon in a bubble of soapy water, 'which had neither reflection nor refraction and yet was diaphanous'. He observed it carefully: colours swirling and changing; bubbles blown about by the air. 'It is pretty hard to imagine,' Hooke told them, 'what curious net or invisible body it is, that should keep the form of the bubble, or what kind of magnetism it is, that should keep the film of water from falling down.' Really, it was hard to know anything at all.

2

H·G·WELLS'
THE War OF THE Worlds
FRANKENSTEIN.
"By the glimmer of the half-extinguished light, I saw the dull, yellow eye of the creature open; it breathed hard, and a convulsive motion agitated its limbs. *** I rushed out of the room."

MARGARET ATWOOD

OF THE MADNESS OF MAD SCIENTISTS: JONATHAN SWIFT'S GRAND ACADEMY

Margaret Atwood is the author of more than thirty volumes of poetry, fiction and non-fiction, and is perhaps best known for her novels, which include *The Edible Woman*, *The Handmaid's Tale*, *The Robber Bride*, *Alias Grace* and *Oryx and Crake*. Her latest book is the novel *The Year of the Flood*. Her work has been published in more than forty languages.

T HOSE EARLY FELLOWS OF THE ROYAL SOCIETY WERE EARNEST SEEKERS AFTER TRUTH AND PILLARS OF THE COMMUNITY. THEY WERE ALSO, FOR SOME, FIGURES OF FUN AND – AS MARGARET ATWOOD EXPLAINS – THE INSPIRATION FOR A MORE SINISTER ARCHETYPE.

In the late 1950s, when I was a university student, there were still B movies. They were inexpensively made and lurid in nature, and you could see them at cheap matinee double bills as a means of escaping from your studies. Alien invasions, mind-altering potions and scientific experiments gone awry featured largely.

Mad scientists were a staple of the B-film double bill. Presented with a clutch of white-coated men wielding test tubes, we viewers knew at once – being children of our times – that at least one of them would prove to be a cunning megalomaniac bent on taking over the world, all the while subjecting blondes to horrific experiments from which only the male lead could rescue them, though not before the mad scientist had revealed his true nature by gibbering and raving. Occasionally the scientists were lone heroes, fighting epidemics and defying superstitious mobs bent on opposing the truth by pulverising the scientist, but the more usual model was the lunatic. When the scientists weren't crazy, they were deluded: their well-meaning inventions were doomed to run out of control, creating havoc, tumult and piles of messy goo, until gunned down or exploded just before the end of the film.

***

Where did the mad scientist stock figure come from? How did the scientist – the imagined kind – become so very deluded and/or demented?

It wasn't always like that. Once upon a time there weren't any scientists, as such, in plays or fictions, because there wasn't any science as such, or not science as we know it today. There were alchemists and dabblers in black magic – sometimes one and the same – and they were depicted, not as lunatics, but as charlatans bent on fleecing the unwary by promising to turn lead into gold, or else as wicked pact-makers with the Devil, hoping – like Dr Faustus – to gain worldly wealth, knowledge and power in exchange for their souls. The too-clever-by-half part of their characters may have descended from Plato's Atlanteans or the builders of Babel – ambitious exceeders of the boundaries set for human being, usually by some god, and destroyed for their presumption. These alchemists and Faustian magicians certainly form part of the mad scientist's ancestral lineage, but they aren't crazy or deluded, just daring and immoral.

It's a considerable leap from them to the excesses of the wild-eyed B-movie scientists. There must be a missing link somewhere, like the walking seal discovered just recently – though postulated by Charles Darwin as a link between a walking canid and a swimming seal. For the mad scientist missing link, I propose Jonathan Swift, acting in synergy with the Royal Society. Without the Royal Society, no *Gulliver's Travels*, or not one with scientists in it; without *Gulliver's Travels*, no mad scientists in books and films. So goes my theory.

***

Opposite:
Jonathan Swift
by Charles Jervas,
circa 1718.

I read Jonathan Swift's *Gulliver's Travels* as a child, before I knew anything about the B-movie scientists. Nobody told me to read it; on the other hand, nobody told me not to. The edition I had was not a child's version, of the kind that dwells on the cute little people and the funny giant people and the talking horses, but dodges any mention of nipples and urination, and downplays the excrement. These truncated versions also leave out most of Part Three – the floating island of Laputa, the Grand Academy of Lagado with its five hundred scientific experiments, the immortal Struldbrugs of Luggnagg – as being incomprehensible to young minds. My edition was unabridged, and I didn't skip any of it, Part Three included. I read the whole thing.

I thought it was pretty good. I didn't yet know that *Gulliver's Travels* was satirical, that Mr Swift's tongue had been rammed very firmly into his cheek while writing it, and that even the name 'Gulliver', so close to 'gullible', was a tip-off. I believed the letters printed at the beginning – the one from Mr Gulliver himself, complaining about the shoddy way in which his book had been published, and the one from his cousin Mr Sympson, so close to 'simpleton', I later realised – testifying to the truthfulness of Mr Gulliver. I did understand that someone called Mr Swift had had something to do with this book, but I didn't think he'd just made all of it up. In early eighteenth-century terms, the book was a 'bite' – a tall tale presented as the straight-faced truth in order to sucker the listener into believing it – and I got bitten.

Thus I first read this book in a practical and straightforward way, much in the way it is written. For instance, when Mr Gulliver pissed on the fire in the royal Lilliputian palace in order to put it out, I didn't find this either a potentially seditious poke at the pretensions of royalty and the unfairness of courts or a hilarious vulgarism. Rather, having been trained myself in the time-honoured woodsman's ways of putting out campfires, I thought Mr Gulliver had displayed an admirable presence of mind.

The miniature people and the giants did hint to me of fairy tales, but

Part Three – the floating island and the scientific establishment – didn't seem to me all that far-fetched. I was then living in what was still the golden or bug-eyed monster age of science fiction – the late forties – so I took spaceships for granted. This was before the disappointing news had come in – No intelligent life on Mars – and also before I'd read H.G. Wells' *The War of the Worlds*, in the light of which any life intelligent enough to build spaceships and come to Earth would be so much smarter than us that we'd be viewed by them as ambulatory kebabs. So I considered it entirely possible that, once I'd grown up, I might fly through space and meet some extraterrestrials, who then as now were considered to be bald, with very large eyes and heads.

Why then couldn't there be a flying island such as Laputa? I thought the method of keeping the thing afloat with magnets was a little cumbersome – hadn't Mr Swift heard of jet propulsion? – but the idea of hovering over a country that was annoying you so they'd be in full shadow and their crops wouldn't grow seemed quite smart. As for dropping stones on to them, it made perfect sense: kids of the immediately post-war generation were well versed in the advisability of air superiority, and knew a lot about bombers.

I didn't understand why these floating-island people had to eat food cut into the shapes of musical instruments, but the flappers who hit them with inflated bladders to snap them out of their thought trances didn't seem out of the question. My father was by that time teaching in the Department of Zoology at the University of Toronto, and growing up among the scientists, and thus being able to observe them at work, I knew they could be like that: the head of the Zoology Department was notorious for setting himself on fire by putting his still-smouldering pipe into his pocket, and could have made excellent use of a flapper.

When I got as far as the Grand Academy of Lagado I felt right at home. In addition to being the golden age of bug-eyed monsters the late forties was also the golden age of dangerous chemistry sets for children – now prohibited, no doubt wisely – and my brother had one. 'Turn water

to blood and astonish your friends!' proclaimed the advertisements, and this was no sooner said than done, with the aid of a desirable crystal named – as I recall – potassium permanganate. There were many other ways in which we could astonish our friends and, short of poisoning them, we did all of them. I doubt that we were the only children to produce hydrogen sulphide ('Make the smell of rotten eggs and astonish your friends!') on the day when our mother's bridge club was scheduled to meet. Through these experiments, we learned the rudiments of the scientific method: any procedure done in the same way with the same materials ought to produce the same results. And ours did, until the potassium permanganate ran out.

These were not the only experiments we performed. I will not catalogue our other adventures in science, which had their casualties – the jars of tadpoles dead from being left by mistake in the Sun, the caterpillars that came to sticky ends – but will pause briefly to note the mould experiment, consisting of various foodstuffs placed in jars – our home-preserving household had a useful supply of jars – to see what might grow on them in the way of mould. Many-coloured and whiskery were the results, which I mention now only to explain why the Grand Academy 'projector' who thought it might be a brilliant idea to inflate a dog through its nether orifice in order to cure it of colic raised neither of my eyebrows. It was a shame that the dog exploded, but this was surely a mistake in the method rather than a flaw in the concept; or that was my opinion.

Indeed, this scene stayed with me as a memory trace that was reactivated the first time I had a colonoscopy, and was myself inflated in this way. You had the right idea, Mr Swift, I mused, but the wrong application. Also, you thought you were being ridiculous. Had you known that the dog-enlarging anal bellows you must have found so amusing would actually appear on Earth 250 years later in order to help doctors run a tiny camera through your intestines so they could see what was going on in there, what would you have said?

And so it is with the majority of the experiments described in the Grand Academy chapters of *Gulliver's Travels*. Swift thought them up as jokes, but many of them have since been done in earnest, though with a twist. For instance, the first 'projector' Gulliver meets is a man who has run himself into poverty through the pursuit of what Swift devised as a nutty-professor chase-a-moonbeam concept: this man wants to extract sunbeams out of cucumbers so he can bottle them for use in the winter, when the supply of sunbeams is limited. Swift must have laughed into his sleeve, but I, the child reader, found nothing extraordinary in this idea, because every morning I was given a spoonful of cod liver oil, bursting with Vitamin D, the 'Sunshine Vitamin'. The projector had simply used the wrong object – cucumbers instead of cod.

Some of the experiments being done by the projectors interested me less, though they have since contributed to Swift's reputation for prescience. The blind man at the Academy who's teaching other blind people to distinguish colours by touch was doubtless intended by Swift to represent yet more foolishness on the part of would-be geniuses, but now there are ongoing experiments involving something called the BrainPort – a device designed to allow blind people to 'see' with their tongues. The machine with many handles that, when turned, causes an array of oddly Chinese-looking words to arrange themselves into an endless number of sequences – thus writing masterpieces eventually, like the well-known infinitely large mob of monkeys with typewriters – is now thought by some to be a forerunner of the computer.

Predicting the future and suggesting the invention of handy new devices was, however, very far from Swift's intention. His 'projectors' – so called because they are absorbed in their projects – are a combination of experimental scientist and entrepreneur; they exist within *Gulliver's Travels* as pearls on his long string of human folly and depravity, midway between the Lilliputians and their tiny fracas and petty intrigues and the brutal, nasty, smelly, ugly and vicious Yahoos of the fourth book, who represent humanity in its bared-to-the-elements Hobbesian basic state.

But Swift's projectors aren't wicked, and they aren't really demented. They're even well meaning: their inventions are intended for the improvement of mankind. All we have to do is give them more money and more time and let them have their way, and everything will get a lot better very soon. It's a likely story, and one we've heard many times since the advent of applied science. Sometimes this story ends well, at least for a while – science did lower the human mortality rate, the automobile did speed up travel, air conditioning did make us cooler in summer, the 'green revolution' did increase the supply of food. But the doctrine of unintended consequences applies quite regularly to the results of scientific 'improvements': agriculture can't keep up with the population explosion with the result that millions are leading lives of poverty and misery, air conditioning contributes to global warming, the automobile promised freedom until – via long commute distances, clogged roads and increased pollution – it delivered servitude. Swift anticipated us: the projectors promise an idyllic future in which one man shall do the work of ten and all fruits shall be available at all times – *pace* automation and the supermarket – but 'The only inconvenience is, that none of these projects are yet brought to perfection, and in the meantime, the whole country lies miserably waste, the houses in ruins, and the people without food or clothes.' Under the influence of the projectors the utopian pie is visible in the sky, but it remains there.

As I've said, the projectors are not intentionally wicked. But they have tunnel vision – much like a present-day scientist quoted recently, who, when asked why he'd created a polio virus from scratch, answered that he'd done it because the polio virus was a simple one, and that next time he'd create a more complex virus. A question most of us would have understood to have meant, 'Why did you do such a potentially dangerous thing?' – a question about ends – was taken by him to be a question about means. Swift's projectors show the same confusion in their understanding of ordinary human desires and fears. Their greatest offence is not against morals: instead they are offenders against common sense – what Swift might have called merely 'sense'. They don't intend to cause harm, but by refusing to admit the adverse consequences of their actions, they cause it anyway.

The Grand Academy of Lagado was recognised by Swift's readers as a satire upon the Royal Society, which even by Swift's time was an august and respected institution. Though English seekers after empirical facts had been meeting since 1640, the group became formalised as the Royal Society under Charles II, and as of 1663 was referred to as 'The Royal Society of London for Improving Natural Knowledge'. The word 'natural' signifies the distinction between such knowledge – based on what you could see and measure, and on the 'scientific method': some combination of observation, hypothesis, deduction and experiment – from 'divine' knowledge, which was thought to be invisible and immeasurable, and of a higher order.

Though these two orders of knowledge were not supposed to be in conflict, they often were, and both kinds might be brought to bear on the same problem, with opposite results. This was especially true during outbreaks of disease: victims and their families would resort both to prayer and to purging, and who could tell which might be the more efficacious? But in the first fifty years of the Royal Society's existence, 'natural knowledge' gained much ground, and the Royal Society acted increasingly as a peer-review body for experiments, fact-gathering and demonstrations of many kinds.

Swift is thought to have begun *Gulliver's Travels* in 1721, which was interestingly enough the year in which a deadly smallpox epidemic broke out, both in London and in Boston, Massachusetts. There had been many such epidemics, but this one saw the eruption of a heated controversy over the practice of inoculation. Divine knowledge had varying views: was inoculation a gift from God, or was smallpox itself a divine visitation and punishment for misbehaviour, with any attempt made to interfere with it being impiety? But practical results rather than theological arguments were being increasingly credited.

In London, inoculation was championed by Lady Mary Wortley Montagu, who had learned of the practice in Turkey when her husband had been Ambassador there; in Boston, its great supporter was, oddly enough, Cotton Mather – he of the Salem witchcraft craze and Wonders of the Invisible World – who had been told of it by an inoculated slave from Africa. Both, though initially vilified, were ultimately successful in their efforts to vindicate the practice. Both acted in concert with medical doctors – Mather with Dr Zabdiel Boylston, who, in 1826, read a paper on the results of his practice-cum-research to the Royal Society, Lady Mary with Dr John Arbuthnot.

You might think Swift would have been opposed to inoculation. After all, the actual practice of inoculation was repulsive and counter-intuitive, involving as it did the introduction of pus from festering victims into the tissues of healthy people. This sounds quite a lot like the exploding dog from the Grand Academy of Lagado and such other Lagadan follies. In fact, Swift took the part of the inoculators. He was an old friend of Dr Arbuthnot, a fellow member of the Martinus Scriblerus Club of 1714, a group that had busied itself with satires on the abuses of learning. And, unlike the ridiculous experiments of the 'projectors' – experiments that may have been invented by Swift with the aid of some insider hints from Dr Arbuthnot – inoculation seemed actually to work, most of the time.

It isn't experimentation as such that's the target of Book Three, but experiments that backfire. Moreover, it's the obsessive nature of the projec-tors: no matter how many dogs they explode, they keep at it, certain that the next time they inflate a dog they'll achieve the proposed result. Although they appear to be acting according to the scientific method, they've got it backwards. They think that because their reasoning tells them the experiment ought to work, they're on the right path; thus they ignore observed experience. Although they don't display the full-blown madness of the truly mad fictional scientists of the mid-twentieth century, they're a definitive step along the way: the Lagadan Grand Academy was the literary mutation that led to the crazed white-coats of those B movies.

***

There were many intermediary forms. Foremost among them was, of course, Mary Shelley's Dr Frankenstein, he of the man-made monster – a good example of an obsessive scientist blind to all else as he seeks to prove his theories by creating a perfect man out of dead bodies. The first to suffer

Mary
Wollstonecraft
Shelley
by Richard
Rothwell,
exhibited 1840.

Frontispiece
engraved for the
1831 edition of
Frankenstein.

from his blindness and single-mindedness is his fiancée, murdered by the creature on Dr Frankenstein's wedding night in revenge for Frankenstein's refusal to love and acknowledge the living being he himself has created. Next came Hawthorne's various obsessed experimenters. There's Dr Rappacini, who feeds poison in small amounts to his daughter, thus making her immune to it though she is poisonous to others, and is thus cut off from life and love. There's also the 'man of science' in 'The Birthmark', who becomes fixated on the blood-coloured, hand-shaped birthmark of his beautiful wife. In an attempt to remove it through his science – thus rendering her perfect – he takes her to his mysterious laboratory and administers a potion that undoes the bonds holding spirit and flesh together, which kills her.

Both of these men – like Dr Frankenstein – prefer their own arcane knowledge and the demonstration of their power to the safety and happiness of those whom they ought to love and cherish. In this way they are selfish and cold, much like the Lagadan projectors who stick to their theories no matter how much destruction and misery they may cause. And both, like Dr Frankenstein, cross the boundaries set for human beings, and dabble in matters that are either a) better left to God, or b) none of their business.

The Lagadan projectors were both ridiculous and destructive, but in the middle of the nineteenth century the mad scientist line splits in two, with the ridiculous branch culminating in the Jerry Lewis 'nutty professor' comic version, and the other leading in a more tragic direction. Even in 'alchemist' tales like the Faustus story, the comic potential was there – Faustus on the stage was a great practical joker – but in darker sagas like *Frankenstein* this vein is not exploited.

In modern times the 'nutty professor' trope can probably trace its origins to Thomas Hughes' extraordinarily popular 1857 novel, *Tom Brown's School Days*. There we meet a boy called Martin, whose nickname is 'Madman'. Madman would rather do chemical experiments and explore biology than parse Latin sentences – a bent the author rather approves than not, as he sees in Madman the coming age:

> If we knew how to use our boys, Martin would have been
> seized upon and educated as a natural philosopher. He had a
> passion for birds, beasts, and insects, and knew more of
> them and their habits than any one in Rugby ... He was also
> an experimental chemist on a small scale, and had made
> unto himself an electric machine, from which it was his
> greatest pleasure and glory to administer small shocks to any
> small boys who were rash enough to venture into his study.
> And this was by no means an adventure free from excite-
> ment; for besides the probability of a snake dropping on to

your head or twining lovingly up your leg, or a rat getting into your breeches-pocket in search of food, there was the animal and chemical odour to be faced, which always hung about the den, and the chance of being blown up in some of the many experiments which Martin was always trying, with the most wondrous results in the shape of explosions and smells that mortal boy ever heard of.

Despite the indulgent tone, the Lagadan comic aspects are in evidence: the chemical experiments that blow up, the stinky substances, the mess, the animal excrement, the obsession.

The tragic or sinister mad scientist evolutionary line runs through R.L. Stevenson's 1886 novel, *Dr Jekyll and Mr Hyde*, in which Dr Jekyll – another of those cross-the-forbidden-liners, with another of those mysterious laboratories – stumbles upon, or possibly inherits from Hawthorne, another of those potions that dissolve the bonds holding spirit and flesh together. But this time the potion doesn't kill the drinker, or not at first. It does dissolve his flesh, but then it alters and re-forms both body and soul. There are now two selves, which share memory, but nothing else except the house keys. Jekyll's potion-induced second self, Hyde, is morally worse but physically stronger, with more pronounced 'instincts'. As this is a post-Darwinian fable, he is also hairier.

Dr Jekyll is then betrayed by the very scientific method he has relied upon. Time after time, the mixing up of the potion and the drinking of it produce the same results; so far, so good-and-bad. But then the original supply of chemicals runs out, and the new batch doesn't work. The boundary-dissolving element is missing, and Dr Jekyll is fatally trapped inside his furry, low-browed, murderous double. There were earlier 'sinister double' stories, but this one – to my knowledge – is the first in which the doubling is produced by a 'scientific' chemical catalyst. As with much else, this kind of transmutation has become a much-used comic book and

Opposite:
'The features seemed to melt and alter'. The transformation from Dr Jekyll to Mr Hyde. Image taken from *The Strange Case of Dr Jekyll and Mr Hyde,* illustrated by S.G. Hulme Beaman. Originally published in 1930.

filmic device. (The Hulk, for instance – the raging, berserk alter ego of reserved physicist Bruce Banner – came by his greenness and bulkiness through exposure to the rays from a 'gamma bomb' trial supervised by Dr Banner himself.)

Next in the line comes H.G. Wells' 1896 Dr Moreau – he of the Island, upon which he attempts, through cruel vivisection experiments, to sculpt animals into people, with appalling and eventually lethal results. Moreau has lost the well-meaning but misguided quality of the projectors: he's possessed by a 'passion for research' that exists for its own sake, simply to satisfy Moreau's own desire to explore the secrets of physiology. Like Frankenstein, he plays God – creating new beings – but like Frankenstein, the results are monstrous. And like so many of the sinister scientists who come after him, he is 'irresponsible, so utterly careless! His curiosity, his mad, aimless investigations, drove him on …'

From Moreau, it's a short step to the Golden Age of mad scientists, who became so numerous in both fiction and film by the mid-twentieth century that everyone recognised the stereotype as soon as it made its appearance.

Its lowest point is reached, quite possibly, in the B-movie called variously *The Head That Wouldn't Die* or *The Brain That Wouldn't Die*. The scientist in it is even more seriously depraved than usual. The head in question is that of his girlfriend; it comes off in a car accident, after which incident most men might have cried. But the mad scientist is building a Frankenstein monster out of body parts filched from a hospital, underestimating as usual the monster's clothing size – why do those monsters' sleeves always end halfway down their arms? – so he wraps the girl's head in his coat and scampers off with it across the fields. Once under a glass bell with wires attached to its neck and its hair in a Bride of Frankenstein frizzle, the head gives itself to thoughts of revenge while the scientist himself haunts strip clubs in search of the perfect body to attach to it.

***

There's another element in Book Three of *Gulliver's Travels* that bears mention here because it so often gets mixed into the alchemist/mad scientist sorts of tales: the theme of immortality. On the island of Luggnagg, the third in Swift's trio of capital-L islands, Gulliver encounters the immortals – children born with a spot on their foreheads that means they will never die. At first, Gulliver longs to meet these 'Struldbrugs', whom he pictures as blessed: surely they will be repositories of knowledge and wisdom. But he soon finds that they are on the contrary cursed, because, like their mythological forebears Tithonus and the Sibyl of Cumae, they do not receive eternal youth along with their eternal life. They simply live on and on, becoming older and older, and also 'opinionated, peevish, covetous, morose, vain … and dead to all natural affection'. Far from being envied, they are despised and hated; they long for death, but cannot achieve it.

Immortality has been one of the constant desires of humanity. The means to it differ – one may receive it through natural means, as in Luggnagg, or from a god, or by drinking an elixir of life, or by passing through a mysterious fire, as in Rider Haggard's novel *She*, or by drinking the blood of a vampire; but there's always a dark side to it.

Luggnagg is Gulliver's last noteworthy Book Three stop. Through his encounter with the Struldbrugs, he's drawing close to the heart of Swift's matter: what it is to be human. In Book Four he plunges all the way in: his final voyage takes him to the land of the rational and moral talking-horse Houyhnhnms, and brings him face to face with an astonishingly Darwinian view of humanity's essence. The filthy apelike beasts called Yahoos he encounters there are viewed by the Houyhnhnms as beasts, and treated as such; and, much to Gulliver's dismay, he is at last forced to recognise that, apart from a few superficial differences such as clothing and language, he too is a Yahoo.

As Swift's friend Alexander Pope wrote shortly after the publication of *Gulliver's Travels*, 'The proper study of Mankind is Man.' In our own age, that study is not only proper, it's more necessary than ever. The botched experiments of Swift's projectors and our own exponentially successful scientific discoveries and inventions are both driven by the same forces: human curiosity and human fears and desires. Since, increasingly, whatever we can imagine we can also enact, it's crucial that we understand what impels us. The mad scientist figure is – to paraphrase Oscar Wilde – our own Caliban's face in the mirror. Are we merely very smart Yahoos, and, if so, will we ultimately destroy ourselves and much else through our own inventions?

Science was just coming into being in the age of Swift. Now it's fully formed, but we're still afraid of it. Partly we fear its Moreau-like coldness, a coldness that is in fact real, for science as such does not have emotions or a system of morality built into it, any more than a toaster does. It's a tool – a tool for actualising what we desire and defending against what we fear – and like any other tool, it can be used for good or ill. You can build a house with a hammer, and you can use the same hammer to murder your neighbour.

Human tool-makers always make tools that will help us get what we want, and what we want hasn't changed for thousands of years, because as far as we can tell the human template hasn't changed either. We still want the purse that will always be filled with gold, and the Fountain of Youth. We want the table that will cover itself with delicious food whenever we say the word, and that will be cleaned up afterwards by invisible servants. We want the Seven-League Boots so we can travel very quickly, and the Hat of Darkness so we can snoop on other people without being seen. We want the weapon that will never miss, and the castle that will keep us safe. We want excitement and adventure; we want routine and security. We want to have a large number of sexually attractive partners, and we also want those we love to love us in return, and to be utterly faithful to us. We want cute, smart children who will treat us with the respect we deserve. We want to be surrounded by music, and by ravishing scents and attractive visual objects. We don't want to be too hot or too cold. We want to dance. We want to speak with the animals. We want to be envied. We want to be immortal. We want to be as gods.

But in addition, we want wisdom and justice. We want hope. We want to be good. Therefore we tell ourselves warning stories that deal with the shadow side of our other wants. Swift's Grand Academy and its projectors, and their descendants the mad scientists, are among those shadows.

***

Last week I came across a 'project' that's a blend of art object and scientific experiment. Suspended in a glass bubble with wires attached to it – something straight out of a fifties B-movie, you'd think – is a strangely eighteenth-century Lilliputian coat. It's made of 'Victimless Leather' – leather made of animal cells growing on a matrix. This leather is 'victimless' because it has never been part of a living animal's skin. Yet the tiny coat is alive – or is it? What do we mean by 'alive'? Can the experiment be terminated without causing 'death'? Heated debates on this subject proliferate on the Internet.

The debate would have been right at home in Swift's Grand Academy: a clever but absurd object that's presented straight but is also a joke; yet not quite a joke, for it forces us to examine our preconceptions about the nature of biological life. Above all, like Swift's exploding dog and the proposal to extract sunshine out of cucumbers, the Victimless Leather garment is a complex creative exercise. If 'What is it to be human?' is the central question of *Gulliver's Travels*, the ability to write such a book is itself part of the answer. We are not only what we do, we are also what we imagine. Perhaps, by imagining mad scientists and then letting them do their worst within the boundaries of our fictions, we hope to keep the real ones sane.

3

MARGARET WERTHEIM

LOST IN SPACE:
THE SPIRITUAL CRISIS OF
NEWTONIAN COSMOLOGY

Margaret Wertheim is an Australian-born science writer, lecturer and broadcaster, now based in Los Angeles. Her books include *Pythagoras' Trousers*, a history of the relationship between physics and religion, and *The Pearly Gates of Cyberspace: A History of Space from Dante to the Internet*. She is the founder, with her twin sister Christine, of the Institute For Figuring, an organisation devoted to the poetic and aesthetic dimensions of science and mathematics. Their projects include a giant model coral reef made using crocheting and hyperbolic geometry that has become the biggest art/science project in the world.

THE MAD SCIENTIST PLOTTING WORLD DOMINATION IS A FICTION. BUT IT IS NO FICTION THAT THE MODERN SCIENCE WHICH WE IDENTIFY WITH THE ROYAL SOCIETY WAS A PROFOUND CHALLENGE TO EXISTING WORLDVIEWS AND SYSTEMS OF MEANING. JUST HOW PROFOUND IS EXPLORED BY MARGARET WERTHEIM, WHO WONDERS WHETHER WE HAVE YET COME TO TERMS WITH THE CHANGE.

STARSHIP DREAMING

The Starship *Enterprise* heads into the void, its warp drive set to maximum, its crew primed 'to boldly go where no man has gone before'. The drive engages, a burst of light flares out from the rear engines and with an indefinable Woosh ingrained in the minds of *Star Trek* fans everywhere, the world's most famous spaceship disappears from our screens and zaps across the universe to a far distant galaxy. As one of those besotted millions, I am not here to quibble about the scientific 'errors' in Gene Roddenberry's masterpiece; as far as I'm concerned 'Beam me up, Scotty' remains the most thrilling line on television. What I wish to discuss here is an underlying premise of the series that has tugged at the back of my consciousness since childhood. The crew of the *Enterprise* take it for granted – as do real-life physicists, astronomers and SETI enthusiasts – that our cosmos is a homogeneous space ruled everywhere by the same physical laws. Such continuity is logically necessary if humans are ever to travel to the stars or communicate extraterrestrially. So essential is the idea of spatial homogeneity to modern science it has been named 'the cosmological principle' and it serves as the foundation of our faith that if indeed we are *not* alone then we will share something meaningful with our alien confrères – the Laws of Nature.

In the realms of both science fiction and science practice the importance of this principle is hard to overstate, for it underpins physicists' confidence that the patterns of behaviour discovered here on Earth will govern distant worlds. Apples, planets, stars, galaxies, black holes and the explosive aftermath of the big bang are all compelled by gravity's unifying force. The *Enterprise* can set its navigation system to any spatial coordinates precisely because the cosmological principle assures its crew that when they arrive the physics they know and trust will still be working. In contrast to biology, whose plasticity *Star Trek* writers gleefully celebrate in a myriad polymorphous modes, the laws of physics remain the same everywhere – they are the Platonic ideal at the core of an otherwise capricious cosmos. It is physics that makes ours a *uni-* rather than a *multi-*verse.

To citizens of the twenty-first century the cosmological principle may seem close to tautological. For us *space* is now an arena to be measured and mapped, 'the final frontier' on which we have imposed a metric of parsecs and light years. Yet the idea of spatial continuity was one of the more contentious propositions of the scientific revolution and its consequences have been far reaching. I want to argue here that adopting this view set the stage for an unbearable tension between science and Christianity and has problematised the very concept of a human 'self'. In essence, concepts of space and concepts of self are inextricably entwined so that when a culture adopts a new conception of space, as Western culture did in the seventeenth century, it impacts our sense of not merely *where* we are but of *what* we are. While Newton's synthesis famously united the heavens and Earth, it tore a hole in our social fabric that we are still struggling to comprehend and whose consequences continue to reverberate in the US 'war' between science and religion.

A SHORT HISTORY OF SPACE

The magnitude of the transformation taking place in the sixteenth and seventeenth centuries was not lost on any of its participants. Copernicus, Kepler, Descartes, Galileo and Newton all understood that what was at stake in the revolution under way was the fate of the Christian soul. Each of these men stood on the side of God and argued that the emerging cosmology supported a case for the divine. What all of them feared was a universe stripped of spirit. They believed in a Holy Spirit whose Love in-*formed* the world and in the immanent spirits of their fellow human beings; in 'the new astronomy' they saw the reflected glory of their Creator, whose presence in the material universe supported their faith in Christianity's promise of the soul's eternal salvation. As Johannes Kepler summed up the case: 'For a long time I wanted to become a theologian ... Now, however, behold how through my effort God is being celebrated in astronomy.'

The literally soul-destroying potential of the new cosmology hung like a cloud over the consciousness of seventeenth-century science and the source of this angst originated in concerns quite apart from its mechanistic tendencies. By the middle of the sixteenth century thoughtful minds had begun to discern that the idea of continuity between the terrestrial and celestial realms threatened the foundation of Christian faith as it had been construed for 1,500 years. By supplanting the geocentric finitude of medieval cosmology, the new science threatened to undo the metaphysical balance between body and soul on which Christian theology relied.

Opposite:
Pleiades star
cluster. Around
440 light years
from Earth.

Contrary to accounts given in many popular science books, medieval cosmology was underpinned by a rigorous logic that attempted to encompass the totality of humans as physical, psychological and spiritual beings. Medieval scholars read the world in an iconic rather than a literalist sense; nature was a rebus in which everything visible to the eye represented multiple layers of meaning within a grand cosmic order. The physical world was the starting point for investigations that ultimately sought to comprehend a spiritual reality beyond the material plane and what is so beautiful here is that the metaphysical duality of body and soul was mirrored in the architecture of the cosmos.

As is well known, the medieval cosmos was finite, with the Earth at the centre surrounded by concentric spheres that carried the Sun, the Moon, the planets and stars revolving around us. Beyond the sphere of the stars was the final sphere of the universe proper, what the medievals, following the Greeks, called the *primum mobile*. Technically this constituted the limit of the universe – here, as Aristotle argued, space and time ended. Critically, because physical space was finite, medieval minds could imagine that 'beyond' the material world there was plenty of 'room' left for some other kind of space. On medieval cosmological diagrams we see it labelled the 'Heavenly Empyrean'. What lay 'beyond' physical space was the spiritual space of God and the soul.

In the final stanzas of *The Divine Comedy* Dante enacts this transition. Having traversed the span of his universe from the depths of Hell in the centre of the Earth, up the purifying mountain of Purgatory and through the celestial layers, Dante pierces the shell of the *primum mobile* and bursts through the skin of the world to come face to face with God, 'the Love that moves the Sun and the other stars'. For medieval thinkers *this* spiritual domain was the primary realm of the Real with the physical realm serving as a secondary and rather pale reflection. Just what it meant to have a 'place' outside physical space was a question that much exercised medieval minds – no scholar of the Middle Ages believed that Heaven lay literally beyond the stars. Yet whatever the philosophical difficulties, scholars of the time insisted that physical space was not the totality of reality but one half of a larger metaphysical whole.

This dualism of body and soul – matter and spirit – was mirrored in a dualism that was believed to exist between the terrestrial and celestial realms. Again, following the Greeks, medieval natural philosophy held that the two regions were qualitatively distinct regions: in the terrestrial realm things were made up from the four mundane elements, earth, air, fire and water; those in the celestial realm (stars, planets, comets and so on) were composed of a fifth element, or quintessence, also known as the æther. Everything in the terrestrial realm was subject to decay and death, those in the heavens were believed to be eternal, prone neither to decay nor change. Subtleties compounded, for the celestial realm was not itself homogeneous.

Fourteenth-century Italian Renaissance poet Dante Alighieri holding his book *The Divine Comedy* against the backdrop of Hell, Purgatory and Paradise. Painting by Domenico Michelino, 1465.

Ascending from the surface of the Earth medieval cosmology posited that in each successive sphere things became more ethereal. In effect, celestial space exhibited a vector of grace: the closer one got to God, the more 'pure' the region was said to be. Within the scheme of medieval cosmology, celestial space thus served as a mediating zone between the purely material realm of the Earth and the purely spiritual realm of the Empyrean. To put it another way, the celestial heaven of the planets and stars stood as a metaphor for and pointer to the religious Heaven of God and the soul, and the whole of medieval thinking rejoiced in this analogy.

But what if terrestrial space and celestial space were not qualitatively distinct? What if the cosmos was a homogeneous domain? Just such an idea began to bubble into European consciousness during the fifteenth century forming the seeds of what would become, in the seventeenth, a full-blown reconfiguration of Western cosmological thinking.

The first person to express this vision in anything like its modern form was a cardinal of the Roman Catholic Church, Nicolas of Cusa, who completed in 1440 a masterpiece of scientific proleptics entitled *On Learned Ignorance*. The universe Cusa proposed had no crystal spheres and no hierarchy of planets; in one daring swoop he abolished the distinction between the 'base' Earth and the 'ethereal' heavens, positing that the stars and planets were also mundane material bodies. Cusa's cosmos was infinite – 'unbounded' is the word he used – a space in which all regions were materially and spiritually on par. He even suggested that other stars were peopled by other physical beings, an idea that would not be broached again for 150 years. Cusa's ideas were too radical for most of his contemporaries, but in the sixteenth century the tectonic plates of the Western psyche began to shift, resulting in the work of Copernicus and all that came after him.

Why was such a shift occurring? After all, the medieval world picture had held stable as a philosophical construct for more than a thousand years. The telescope had not yet been invented, astronomical observations were not qualitatively better, the Ptolemaic model continued to yield reasonable results. Cosmological technologies were not perceived to be failing. So what *was* going on? Astronomy wasn't undergoing a crisis, nonetheless underlying conceptions of how reality might be were beginning to change. We know this primarily not from what scientifically minded thinkers were saying but from what painters were doing. Long before the rise of science a new Western attitude to space was apparent in the realm of art, and to understand the cosmological transformation wrought by Galileo and Newton in the seventeenth century it is instructive to turn first to the frescoes of Giotto. Here we can see explicitly the spiritual stakes that were coming into play as Europeans began to feel their way out of a medieval world.

Along with medieval philosophy, medieval art focused on the numinous realm of the soul. Art also was iconic, aiming to represent the spiritual order beyond the material world. One way of conveying that order was through scale; thus Christ would be the largest figure in a painting, with angels next in size, followed by saints and martyrs, then ordinary human beings. Backgrounds too were iconic; gold and azure represented Heaven, whose value was viscerally present in exorbitantly expensive gold leaf and lapis lazuli pigments. Depth was almost absent from these images. But in the late twelfth century representation began to undergo a subtle transformation with a gradual interest in three-dimensionality starting to emerge. This new style reached a crescendo with Giotto's work in the Arena Chapel in Padua in which he depicted a sequence of near-life-sized images recounting the life of Christ. What is immediately startling about these Christ Cycle frescoes is their sense of physical presence. Figures look solid and are anchored to the ground as if compelled by gravity. We are clearly no longer in Heaven but on Earth. Everyone appears at the same scale: Christ and humans and angels. Flat blue and gold backgrounds are replaced by attempts at genuine landscapes; there are mountains, trees and carefully observed studies of animals. Buildings seem to be leaping out of the surface. True, they are not entirely convincing, but one feels here that the artist is striving to convey three-dimensional space.

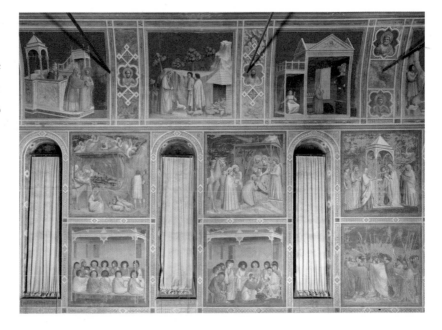

All this was in keeping with a revivified interest in the natural world. After the hiatus of the early Middle Ages, scholars had begun to recover the science and mathematics of ancient Greece, and during the thirteenth century the study of nature underwent a renaissance. With his careful attention to empirical detail, Giotto reflected this novel scientific bent. For artists and their patrons (many of whom were leaders of the Catholic Church) the observations of the outer eye were becoming more interesting than the revelations of the inner eye. In short, visual attention was shifting towards the material realm.

Paradoxically, this refocusing from spirit to matter was given credence by a novel theological development and it is here that science and religion intersect in a uniquely Western way. As Europeans recovered the heritage of the Greeks one thinker they increasingly encountered was Pythagoras, a mathematician and mystic who had dreamed the dream that would become modern physics. In the fifth century BCE Pythagoras posited that the structure of the world was determined by mathematics: 'All is number', he famously declared. A small band of medieval thinkers took Pythagorean precepts and transformed them into a Christian context, giving rise to the then-novel idea that God had created the material world according to mathematical rules.

Among God's primary tools was Euclidean geometry and in 1267 the Franciscan friar Roger Bacon argued in a treatise to Pope Clement IV that artists ought to follow their Creator and construct images accordingly with

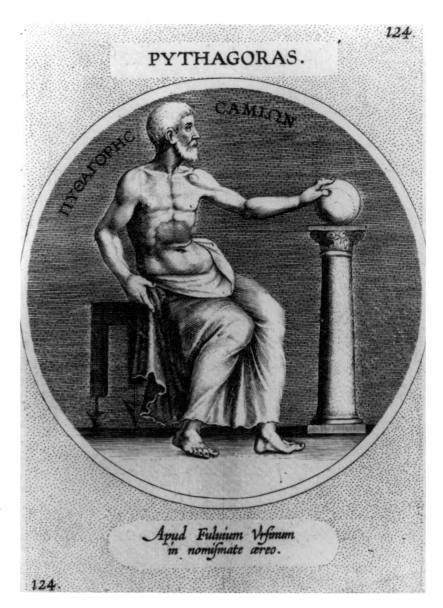

The Greek philosopher, mathematician and scientist Pythagoras.

geometric relationships. Bacon called the new style 'geometric figuring' and he proposed that the Church encourage painters to adopt it as a matter of principle. Artists who did so would not just be rendering Creation truthfully, Bacon said, they could also serve a powerful propaganda purpose, for according to him the techniques of three-dimensional verisimilitude were so psychologically powerful that viewers beholding such images would believe they were actually witnessing the scenes depicted. They would believe they were really *seeing*, for example, Christ raising Lazarus in front of them. To put this into current parlance, Bacon was suggesting that 'geometric figuring' acted as a kind of virtual reality and, as he saw it, this medieval VR would have the power to convert unbelievers to the Christian faith.

From the fourteenth through sixteenth centuries artists elaborated Bacon's vision with ever-greater finesse, a movement that culminated in the formalisms of 'linear perspective'. The consequences of this representational revolution reached far beyond the painted surfaces of the churches from which it began. Art historian Samuel Edgerton has argued that 'geometric figuring' retrained European minds to see space in a Euclidean sense and that in this respect Renaissance artists from Giotto through Raphael paved the way for the physicists who came after them. Edgerton's thesis helps make sense of a historical conundrum, for following Aristotle most Western thinkers pointedly rejected a Euclidean view of space. As physicist and science historian Max Jammer has stressed, such a view of space was not 'thought reasonable until the seventeenth century'. No other culture we know of has conceived of its cosmic scheme in this mathematical manner, and even in the West most learned people rebelled against the idea for several hundred years. Perspectival painting served to introduce the concept as a visceral *experience*, subverting intellectual objections by giving viewers a powerful psychological illusion that the painted scenes they were looking at were actually there.

By the mid-sixteenth century, educated Europeans were coming to believe that the space around them here on Earth *was* a Euclidean realm.

But that raised an uneasy question: How far out does this space extend? Does it extend to the Moon? To Mars? To the Sun and stars? Though not articulated in quite this form during the Renaissance, the question assumed immense importance because it challenged the medieval distinction between the terrestrial and celestial realms. If Euclidean space proceeds beyond the Earth then that suggests that similar laws and similar things should be found in both regions.

The unification of the two domains was of course cemented by Isaac Newton and in some ways it remains his most profound legacy. Newton showed that the same force of gravity that makes an apple fall to the ground also operates to keep the Moon revolving around the Earth and the planets orbiting the Sun. Newton's law (a Pythagorean triumph if ever there was one) demonstrated an essential continuity, for if gravity operates between celestial bodies then they *too* must be mundane matter like the pebble that rolls down a hill. Moreover, once astronomers abandoned the medieval distinction between earthly and celestial space, there was no longer any reason to imagine a limit to the physical world. Why should physical space not go on for ever? By the end of the eighteenth century, that view had become scientific orthodoxy.

WHERE IS HEAVEN?

This new cosmology had profound theological consequences, for with physical space extended to infinity there was literally no room left for Heaven. One could say, as liberal theologians *do*, that the realm of the soul is simply beyond the material plane and leave it enigmatically at that, yet with physical space infinitised the whole question of what a 'beyond' might constitute became increasingly problematic. For better or worse, one of the consequences of the scientific revolution was to write out of Western cosmology any sense of spiritual space as a legitimate aspect of the Real.

Newton himself was concerned about the matter and tried hard to rescue the situation by associating space with God. Picking up on a tradition that originates in Judaism, he posited space as the medium through which the deity's presence permeates the world. Space, he said, was God's *sensorium*, the substrate through which He sees all, feels all, knows all. Space was indeed synonymous with divine Knowing. As President of the Royal Society Newton understood that the new science had to do much more than make empirical predictions – it had to be acceptable to reasonable society. Galileo and Descartes had both run afoul of such expectations about what a cosmology should deliver and Newton was determined not to make the same deistic mistake. As Britain's leading representative for science, he comprehended that neither the people nor the patrons would support the endeavour if it was seen to be in conflict with wider spiritual needs. The Royal Society stood on the side of reason, but it also allied itself with the state, the King and God. All this wasn't just a propaganda exercise, for psychologically speaking, Newton needed reasons to accept the new space himself – God made the void 'reasonable' to *him*.

Newton had good cause to worry, for soon after his death less religious minds stripped the theological embellishments from his system leaving humans alone in the void. Increasingly in the age of science we have confronted the dilemma that if we want to claim something is real, we have to posit its position in physical space. If one can't point to coordinates on a map, then more and more one invites the accusation that whatever it is, is not real at all. Hence the liberal theological dilemma about Heaven. Where is it? Both Hell and Purgatory could easily be abandoned, but Heaven – the domain of human salvation – is critical to Christian integrity. The soul also became collateral damage as 'Man' was transformed into 'an atomic machine'. Without its own *place* in the cosmic scheme, the spirit was disenfranchised. Humans became mere bodies, flecks of dust residing on a chunk of rock orbiting a small and insignificant star in the outer suburbs of a very mundane galaxy. We moderns are not only not at the centre of the universe, as spiritual beings we actually don't exist in this world.

POST-NEWTONIAN SPACE

During the twentieth century physicists developed a post-Newtonian vision of space beginning with Einstein's relativity theories and proceeding to so-called 'hyperspace' theories. How have these ideas impacted on the discussion above? Relativity compounds the problem in a truly fascinating way. General relativity, which is the cosmological version of Einstein's ideas, replaced the three-dimensional Euclidean void of Newton's cosmology with a four-dimensional Minkowskian void that now includes *time* as part of the spatial matrix. Physicists call it *spacetime*, and treat time as effectively another dimension of space. From a theological perspective the consequences here are non-trivial because in a purely relativistic cosmos nothing really 'happens'. Time unwinds itself in a manner predetermined by the tensor equations; nothing evolves or comes into being that wasn't already inherent at the start. In a purely relativistic cosmos (where there are no quantum effects) time is thereby neutered: there is no *happening* whatever. From a four-dimensional perspective the universe *just is*, complete and whole as a pre-set form. If this cosmos is a thought in the mind of God, it is one that is effectively static. Now that might be OK for God – who has always been said to see time whole – but it is not OK for human souls whose destiny cannot be pre-ordained. Christian theology demands that time be open so that individuals truly have a choice about what decisions they make. As moral beings our 'worldlines' cannot be set by analytic equations; for Heaven to mean anything, we must be able to act on our own *volition*. In short, the Christian concept of salvation requires a concept of spacetime that is more dynamic and incomplete than relativity allows.

Hyperspace theories add further complications. These theories extend Einstein's concept of space from four dimensions to ten or eleven. Where Einstein folded time into the spatial matrix, hyperspace theories aim to fold in everything. Here matter itself becomes a by-product of the shape of space. In hyperspace theories there is actually nothing but space curled up into patterns – everything that exists from protons and petunias to planets and people is at core complex enfoldings of space. The English physicist Paul Davies has called this 'structured nothingness'. We may think of it as a kind of cosmic origami. At the start of our universe, space had *no* structure – it was simple and unformed like a blank sheet of paper, then as time proceeded the 'paper' crinkled up into ever more elaborate structures, eventually giving rise to the complexities we see today.

Where does this take us theologically? Unlike relativity's God, the God of hyperspace theory is an active and dynamic Creator. As a fan of origami it thrills me to think of Him whiling away the tedium of eternity folding space into increasingly subtle forms. He is an architectonic genius, a veritable master of structure. A standing ovation for origami God, I say. But where do *we* stand in this picture? Is there a place in the hyperspace cosmos for humans as spiritual beings? It seems to me there is not – at least not in a way that I believe was a central aspect of the medieval world picture. In the hyperspace vision of cosmology, space becomes not just the *arena* of reality, as it was for Newton and Einstein, but reality itself. Here, there is actually nothing but structured space. This is an extraordinary philosophical move. Newton's cosmos contained three fundamental things: matter, space and force (epitomised by gravity). With hyperspace theories there is now just one fundamental thing – space – everything else being a by-product of this fundamental 'stuff'. What we have here is literally a post-material account of the world, for matter has now been relegated to secondary status. At first glance that might seem like a good thing for the spiritualists, and some people have tried to read it that way. Western culture has a long tradition of opposing matter and spirit, so something that is *not* matter can easily be read within this tradition as *ipso facto* spiritual. I believe such optimism will prove to be as historically futile as Newton's hope that space would be read as God's *sensorium*.

The problem is that in hyperspace theories everything is reduced to a seamless monism. Everything is collapsed into a single category. This is precisely the mistake that Descartes sought to avoid with his infamous dualism. As a man of science Descartes wanted to articulate what the new science could do, but as a devout Catholic he also wanted to preserve the gift of Christian salvation. His answer was to postulate two distinct 'realms' of experience: the *res extensa* or extended realm of matter in motion, and the *res cogitans*, the 'realm' of thoughts, feelings, morality and spiritual consequence. The new science would tell us about the former, but for Descartes science would have nothing to say about the latter. In effect, Descartes tried to preserve the dualism inherent in medieval thinking while also opening up the possibilities he so boldly saw in the emerging science. As a Catholic, he understood that the Christian soul could not be bound by mathematical laws, and since he believed that mathematics *was* the language of the material world there had to be some 'realm' apart from those laws.

Descartes failed in the same sense that Newton failed; his theological trappings were stripped away by later generations who took what he had done and used it to promulgate a purely secular cosmology. Since the Enlightenment we have come to use the word 'cosmos' to mean the purely physical world and 'cosmology' to mean our concept of the material domain alone. We have forgotten the wider picture in which 'the cosmos' encompassed multiple levels of being; we tell ourselves that older cosmologies are childish tales and that we moderns supposedly have outgrown these stories and faced reality 'squarely' to work out where we 'truly' are.

SPACE AND SELF

In discussions about science and religion it is often noted how corrosive a mechanistic philosophy was to the Christian idea of a soul; what is not widely understood is how important a role our conception of space has played in this story. Eighteenth-century natural philosophy was premised on a neutral, homogeneous, infinite and passive space. The very qualities of Euclid's ideal that made it such a fruitful foundation for the development of physical science are just the qualities that have become so problematic for those who wish to assert the reality of a 'spiritual' plane of being. For medieval Christians, a dualistic conception of the human person went hand in hand with a dualistic spatial scheme; with the advent of a purely physicalist world picture it has become increasingly difficult to argue for the reality of *any* kind of non-physical dimension to human existence.

Christians are not the only ones who might be troubled by this development. Secularists can be concerned too, for the equation of physical space with the totality of 'reality' also problematises the idea of a human *self*. What does it mean to say that the self exists if we cannot locate it on a map? In talks I give about this subject I am sometimes asked during question time to 'prove that the self exists'. It is always a young man who makes this demand and he is usually a student of physics or philosophy. He is well read and he means his question in earnest. He does not believe that the self exists and he wants me to prove it does. At first I was puzzled by this demand, then I realised how I should answer: If the self does not exist, I say, then *his* self doesn't, so I shall move right along to a question from someone who does. I assume there are some selves in the room who do exist.

But are there? In the mathematically defined space of modern cosmology do any of us exist?

A SCIENCE OF MIND

In the early eighteenth century, the philosopher John Locke claimed that it wasn't stable for a society to have only a science of body. According to Locke, we would eventually need to develop a complementary science of mind, which is what Freud attempted in the late nineteenth century. The psychoanalytic tradition of the past century may be read, in part, as one

JOHN LOCKE ESQ.ʳ F. R. S.

reaction to the cosmological shift that took place two hundred years earlier. Freudian psychoanalytics and its many descendants are attempts to make sense of the self in a non-spatial framework and in a very real way to get beyond the metaphysical dualism of our Christian and Greek heritage. Personally I find myself greatly in sympathy with the whole exercise and although I think its therapeutic effects are easily overstated, I do believe the psychoanalytic stream of theory and practice is a powerful response to what remains within our society a cosmologically inspired trauma.

I do not mean to propose here that every individual is personally feeling this rent; but it is clear that a great many of us are. For all of the immense practical and epistemic triumphs of modern scientific metaphysics, which is premised on a homogeneous continuous conception of space, it is manifestly not being accepted by huge slabs of our population. Reactions against it have been vast and varied from William Blake's scathing poetic critiques (that science would 'conquer by rule and line' and 'unweave the rainbow'), to Alfred Whitehead's enigmatically difficult 'process philosophy', which attempts to articulate a reality in which neither matter nor mind take precedence, rather both are artefacts of a fundamentally procedural world. Intellectual alternatives to pure physicalism are myriad: Teilhard de Chardin, Loren Eiseley, Mircea Eliade and Rupert Sheldrake may all be read as responses, to say nothing of the exponentially expanding volume of New Age literature. To the continuing horror of many champions of science, belief in astral planes, psychic channelling, reincarnation and past lives seems to be growing stronger.

In part I believe what this represents is a widespread social refusal of spatial monism. Whole sectors of our society are just not buying it! More than twenty million people bought *The Celestine Prophecy* (it is one of the most successful books of all time), which posits that when we become the beings we ought to be our souls 'cross over' (via some processes of quantum mechanics) to a higher spatial plane. In the age of science, one of the most pervasive fantasies is indeed the existence of other spaces of being: from the

Opposite:
Portrait of John
Locke, after Sir
Godfrey Kneller.

X-Files and *Buffy the Vampire Slayer* to *Lost* and *Battlestar Galactica,* our television screens offer a steady diet of realities in which multiple spaces and planes of being co-exist. (Cyberfiction offers yet another response – the fantasy of downloading one's mind into a computer to live for ever in a virtual world is nothing more, though a good deal less, than a technological version of Heaven.) One of the great philosophical projects of the post-Enlightenment era has been to articulate non-spatialised conceptions of the self in relation to the cosmos; yet judging by the evidence of the most pervasive medium on our planet the enterprise has met with little success in a sociological sense. Even science fiction writers – Carl Sagan, no less – keep on inventing wormholes through the physio-spatial matrix to *other,* suspiciously spiritualised, places of being.

Those of us who love science may choose to interpret all this as a kind of play, and in some sense it is, but the refusal to accept spatial monism is also in part fuelling the rise of Creationism and other fundamentalist brands of Christianity. At the same time that spatial monism erased the division between earthly and heavenly space, it also provided a platform for erasing any fundamental distinction between living and non-living things. In the new era of science, *continuity* itself became the epistemic model – the continuity of the laws of nature, the continuity of space, the continuity of matter, the continuity of life. No body is special, because no thing is special, because no place is special. Humans are related to apes because, in the end, we are all just inert matter floating in a homogeneous void. The fundamentalist rebellion against Darwinism is not just a rejection of the continuity proposed by biology but in a wider, and less obvious way a rejection of the very premise of totalised cosmic continuity. Christians who insist on a space for the soul wish to reclaim that part of the medieval world picture that literally gave a *place* to moral human agents. Though I do not endorse their specific responses, I believe that in this respect the religious right point us to a deep and abiding sociological problem that will not be easily resolved and which ought not be so readily dismissed.

CONCLUSION

At all times in *The Divine Comedy* Dante knew where he was. He was embedded in a cosmos that gave him a position physically, spiritually and psychologically. One of the many strengths of the *Comedy* is that it gives a concrete landscape to both soul *and* psyche. While the book must be read as the journey of a Christian soul through Hell and Purgation towards Paradise, it can also be read as a journey of psychological self-examination and healing. The descent into Hell is a literal depiction of human psychic suffering; the trip up Mount Purgatory is the therapeutic path. We can gauge Dante's progress by the state of his surroundings – we feel the anguish as we slog with him through the ditches of the Malebolge, we rejoice with relief as he trots up the marble ramps of the mountain. Dante may be a sinner, but he is never lost – his cosmos tells him in the very texture of his surroundings where he stands as a material body, as a Christian soul and as a human self.

Several years ago I gave a lecture at a small university in the American South. After the lecture I was taken aside by a professor at the school, an anthropologist who had done field work in Namibia with the Himba tribe. One day, he told me, he was approached by a Himba man who asked him a question: 'Do you Westerners really see the space between you as empty?' 'Yes,' my American interlocutor replied, 'that is the way our science tells us to see the world.' The Himba man went on to explain that, in his culture, people saw the world in a different way. According to their worldview, each person is surrounded by a kind of self-space which extends out around the individual. Going about their daily business, he and his fellow villagers found their self-spaces continually intersecting. They rarely found themselves 'alone' – their 'selves' being continually in *touch* with others. Having explained this way of seeing, the Namibian man asked the American professor a second question: 'If you people really see yourselves as isolated points alone in empty space, how do you bear it?'

It seems to me that as a society we are not bearing it. Unlike Dante, we are lost in space.

D. ISAACVS NEWTON EQVES
REG. SOCIETATIS PRÆSES. AN°. 1703

4

J. Newto

NEAL STEPHENSON

ATOMS OF COGNITION: METAPHYSICS IN THE ROYAL SOCIETY, 1715–2010

Neal Stephenson is the author of the three-volume historical epic *The Baroque Cycle* (*Quicksilver*, *The Confusion* and *The System of the World*) and the novels *Cryptonomicon*, *The Diamond Age*, *Snow Crash* and *Zodiac*. He lives in Seattle, Washington. His latest novel is the alternate reality epic *Anathem*.

T HE LOOSENING OF THE MOORINGS OF THOUGHT WHICH CAN BE SENSED IN THESE EARLY DAYS OF THE SOCIETY HELPED SUSTAIN SOME VAST INTELLECTUAL DISPUTES. AS NEAL STEPHENSON EXPLAINS, ONE OF THE SHARPEST, BETWEEN THE TWO GIANTS NEWTON AND LEIBNIZ, INVOLVED SOME VERY STRANGE METAPHYSICS – NEARLY AS STRANGE, IT TURNS OUT, AS TWENTY-FIRST-CENTURY PHYSICS.

> This philosophy is a gift of God to this old world, to serve as the only plank, as it were, which pious and prudent people may use to escape the shipwreck of atheism which now threatens us.
> – Leibniz, in a 1669 letter to Thomasius

Isaac Newton was slow to join the Royal Society – in the Charter Book that lives in the Society's vault, his signature does not appear until the ninth page – but by the second decade of the eighteenth century he had become its President. His unquestioned status as the greatest mind of his generation, combined with his political connections as Master of the Mint and his ruthlessness toward those he perceived as rivals, had given him an unusual degree of power. This he brought to bear against the only living person who could even hope to challenge his intellectual supremacy: Gottfried Wilhelm Leibniz, who despite being a foreigner (he was Hanoverian) had been made a Fellow of the Royal Society in 1673, largely in recognition for his invention of the Stepped Reckoner, a mechanical computer.

The contrasts between Newton and Leibniz were lavish. Newton seems to have had an entirely accurate sense of just how he compared to his contemporaries, and acted accordingly without concern for dusty precedents or the personal feelings of those who clung to them. When confronted with anything less than uncritical acceptance of his work, he lashed out and then secluded himself. He published rarely but ex cathedra, handing down nearly flawless treatises over which he had toiled for years or decades, perfectly organised into definitions, axioms, lemmas and laws, framing a mathematical physics that could be used to explain past observations and to make verifiable predictions.

Isaac Newton's signature.

Leibniz was an accomplished courtier who maintained long friendships with the Electress of Hanover, the first Queen of Prussia, the sister-in-law of Louis XIV, and the future Queen Consort of England, while moonlighting, late in his career, for Peter the Great. He corresponded so heavily that scholars are still sorting through his unpublished papers. In his philosophy he practised an ecumenicism that in a lesser mind would strike us as suspicious or even craven. Leibniz seems never to have met a philosopher or a theologian he didn't like, and his metaphysics developed out of an effort to harmonise the ancient thinking of (both) Plato and Aristotle with tenets of

Christian and Jewish theology and with the 'mechanical philosophy' the Royal Society had been created to champion. It is impossible to know precisely what he was thinking without perusing his vast legacy of papers. In effect, Leibniz's philosophy ceased to exist at the moment he died. Since then, anyone who has wanted to know it has first had to reconstruct it, which is only possible for forensically inclined scholars, fluent in Latin, French and German, and well versed in the history of Western philosophy, Christian theology and Enlightenment science.

Given Leibniz's stature as one of the great thinkers of Western history, one might expect that, as of the 350th anniversary of the founding of the Royal Society, all of his writings would long since have been published, and that everything would be known about his philosophy. But the question of 'what did Leibniz believe, and when did he believe it?' is unsettled and is the topic of current research and debate.

A squalid row over the origins of the calculus, which these two men had independently invented decades earlier, became the public face of the conflict, which is regrettable since it is not very interesting and since it reflects dreadfully on the combatants. Much more significant in the long run was a debate on topics that reach so deeply into the foundations of science that they are still discussed in our times. This broke the surface in the last year of Leibniz's life, in an exchange of letters that has come to be known as the Leibniz–Clarke correspondence.

The year was 1715, and because of two royal deaths (in England, Queen Anne; in Hanover, Electress Sophie), Princess Caroline of Brandenburg-Ansbach had just become the Princess of Wales. To the modern reader, Caroline seems less like a real historical personage than a plucky, clever, independent-minded heroine from some post-feminist historical novel. A noble but poor orphan, raised as a ward of the Prussian court, she was conversant with scientific topics of the day, largely because she had been tutored in them by Leibniz. She had married into the Hanoverian dynasty and had moved with it to London, where her father-

D. ISAACVS NEWTON EQVES
REG. SOCIETATIS PRÆSES. ANº. 1703.

in-law had been crowned King George I. The sixty-nine-year-old Leibniz, who had become unfashionable and, because of the dispute over the calculus, something of a political problem, had been left behind in Germany. He wrote a short letter to Caroline, warning her that religion was declining in England; that John Locke did not believe in the immortality of the soul; and that Sir Isaac Newton held to some strange views about the relationship between God and the physical universe.

Anyone who has blithely forwarded a private email to a corporate mailing list, with incalculable consequences, will recognise what happened next: Caroline made Leibniz's letter known, and one Samuel Clarke stepped forward to rebut Leibniz's charges. The result was a series of letters (five each by Leibniz and Clarke) over the course of a year, at which point Leibniz died. Clarke, though he had serious credentials in his own right both as theologian and scientist, was acting as a spokesman for Newton, and so the correspondence can fairly be read as a debate between Leibniz and Newton.

In the opening round, the combatants practically trip over each other in their eagerness to remind the Princess that atheism is bad and that true natural philosophy in no way conflicts with religion. There is no reason to think that either of them is being disingenuous. The scientific revolution had created doubts about the existence of God, or at least the veracity of religious dogma, in the minds of many; but not Newton or Leibniz.

These concerns are dispensed with in a few paragraphs. The bulk of the correspondence, which runs to about eighty pages, resembles an email exchange that devolves, as it goes on, into several distinct threads, each concerning a specific sub-topic. The correspondents begin to number their paragraphs (Leibniz's fifth letter contains 130 of them), the better to keep track of all the rebuttals and counter-rebuttals. The over-arching theme is the relationship of God to the universe, and more specifically the universe as perceived, measured and understood by scientists. Leibniz, in the universal manner of authors promoting their latest work, finds frequent occasion to mention his books *Theodicy* and *Monadology*. Even when he isn't mentioning them by name, he is presenting arguments, and using terminology, derived from them.

My theme is the legacy of Leibniz's metaphysics from the time of his death down to the present day, and so a direct summary of that system, based on the scholarship of latter-day researchers, will do better service than any attempt to untangle the points and counter-points in the correspondence. The account presented below is patterned after the work of Christia Mercer of Columbia University. Her book *Leibniz's Metaphysics: Its Origins and Development*, published in 2001 by Cambridge University Press, is a formidable work of forensic scholarship that can in no way be improved by my attempts to summarise it.

In 1661, at the age of fourteen, Leibniz had formed a resolution to embrace the new mechanical philosophy. For most natural philosophers of the era, this meant rejecting the Aristotelian worldview of the medieval schoolmen. As mentioned, though, Leibniz was an ecumenicist and a conciliator, and so for him it meant, rather, the beginning of a lifelong quest to reconcile certain select, precisely defined tenets of Aristotelian and Platonic thought with modern science.

In his metaphysical reasoning, Leibniz is at least as meticulous as is Newton in his mathematical physics. Bertrand Russell called Leibniz's system 'profound, coherent, largely Spinozistic, and amazingly logical'. Newton, however, can verify his results by comparing them to observations, while Leibniz is beholden to no one except Leibniz. By pure thinking, Leibniz fabricated a metaphysical system that could hardly be more at odds with that of Newton, or indeed any other person who attempts to think in a commonsensical way about how the world might work.

Where Newton's work is grounded in Euclidean geometry, Leibniz begins with certain precepts that he takes to be axiomatic, such as the Principle of Sufficient Reason (nothing exists without a reason; there is no effect without a cause) and the Identity of Indiscernibles (two individual things cannot differ in number alone; it must be possible to explain why they are distinct based on some intrinsic difference). Newton developed calculus because it enabled him to solve problems in his theory of gravitation; Leibniz developed it as an outgrowth of his fascination with the problem of the Continuum, which asks how a line can be made up out of points, a span of time from instants, or a thought from the minute perceptions and endeavours of a mind. Just as Newton would not bother developing a physics that could not explain the fact that planets move in elliptical orbits, Leibniz had no time for any metaphysics that was incompatible with the transubstantiation of the Eucharist (both the Protestant and the Catholic versions!) and the incarnation of God in Christ. Much of the pick-and-shovel work of his *Monadology* came from a 1671 tract about the Incarnation of God.*

The modern reader, following the development of Leibniz's ideas over the years between 1661 and his death in 1716, veers between finding it all quite reasonable and feeling as though it must have come from an alien planet. Just when one is about to judge Leibniz as having the strangest mind of anyone who ever lived, one remembers Newton and his lifelong obsession with alchemy and his strenuous efforts to predict the exact date of the End Times by ransacking the Book of Revelation for encrypted clues.

It takes an entire book such as Mercer's to explain Leibniz's full chain of reasoning, so there is not room here to attempt any such thing. The end point – Leibniz's mature system, as described in *Monadology* – may be summarised as follows:

Matter, assumed by most to be the primary stuff of the universe, extended in space and time, is, in fact, unreal. Atomism in its conventional form – the idea that physical objects can be divided and subdivided up to a certain point, but (for some, usually unspecified, reason) no further, and that the result is a collection of tiny indivisible matter-bits moving around in empty space and banging into one another – is all wrong. The true atoms – the fundamental, indivisible units that make up the universe

* This perennial theological chestnut seems to have occasioned some soul-searching for Newton as well, since he risked serious trouble by semi-openly espousing the Arian heresy, which denies the Trinity.

– are not spatiotemporal and so are not bound by spatial and temporal constraints; rather, space and time are epiphenomena of their activities, which are mental (today we might say computational) rather than physical. Leibniz calls these mind-atoms by the name of monads.

Use of 'mind' and 'mental' is apt to give modern readers the wrong idea. Many translators of Leibniz (including Russell) choose the word 'soul' instead of 'mind', which is even more confusing. A word about those words is, therefore, in order. Extension (occupying physical space) and duration (persisting through time) are obvious properties of matter that had long been of interest to natural philosophers. Beginning around 1671, Leibniz added a third element, namely *cognitio*, which can be translated as 'thought' or 'knowledge'. In his metaphysics, *cognitio* is a property that things can possess and that makes them different from inert matter. Early in his career, it is as fundamental as extension. Later, it becomes more so. Previously, he had admitted God and the human mind as the only two incorporeal principles in his system; the key move he now made was to admit the possibility of cogitating entities ('minds' or 'souls') that were neither divine nor human, and to make them and 'endeavour' – the smallest possible unit of cogitation, which is to *cognitio* as a point is to a line or an instant is to time – as fundamental as space and time. Later, he goes on to deny the primary reality of space and time altogether and to assert that the created world consists entirely of these unextended monads and that the universe is created from moment to moment as a result of their cognition. In this he breaks from the metaphysics assumed by Newton (and almost anyone else who has thought in a commonsensical way about space, time and atoms) in which space and time have an absolute reality, and form a sort of lattice on which the laws of physics are enacted, and, indeed, without which they cannot even be written down.

Because the monads do not exist in space and time, they are free to take on certain powers and properties that would otherwise be implausible: (1) each monad perceives the state of every other monad in the universe, and

(2) each exists in a certain state, and is capable of changing that state. This process of continual internal state-change is the cogitation that is the raison d'être of the monad and the fundamental process of the universe.

Internal and intrinsic to each monad is a rule (dubbed by Mercer the Production Rule) that governs how it changes its state in response to its current state and the perceived state of all of the other monads. And just as the constraints of space and time are inapplicable to monads, so cause and effect work differently, for each monad is causally independent of all other monads. It makes its own decisions by its own lights, obeying its intrinsic rule.

This raises the obvious objection that if the states of the other monads serve as inputs to the production rule, then there would seem to be a cause-and-effect relationship at work, but Leibniz doggedly maintains that no such relationship exists and that coordination among monads comes about, not through causal linkages, but as the result of a divinely ordained pre-established harmony that brings all of the monads into a kind of synchronisation without encroaching on their independence. For minds and cogitation are, to Leibniz, the ultimate reality, and unless the minds have free will, they are not minds at all but physical mechanisms numbly obeying deterministic rules.

This is the one feature of the Monadology that might (I speculate) have aroused some competitive anxiety in Newton's mind. The Leibniz–Clarke correspondence probably would not have drawn the attention of so many important people were it not that traditional (spatiotemporal) atomism, combined with the then-new science of mathematical physics, seems to lead ineluctably to what was later called Laplacian determinism. If the behaviour of all objects can be explained in terms of spatiotemporal atoms, and if the atoms' behaviour, in turn, is subject to Newton's deterministic mathematical laws, then there is no room for free will. Humans are robots and religion is a fraud.

Newton was aware of this problem. He had no intention of promulgating a philosophy that stripped humans of free will. He seems to have got around it by positing supernatural intervention, i.e., by recourse to entities and powers that lay outside the system described by his science. Leibniz's approach, bizarre as it might be in many respects, was, in a sense, more scientific; free will was no longer a problem that needed to be explained away, but an intrinsic feature of every monad.

Monadology spent the next two centuries on the ash-heap of intellectual history. After Leibniz's death, a faulty version was published by one of his disciples, and its errors laid at Leibniz's feet. Then it swam into the gunsights of Immanuel Kant. In his *Critique of Pure Reason*, Kant begins by saying a few complimentary things about Leibniz. Three hundred pages later, having carefully set his pieces out on the board, he annihilates Leibniz's metaphysics in a few sentences. According to Kant's philosophy, Leibniz is correct in thinking that space and time, cause and effect, are not ultimate realities, but rather constructs of mental activity. But by the same token, Kant says, the human mind is powerless to think in any useful or productive way about anything that is outside space and time, cause and effect, and so Leibniz's entire Monadology – or any thinking that attempts to transcend spatiotemporality – is rubbish.

In the day of Newton and Leibniz, metaphysics had been as respectable as mathematics, but the hard-headed empiricists of the scientific world began to kick dirt on it during the nineteenth century and, in the first half of the twentieth, the logical positivists buried it. And indeed, Leibniz's work seems unsound at best, ludicrous at worst, by the scientific standards of the era before relativity, quantum mechanics and Gödel's proof.

Today, metaphysics in general has regained much of its former respectability among philosophers. For almost everyone else, though, it retains the connotations of woolliness that it picked up during that century or so of rough treatment at the hands of empiricists and positivists. Many hard scientists still use 'metaphysics' as a byword for undisciplined, conjectural thinking. Nevertheless, metaphysics is still being practised today: by philosophers openly, by physicists under other names.

A straightforward way of defining metaphysics is as the set of assumptions and practices present in the scientist's mind before he or she begins to do science. There is nothing wrong with making such assumptions, as it is not possible to do science without them. The lepidopterist who records in her notebook that a butterfly is blue may not stop to consider that this is true only because the giant ball of nuclear fuel ninety-three million miles away happens to maintain a surface temperature just right for shedding certain wavelengths of electromagnetic radiation on the Earth; that the eyes of humans have evolved to be sensitive to those wavelengths; that the eye can discriminate slightly different wavelengths as colours; that one of those colours has, by cultural consensus, been defined as 'blue', and so on. Nevertheless, science benefits from the lepidopterist's note that the butterfly is blue.

Even the hardest of hard sciences is replete with assumptions that may fairly be classified as metaphysical. Almost all mathematicians, for example, presume that they are discovering, rather than creating, mathematical truths. Ask a roomful of mathematicians whether three was a prime number a billion years ago (i.e. before there were humans to define it as such) and every hand will go up. And yet to say so is to espouse the metaphysical position that primeness and all the other subject matter of mathematics have a reality independent of the human mind. This assumption goes under various names, one of which is Mathematical Platonism. Likewise, physicists can hardly go about their work without assuming that the physical world answers to laws that may be expressed and proved mathematically – an assumption for which there is plenty of empirical evidence, dating back (at least) to Galileo, but no proof as such.

The revival of Leibniz's fortunes may be dated to approximately 1900, when Bertrand Russell began to publish his studies of Leibniz's unpublished work. While unsparing in his criticisms of Leibniz's character and of his more popular writings, Russell had a high opinion of Leibniz's work on mathematical logic and was fascinated by some of the ramifications of the

Monadology. In his *History of Western Philosophy* (1945) he ends his chapter on Leibniz as follows: 'What I … think best in his theory of monads is his two kinds of space, one subjective, in the perceptions of each monad, and one objective, consisting of the assemblage of points of view of the various monads. This, I believe, is still useful in relating perception to physics.'

Leibniz then came to the attention of a wide range of thinkers. To tell the story in chronological order, including all of the requisite details about those who have knowingly or unknowingly echoed Leibniz's views, would require a substantial book in itself, of which the following might serve as a brief sketch or outline.

1. The debate on free will vs. determinism is no more settled today than it was at the time of the Leibniz–Clarke correspondence, and so in that sense (at least) Monadology is still interesting as a gambit, which different observers might see as heroic, ingenious, or desperate, to cut that Gordian knot by making free minds or souls into the fundamental components of the universe.

2. Leibniz's interpreters made use of the vocabulary at their disposal to translate his terminology into words such as 'mind', 'soul', 'cognition', 'endeavour', etc. This, however, was before the era of information theory, Turing machines and digital computers, which have supplied us with a new set of concepts, a lexicon, and a rigorous science pertaining to things that, like monads, perform a sort of cogitation but are neither divine nor human. A translator of Leibniz's work, beginning in AD 2010 from a blank sheet of paper, would, I submit, be more likely to use words like 'computer' and 'computation' than 'soul' and 'cognition'. During Leibniz's era, the only person who had thought seriously about such machines was Leibniz himself; building on earlier work by Blaise Pascal, he designed, and caused to be built, a mechanical computer, and envisioned coupling it to a formal logical system called the *Characteristica Universalis*. He invented binary arithmetic, and, according to no less an authority than Norbert Wiener, pioneered the idea of feedback.

3. In particular, the monads' production rule scheme clearly presages the modern concept of cellular automata. Quoting from Mercer's work:

The Production Rule of F is a rule for the continuous production of the discrete states of F so that it instructs F about exactly what to think at every moment of F's existence. Following Leibniz's suggestion, if F exists from t_1 to t_n and has a different thought at each moment of its existence, then at every moment, there will be an instruction about what to think next. The present thought occurring at t_1, together with the Production Rule, will determine what F will think at t_2.

Combined with the monadic property of being able to perceive the states of all other monads, this comes close to being a mathematically formal definition of cellular automata, a branch of mathematics generally agreed to have been invented by Stanislaw Ulam and John von Neumann during the 1940s as an outgrowth of work at Los Alamos. The impressive capabilities of such systems have, in subsequent decades, drawn the attention of many luminaries from the worlds of mathematics and physics, some of whom have proposed that the physical universe might, in fact, consist of cellular automata carrying out a calculation – a hypothesis known as Digital Physics, or It from Bit.

4. Leibniz insisted that each monad perceived the states of all of the others, a premise that runs counter to intuition, given that this would seem to require that an infinite amount of information be transmitted to and stored in each monad. Of all the claims of Monadology, this must have seemed the easiest to refute a hundred years ago. Since then, however, it has been given a new lease on life by quantum mechanics. Consider, for example, the Pauli exclusion principle, which states (for example) that in a helium atom with two electrons in the same orbital, the two must have opposite spins. It is not possible for both of them to possess exactly the same state. Each of the two electrons somehow 'knows' the direction of the other's spin and 'obeys' the rule that its spin must be different. The Pauli exclusion principle is Leibniz's identity of indiscernibles principle translated directly

into physics. Moreover, the ability of an electron to 'know' the state of another electron, without any physical explanation as to how this information is transmitted and stored, is strongly reminiscent of Monadology. Elementary descriptions of quantum mechanics tend to limit themselves to extremely simple systems, such as individual particles or atoms, since beyond there the mathematics becomes intractable. But the same principles apply, albeit in vastly more complex form, in larger systems: the quantum state of each particle is dependent upon the states of all the other particles in the system.

5. Leibniz's notion that the ultimate entities in the universe were non-spatiotemporal received a kind of weak boost from general relativity, which called into question the idea of absolute space and time as a fixed lattice on which the laws of physics were enacted. More recently, absolute space and time have come under more concerted attack as some physicists have sought to develop so-called background-independent theories. The idea of background independence is explained in more detail in Lee Smolin's *The Trouble with Physics*, and the history of the concept of absolute space and time, from the Babylonians forwards, is told by Julian Barbour in his magisterial *The Discovery of Dynamics*. That space and time have an absolute reality, and that the laws of physics must be hung on a fixed spatiotemporal lattice, are metaphysical assumptions. Very reasonable, empirically grounded assumptions to be sure, but assumptions nonetheless. Resulting theories are called background-dependent. Various efforts have been made to derive background-independent theories that make no assumptions as to the fundamental reality of space and time. Barbour in particular has done seminal work along these lines, showing that general relativity is a realisation of a relational, i.e. Leibnizian, view of space and time. More recently, other researchers, notably Smolin, have sought to unify Barbour's formulation of general relativity with quantum mechanics, the aim being to develop a background-independent theory of quantum gravity according to which space

and time are emergent properties resulting from interactions of more fundamental entities joined together in a graph of connections. This theory, which is called loop quantum gravity, is proposed as an alternative to string theory, which is background-dependent.

6. The Leibnizian concept of pre-established harmony was viciously mocked by Voltaire in *Candide*, and has become no easier for sophisticated people to accept since then. Stripped of its theological overtones and saccharine connotations, though, the concept has a reasonably clear analogue in modern physics.

a) Newtonian mechanics exactly describes the behaviour of individual bodies (provided, as Einstein later discovered, that they are reasonably large and slow-moving). Its laws are expressed in terms of individual particles: a particle moves in a straight line unless acted upon by a force. The force acting on a particle is equal to the product of its mass and acceleration ($F = ma$). As any first-year physics student learns the hard way, naïvely using the $F = ma$ approach to describe systems comprising many independent parts soon becomes mathematically intractable.

b) Leibniz is credited with having written down the law now known as conservation of energy (which he denoted *vis viva*). In any system of particles, the product of the mass and the square of the velocity of each particle, summed over all of the particles in the system, remains constant. When this, and the law of conservation of momentum, are imposed as constraints on a system, the mathematics frequently gets easier, to the point where it becomes possible to produce results not obtainable otherwise. Conservation of energy does not contradict Newton's laws, and, in fact, is derivable from them, and so from a strictly mathematical point of view it adds nothing to Newtonian physics. It does, however, introduce a different way of thinking about physical systems. The naïve reductionist strategy of the first-year physics student gives way to a global approach in which the system as a whole must obey certain rules, to which the detailed movements and interactions of its components are seen as subordinate.

c) The physicists of the late eighteenth and early nineteenth century developed new tools based on the notion of state or configuration spaces framed not of spatial dimensions but of all the generalised coordinates and momenta needed to specify the state of the system. Any possible state can be represented as a point in that space, and its evolution over time as a trajectory. The behaviour of such trajectories is governed by an 'action principle' that encodes all of the applicable physical laws, such as conservation of momentum and of energy. Action principles in classical state space are a mathematical reformulation of Newton's laws, not an alternative to them. The change in point of view from physical trajectories in Cartesian space to action in state space is nonetheless significant. It is a further step away from the reductionist and toward the global approach. It seems to inject a teleological aspect that is not present in the older formulation, and so has occasioned some introspection among philosophically inclined scientists. In his *Lectures on Physics*, Richard Feynman interpolated a single, anomalous chapter on the topic, simply because of his abiding fascination with it. It allows the physicist to predict the behaviour of a complex system without having to work out the detailed interactions among its physical atoms. It leads to important results from thermodynamics and it is directly applicable to quantum mechanics. It is a way of thinking, systematically and rigorously, about compossibility, a concept important to Leibniz. Many possible states of affairs might exist or, to put it another way, there are an infinite number of possible worlds. But not all states of affairs are *compossible*; some are mutually contradictory, and while it is possible to imagine a universe in which contradictory states of affairs coexist, it is not possible for such a universe to come into practical being. The configuration space that describes the universe contains an infinity of points, each of which represents a different state of affairs, but most of these are incoherent. Only certain points – certain universes – make sense internally, and those points lie on trajectories that describe the logical evolution, according to physical law, of those universes over time. If one adopts this frame of reference for considering Leibniz's concept of the pre-established harmony, and excludes (or at least adopts an agnostic stance toward) the notion that it was all set up at the beginning by God, it is easier to come to grips with Leibniz's idea that the monads act in a coherent way somehow transcending detailed cause-and-effect interactions.

d) That much is true of classical (i.e. pre-quantum) state space theory, even though it adds nothing beyond Newton's original laws. The quantum version of the theory, on the other hand, requires that actions over all possible worlds be brought together in a calculation yielding the probability that any one state of affairs will eventuate. As Feynman puts it, 'It isn't that a particle takes the path of least action but that it smells all the paths in the neighbourhood and chooses the one that has the least action …' The picture is reminiscent of Leibniz's 'best of all possible worlds'.

7. Possible-world theory has come in for serious study in recent decades both by philosophers and physicists. For impressively technical reasons that are likely to leave lay readers nonplussed, David Lewis (*Plurality of Worlds*) posited that all possible worlds really exist and are no less real than the one we live in. Such notions are the subject of current philosophical research, under the rubrics of modal realism and actualist realism. Among physicists, Hugh Everett launched the many-worlds interpretation of quantum

mechanics in the late 1950s, since which time it has slowly but steadily garnered support. A particularly eloquent latter-day treatment can be found in David Deutsch's *The Fabric of Reality*.

8. Kurt Gödel (1906–1978) who early in his life became known as 'the greatest logician since Aristotle' because of his astonishingly original work on the foundations of mathematics, devoted much of the second half of his life to the development of a rigorous metaphysical system that was to be based upon the work of Leibniz, with whom he had a fascination that became notorious.

Kurt Gödel.

Gödel was a strong mathematical Platonist who thought in a serious way about the notion that the entities that are the subject matter of mathematics really exist, though not in our physical universe, and that when we do mathematics we in some sense perceive those entities. An almost painfully meticulous scholar, he was well aware of Kant's objections to Leibniz's metaphysics, and understood that those objections would have to be dealt with in order for him to make any progress. According to his friend and biographer Hao Wang, Gödel discovered the works of Edmund Husserl (1859–1938) in the late 1950s and devoted much of the remainder of his life to studying them. He felt that Husserl had solved many, if not all, of the metaphysical problems that Gödel had set for himself, including doing away with Kant's objections to Leibniz's work. Husserl is prolix, prolific and infamously difficult to read (even Gödel complained of this) and so a reader of sub-Gödel IQ, eyeing a heap of Husserl translations on a table, might despair of ever putting his finger on the passages that Gödel is thinking of. Fortunately, Hao Wang did us the favour of listing the specific Husserl books that Gödel most admired. One of them is *Cartesian Meditations*, based on a series of lectures that Husserl delivered late in his career. In the fifth and last of these, Husserl gets around to mentioning, in an approving way, Leibniz and monads. Husserl has come round to Leibniz's way of thinking, but he has got there by taking a different route, pioneered by Husserl, through phenomenology – the premises and development of which I'll spare the reader. Since Gödel's death, mathematical Platonism has come in for serious study both by philosophers such as Edward N. Zalta, a metaphysician at Stanford University, and scientists such as Max Tegmark, an MIT cosmologist. Zalta and Tegmark (like Deutsch) have been influenced by David Lewis' work on modal realism. Beginning from different premises, they have arrived at markedly similar approaches.

***

None of these latter-day echoes of Leibnizian thinking has generated traceable, exact results in the same way that, for example, Newtonian mechanics was able to predict the orbit of the Moon. If such a thing happens in the future – if, for example, the practitioners of loop quantum gravity use their theory to make predictions that are verified by experiment – then credit will have to go to them and not to Leibniz, who could never have imagined such a science. It's not the point of this chapter, in other words, to argue that Leibniz was right, much less that Newton was wrong. Leibniz was not even doing science as we now define the term. My conclusions are two. First of all, that the infamous duel between Newton and Leibniz – which was only superficially about who had invented the calculus – came back from the dead a hundred years ago to exert remarkable influence over the course of modern science. Secondly, that Leibniz's most fundamental assumption, namely that the universe makes sense and that the human has the power to make sense of it and that, consequently, pure metaphysics is no waste of time, remains perhaps the central question of all science. In 1960, Eugene Wigner wrote a paper, *The Unreasonable Effectiveness of Mathematics in the Physical Sciences*, in which he addressed the nearly miraculous way in which pure mathematics – seemingly a product of human cognition, and nothing else – predicts the behaviour of the physical world. The examples cited by Wigner would have made sense to Leibniz. Leibniz, however, would have been baffled by Wigner's use of the adjective 'unreasonable' in the title of his paper. Wigner was a modern: a product of a sceptical age. He was uneasy (or felt obliged to pretend to be uneasy) with the philosophical implications of the way in which the physical world answered to mathematics. This unease could not have been more alien to Leibniz, who, during his long philosophical career, questioned many things that would have been easier to leave alone, but believed, with a kind of medieval serenity, in the reasonableness of Creation.

REBECCA NEWBERGER GOLDSTEIN

What's in a Name? Rivalries and the Birth of Modern Science

Rebecca Newberger Goldstein is a philosopher and a novelist. Her work, both literary and scholarly, has received numerous prizes, including a MacArthur 'genius' award. Her non-fiction works are *Incompleteness: The Proof and Paradox of Kurt Godel* and *Betraying Spinoza: The Renegade Jew Who Gave Us Modernity.* Her fiction includes the novels *Properties of Light: A Novel of Love, Betrayal and Quantum Physics, Mazel, The Dark Sister* and *The Mind-Body Problem,* as well as a volume of stories, *Strange Attractors.* Her latest book is *36 Arguments for the Existence of God: A Work of Fiction.*

A MID THE DIVISIONS IN THOUGHT WHICH MARKED THE SCIENTIFIC REVOLUTION, THE FOUNDERS OF THE ROYAL SOCIETY INSISTED ON BINDING TOGETHER TWO CONTENDERS FOR THE BASIS OF NATURAL EXPLANATION. AS REBECCA GOLDSTEIN EXPLAINS, THERE WERE DEEP COMMITMENTS TO THE PRIMACY OF EXPERIMENTAL RESULTS, AND TO RECOGNISING UNDERLYING MATHEMATICAL PATTERNS. BUT THE REALLY POWERFUL TRICK, THEN AS NOW, LAY IN FINDING HOW TO BRING THEM TOGETHER.

After a lecture given by Christopher Wren, then the Gresham College Professor of Astronomy, twelve prominent gentlemen, deciding that they would meet weekly to discuss science and perform experiments, recorded their intention to form a 'Colledge for the Promoting of Physico-Mathematicall Experimentall Learning'.

It might not have been the most elegant of designations, but it did, in its very wordiness, portend great things. It gave notice to the hope – because it was still, in 1660, only a hope – that two distinct orientations, one mathematical, the other experimental, would be pounded together into one coherent scientific method. The hope paid off, and it was from

A colour engraving
of Gresham
College, home of
the Royal Society
from 1660 to
1710.

within the ranks of the Royal Society that the new compound emerged.
Two cognitive stances that had seemed to have little to do with one another, except in their opposition to the system of natural philosophy dominant
for centuries, were rendered equally necessary in the explanation of physical phenomena.

It was a time of epistemological urgency. A grandly unifying cathedral
of thought was crumbling.[1] The all-inclusive view of the cosmos, laid down
by Aristotle and buttressed by the medical theories of Galen, the astronomy of Ptolemy, and the theology of Christianity, had offered a way of
explaining … absolutely everything. From the falling of objects to the
rising of smoke; from generation and decay to the four basic personality
types; from the relation between body and soul to the pathways of the planets; the supposed nature and reason for every aspect of the world could be
extracted from an interlocking system that employed a homogeneous form
of explanation throughout.

1 There were, of course, political and sociological dimensions to this process, since the grandly unifying system of thought was
not only scientific (or proto-scientific) but also religious and political, making challenges to the system ipso facto religious and
political challenges. I will focus on the scientific aspects of the process, but it is of course naïve to think that this constitutes
the whole story. The history of ideas is hardly hermetically sealed against all but questions of validity and falsification.

The form of explanation had been purpose-driven, or teleological, and its scaffolding was the metaphor of human action. We explain human actions by citing the end state that the agent has in mind in undertaking it. The old system took this familiar model of explanation and expanded it to apply to the world at large. 'To be ignorant of motion is to be ignorant of nature,' Aristotle had written, but by motions he meant not just displacements of bodies but such processes as becoming a parent, gaining knowledge, growing older. All were subsumed under the same conception: a striving to actualise an end state that was implicit in the motion and provided the explanation, the final cause, for the course that the motion took. The explanatory logic of human actions – based on intentions – was one with the explanatory logic of the cosmos.

The working hypothesis behind teleology was, of course, that all natural phenomena and processes do in fact have goals, allowing them to be viewed as potentialities on the way to being actualised. But every form of explanation makes use of some working hypothesis or other, ascribing to nature the features that allow such explanation to work. The mode of physical explanation that was to supplant teleology, making essential use of mathematics, also staked its claim on the world's being a certain way.

We are today understandably prepared to believe that the only reasons anyone might have had to cling to the old crumbling teleological cathedral, in light of the superior science battering it, were speciously theological; and, in fact, such reasons probably did motivate most of those who clung to the old system. Still, there was nothing a priori fallacious about the old system's assumptions about reality, just as there was nothing a priori true about the assumptions that would replace them.

The grand old system was crumbling, and it made for a capacious space into which genius could expand. When foundations fall, everything can and must be rethought. The exhilaration on display in the writings of the new scientists bears witness to how bracingly liberating such possibilities can be, at least for those with the intellectual imagination and bravado to take advantage of them. 'You cannot help it, Signor Sarsi,' Galileo exults in *The Assayer*, written in the form of a letter to a friend, 'that it was granted to me alone to discover all the new phenomena in the sky and nothing to anybody else.'

EXPLANATION RE-EXPLAINED

And what question is more foundational than the question of what counts as a good explanation? All the great men whom we now associate with the formation of modern science – Copernicus (1473–1543), William Gilbert (1544–1603), Francis Bacon (1561–1626), Galileo Galilei (1564–1642), Johannes Kepler (1571–1630), William Harvey (1578–1657), René Descartes (1596–1650), Robert Boyle (1627–1691), John Locke (1632–1704) and Isaac Newton (1643–1727) – were intensely involved with the question of what form explanation ought to take, if teleology was truly to be abandoned, and there was by no means a consensus among them. Two different orientations emerged: one rationalist, stressing abstract reason, the other empiricist, stressing experience.

In some sense, this cognitive split was nothing new. It had made itself felt in the ancient world, in the distinction between the Platonists and the Aristotelians. It is probably as old as thought itself, shadowing two distinct intellectual temperaments. But the new rationalist and empiricist orientations were not like the old. The rationalist orientation looked to mathematics to provide the new mode of explanation. The empiricists saw the new scientific method as emerging out of experimentation. In responding to the need for a new mode of explanation to take the place of teleology, they became epistemological rivals, offering competing models to take the place of the old system's final causes.

The men who met in Gresham College, London,[2] had given notice, in their self-baptism, that the mathematical and experimental approaches were not only compatible but collaborative; even, as it were, *one*. There is

---

2 Many of these same scientifically inclined men had begun meeting earlier, in Oxford, at the end of the 1640s, during the Parliamentary Interregnum between the reigns of Charles I and Charles II, laying the foundations for what would become, with the Restoration, the Royal Society, their ranks now swelled not only by Royalists, but the King himself.

Nicolaus Copernicus

Francis Bacon

Galileo Galilei

Johannes Kepler

William Harvey

René Descartes

Robert Boyle

John Locke

Isaac Newton

Opposite:
Some of the great
men associated
with the formation
of modern science.

an important epistemological claim implicit in their stated intention to promote 'physico-mathematicall experimentall learning', and the claim was by no means demonstrable in 1660. The thinkers whose work inspired them could be divided into those whose stance was slanted toward the new rationalist understanding of physical explanation – Copernicus, Kepler, Galileo, Descartes – and those who espoused the experimental understanding of physical explanation – Francis Bacon, William Gilbert and William Harvey. This list suggests a geographical divide, with the rationalists on the Continent, the empiricists in England, which makes the ecumenicalism of the sources of inspiration all the more noteworthy.

The temperamental distinction between the mathematical rationalists and experimental empiricists could be, in fact, so marked that we can well wonder how these scientific founders made common cause with one another against the old system. How can such different scientific temperaments, proffering such different answers as to what a scientific explanation ought to look like, have conspired to hammer out the new methodology?

William Gilbert, for example, a luminary of the experimental approach, is acknowledged as the founder of the science of magnetism, and his experiments had been ingenious. He had carved out of a lodestone – a piece of naturally magnetic mineral – a scale model of the Earth he called his terrella, or little Earth, and with it he had been able to explain a phenomenon that had been known for centuries. A freely suspended compass needle pointed North, but later observations had revealed that the direction deviated somewhat from true North, and Robert Norman had published his finding in 1581 that the force on a magnetic needle was not horizontal but slanted into the Earth. Passing a small compass over his terrella, Gilbert demonstrated that a horizontal compass would point toward the magnetic pole, while a dip needle, balanced on a horizontal axis perpendicular to the magnetic one, indicated the proper 'magnetic inclination' between the magnetic force and the horizontal direction. The experiments convinced him that the Earth itself was a giant magnet. Galileo, his contemporary, commends his work, but criticises him for not being well-grounded in mathematics, especially geometry.

Galileo, for his part, could be high-handed in regard to experimentation, writing, for example, that it was only the need to convince his ignorant opponents that made him resort to 'a variety of experiments, though to satisfy his own mind alone he had never felt it necessary to make any'.[3] As one historian of science has written, 'If this was seriously meant, it was extremely important for the advance of science that Galileo had strong opponents, and in fact there are other passages in his works which show that his confident belief in the mathematical structure of the world emancipated him from the necessity of close dependence on experiment.'[4]

The two orientations, rationalist and empiricist, were partly defining themselves in opposition to one another, becoming far more adversarial now that the old system was crumbling. That system had blended together both a priori reason and empirical observation, conceiving both as co-dependently involved in scientific explanation. Aristotle had been a biologist, much given to observing the natural world, and the system that had grown up on Aristotelian foundations had always striven to take account of observable facts. So, for example, as more precise observations of the 'wandering' planets were made, a vast complexity of interacting celestial gears, the ever more torturous epicycles and eccentrics, was sketched to accommodate them into the geocentric picture which was an essential part of the old system's teleology. In *Paradise Lost*, John Milton speaks of 'Sphere/With Centric and Eccentric scribbled o'er,/Cycle and Epicycle,

3 *The Scientific Works of Galileo* (Singer, Vol. II, p. 252).

4 E.A. Burtt, *The Metaphysical Foundations of Modern Physical Science* (NY, Prometheus Books, 1999), p. 76. Galileo's rationalist attitude has been echoed by various modern physicists. Paul Dirac, for example, said: 'It is more important to have beauty in one's equations than to have them fit experiments,' and Einstein, too, made such remarks, for example telling Hans Reichenbach that he had been convinced before the 1919 solar eclipse gave confirming evidence that his theory of general relativity was true because of its mathematical beauty. In our day, the hegemony of mathematics has been claimed most insistently by champions of string theory, which has as yet been unable to produce any testable predictions. 'I don't think it's ever happened that a theory that has the kind of mathematical appeal that string theory has has turned out to be entirely wrong,' Nobel laureate Steven Weinberg has said. 'There have been theories that turned out to be right in a different context than the context for which they were invented. But I would find it hard to believe that that much elegance and mathematical beauty would simply be wasted.' (Quoted on *Nova, The Elegant Universe*. http://www.pbs.org/wgbh/nova/elegant/view-weinberg.html.) String theory has been criticised by more empirically inclined physicists, some going so far as to claim the theory does not even qualify as scientific. Thus the schism between scientific rationalists and empiricists continues into our own day.

Orb in Orb'. Such complexity was demanded because of ongoing observation. Aristotelians were not given to ignoring the observable facts. Quite the contrary: they observed processes so as to be able to read out of them the narratives of potentiality actualised.

Then again, Aristotle was also a logician, who had laid down the laws of the syllogism. According to Aristotle, logical demonstration, by way of the syllogism, was a necessary component of *epistêmê*, or scientific knowledge. In his *Posterior Analytics*, he says that scientific knowledge requires that we know the cause 'of why the thing is', and also know that it could not have been otherwise. In other words, scientific knowledge not only must discover causes but demonstrate that they are necessarily the causes, and it is the abstract science of the syllogism that is assigned the latter demonstrative role.

However, both rationalism and empiricism, as they emerged in the seventeenth century, were of an entirely different kind from their counterparts in the old system. The scientific rationalism of Copernicus, Galileo, Kepler, and Descartes had little use for the Aristotelian syllogism, which, so they argued, cannot expand our knowledge but merely rearrange it to set off implicit logical relations. Logic may be perfect, but it is also perfectly inert, incapable of moving substantive discovery forward. For the new scientific rationalists, it is not syllogistic logic but rather mathematics that holds an incomparable active power, capable of generating new knowledge. 'We do not learn to demonstrate from the manuals of logic,' Galileo wrote, 'but from the books which are full of demonstrations, which are mathematical, not logical.' A priori reason in the form of mathematics provides a methodology for discovery. As Galileo was to put it ringingly in *The Assayer*:

> Philosophy is written in this vast book, which continuously lies upon before our eyes (I mean the universe). But it cannot be understood unless you have first learned to understand the language and recognise the characters in which it is written. It is written in the language of mathematics, and the characters are triangles, circles,

and other geometrical figures. Without such means, it is impossible
for us humans to understand a word of it, and to be without them is
to wander around in vain through a dark labyrinth.

It was, more than anything else, the new mathematical conception of the
physical universe that had hastened the crumbling of the old explanatory
system. Copernicus had urged his heliocentric model of the solar system
not on the basis of its empirical superiority – both the geocentric and the
heliocentric pictures could accommodate the data – but on the basis of its
mathematical superiority:

> Nor do I doubt that skilled and scholarly mathematicians will agree
> with me if, what philosophy requires from the beginning, they will
> examine and judge, not casually but deeply, what I have gathered
> together in this book to prove these things … Mathematics is writ-
> ten for mathematicians, to whom these my labours, if I am not
> mistaken, will appear to contribute something.[5]

Under Galileo, the mathematical conceptualising of nature was radically advanced. He took the concept of motion, agreeing with Aristotle that it is the object of scientific explanation, and he reconfigured it into terms that can be expressed precisely in numbers. Distance travelled is quantifiable, as is time elapsed; and, from Galileo onward, motion is conceived of as a comparison between these two factors, the change of distance and the passing of time. Once motion itself had been reconfigured as a mathematical concept, other concepts, which are functions of motion, can be mathematically defined, so that, by developing the equations between the various functions of mathematical motions, new properties can be uncovered. The mathematical expression of the physical allows for what logic could never accomplish: the generation of new descriptions, going beyond the observable. It is the relations between these mathematical properties which, expressed as equations, remain constant between instances, yielding universal laws of nature. And it is these laws that supplant teleology in the new conception of explanation.

A priori mathematics, according to Galileo, does not entirely obviate the need for observation (only the most extreme of rationalists, Spinoza and Leibniz, were to argue the expendability, at least in principle, of all empirical knowledge, claiming that all could be a priori deduced from first principles[6]); but mathematics *does* allow us to deduce unobservable properties and thus to penetrate into the structure of nature.

Of course, this meant that not all of the processes conceived of as motions by Aristotle were Galilean motions. Only motions susceptible to mathematical translation came under the purview of science; the rest were expelled from the possibility of physical explanation. Even more than this, Galileo, and those who followed him, defined physical nature itself in terms of mathematics. It was Galileo who first drew the distinction between primary and secondary qualities. If all aspects of physical reality are mathematically expressible, and if not all aspects of our experience are susceptible to mathematical treatment, the implication is that

5 From his Letter to Pope Paul III, in the *De Revolutionibus*.
6 See footnote on page 119.

not all aspects of our experience are physically real. Our minds contribute to what we seem to see out there in the world. Our experience is not transparent; there is a gauzy veil of subjectivity hung between us and the objective physical world of mathematical bodies, compounded out of mathematically arranged mathematical constituents, mathematically moving through mathematical space over the course of mathematical time. All those aspects of our experience that can be rendered in mathematical language are 'primary' and correlated with what is out there; the rest are 'secondary' qualities, features of our subjective experience, caused by the interaction between the primary qualities out there and our own sensory organs. This distinction was widely accepted, not only by rationalists like Galileo and Descartes, but empiricists like John Locke. The portions of *res cogitans* lurking in our cerebral hemispheres provide a sanctuary for the otherwise inexplicable flotsam and jetsam of perception.

Scientific rationalism, then, as it emerged to challenge the old system, placed its hopes not in logic but in mathematics. Whereas the old system's working hypothesis had been that all physical processes are striving toward an end they seek to accomplish, the working hypothesis of the new rationalists was that all physical processes have a quantitative structure, and it is this abstract structure that distils the laws of nature that provide their explanation. As the über-rationalist Spinoza was to express it:

> Thus the prejudice developed into superstitions, and took deep root in the human mind; and for this reason everyone strove most zealously to understand and explain the final causes of things; but in their endeavour to show that nature does nothing in vain, i.e. nothing which is useless to man, they only seem to have demonstrated that nature, the gods, and men are all mad together … Such a doctrine might well have sufficed to conceal the truth from the human race for all eternity if mathematics had not furnished another

standard of verity in considering solely the essence and properties of
figures without regard to their final causes.[7]

But what of the new empiricism? How was it in opposition to the old
system? Aristotle may not himself have thought much of mathematics, but
he was himself an empiricist, who took observation, most especially of
biological organisms, very seriously; it was his mathematical-maniacal
teacher, Plato, who dismissed sense-data (and many of those in the
Copernicus–Kepler–Galileo camp were neo-Platonists). But Aristotle and
the grand cathedral of thought that was erected around him advocated a
passive form of observation. Nature, working always with its own ends in
view, the very ends which provide the explanation in terms of final causes,
was not to be interfered with. Teleology trumped technology. The very
windingness of the roads of Europe's medieval cities testifies to the old
system's hands-off approach toward nature. These roads were laid out on
paths the rain took as it rolled down inclines. To transpose our own pathways over nature's choices was a violation of the fundamental assumption
of the old system. One must respectfully observe the motions of nature,
since their course had been plotted by their implicit end states, and it is in
the hands-off observation that the explanation emerges.

The new empiricism, in seeking its non-teleological form of explanation, took an aggressively interventionist attitude toward observation. In
doing so it not only asserted its rejection of Aristotelianism, of the teleology
that dictated passive observation; its new active observation, in the form of
experimentalism, claimed to present a new science, a scientia operativa,
that could supplant the old.

7 *The Ethics*, I, Appendix. Some of the new rationalists, such as Descartes, Spinoza and Leibniz, argued that what was generative in mathematical reasoning need not be confined to the quantitative, but could range beyond, and thus give us a form of
explanation so powerful as to obviate any need for observation at all. This belief caused them to attribute unlimited potency to
a priori reason, and explains why they are now more characteristically classified as philosophers rather than scientists. But in
their day there was no segregation between the two types of thinkers, philosophers all, and they all saw themselves as engaged
in the same project of finding the mode of explanation to supplant teleology. A rationalist extremist like Spinoza was as engaged
as any in the scientific project; indeed, he was in close communication with the Fellows of the Royal Society, through his
communications with the indefatigably gregarious first secretary, Henry Oldenburg, and even offered, through Oldenburg, his
critique of some of Boyle's ideas, in several instances not finding them sufficiently scientific. So, for example, in *De Fluditate*
19, Boyle wrote of animals that 'Nature has designed them both for flying and swimming,' which provoked from Spinoza the
response, 'He seeks the cause from purpose' (*causam a fine petit*), which is, of course, a relapse to the old system.

The empiricist Bacon, just like the rationalist Galileo, believed that the experience we are presented with does not reflect nature as it is: 'For the mind of man is far from the nature of a clear and equal glass, wherein the beams of things should reflect according to their true incidence; nay, it is rather like an enchanted glass, full of superstition and imposture, if it be not delivered and reduced. For this purpose, let us consider the false appearances that are imposed upon us by the general nature of the mind …'

Bacon's solution to how to circumvent these false appearances, which he called the 'idols of the cave', lay in his empirical activism. We are not to stand passively by as submissive observers of what nature might offer of itself, but assert ourselves in the gathering of facts through experiment. This assertion is what transforms sense-data, subject to illusion, into facts. The keen but passive gazing that makes sense under the assumptions of teleology made no sense to Francis Bacon.

The Lord Chancellor's metaphors are telling. Nature should be looked on as an uncooperative witness in a courtroom, who must be interrogated and even tortured in order that the information be extracted. Nature should be treated as a slave who must be 'constrained' and 'moulded' and compelled to serve man. We must 'shake her to her foundations'. In short, we force the sense-data to yield up the factual data that nature is actively keeping from us by asserting our own active power over nature in controlled experiments.[8] (Although sometimes these experiments end in nature asserting its power over us: the legend is that Francis Bacon died after contracting pneumonia while undertaking some experiments in the dead of winter on the preservation of meat by freezing.)

Thus for both the new rationalists and the new empiricists there was a veil of subjectivity separating the observer from the observed. In this way the two orientations, no matter how distinct their intellectual temperaments, shared a central attitude that went beyond their mere opposition to the old system and explains why they were, even if rivals, also potential

8 The metaphors of Francis Bacon are a feasting ground for feminist readings of the history of science.

allies. Both insisted, against the old system, on more assertiveness. Mathematics, as opposed to inert logic, inserted a generative power into physical description. Experiments, as opposed to passive observation, allow us to wrest the physical facts from illusory experience.

The old system had seen nature as eminently readable by us. The form of explanation spread throughout the cosmos was one which was familiar and natural to us; after all, it was an essentially human form of explanation, taking the sort of explication that applies to human actions and generalising it. The old system saw us as *of* the universe. There was no reason to suspect our experience, and Aristotle was an unguarded empiricist, an observer who never seemed to worry about what his own mind might be contributing to perception. But not so the post-teleology Baconian empiricist, no more than the post-teleology Galilean rationalist. For both, the experience we have of the world has to be subjected to special treatment in order for reliable information to be extracted.

OF ENDS AND MEANS

The activist empiricism of Bacon was correlated with a practical stance toward scientific knowledge, which blazed forth into utopian zeal:

> I humbly pray … that knowledge being now discharged of that venom which the serpent infused into it, and which makes the mind of man to swell, we may not be wise above measure and sobriety, but cultivate truth in charity … Lastly, I would address one general admonition to all; that they consider what are the true ends of knowledge, and that they seek it not either for pleasure of the mind, or for contention, or for superiority to others, or for profit, or fame, or power, or any of these inferior things; but for the benefit and use of life; and that they perfect and govern it in charity. For it was from the lust of power that the angels fell, from lust of knowledge that

man fell; but of charity there can be no excess, neither did angel or man ever come in danger by it.[9]

Here, too, on this question of the 'true end of knowledge' a temperamental difference parts the new rationalists and empiricists. A Galileo or Descartes would not have been as inclined to archly dismiss 'pleasure of the mind' or 'lust of knowledge' as Bacon had been. Though the scientific rationalists and scientific empiricists might share the belief that experience must be subjected to special treatment to be rendered profitable for science, they had differing views on the profit of science. The experimental/empiricists (Gilbert, Harvey) tended to agree with Bacon's practical goals. As men must experimentally assert their power over nature, so, too, the value of possessing nature's secrets was that they be utilised for the practical improvement of men's lives. For the mathematical/rationalists the knowledge was sufficient unto itself, a thing deserving to be desired, whether it yielded practical improvements or not.

By 1660, the mathematical understanding of physical explanation could not be ignored, not with the work of people like Copernicus, Galileo and Descartes; and the men who came together to form a Colledge for the Promoting of Physico-Mathematicall Experimentall Learning acknowledged the mathematical conception of the physical in their self-designation. Nevertheless by temperament these early men of the Royal Society were more allied with Bacon, Gilbert and Harvey than with Galileo and Descartes. It was the 'experimentall learning' that most engaged them, and so, too, they were inclined to embrace the practical humanitarian goals of science that Bacon had linked with his experimentalism.

Christopher Wren gave the inaugural lecture at Gresham College, after the Royal Society had been officially formed in 1662, and in his address he spoke passionately of the manner in which the new thinking had thrown off the tyranny of the old system of thought, bringing in its stead the freedom of scientific investigation. In the course of his celebratory advocacy he

9 Preface, *The Instauration Magna*, in Bacon, Francis, *The Works*, ed. by J. spedding, R.L. Ellis and D.D. Heath (Houghton Mifflin, 1901), volume IV, 20f.

extolled William Gilbert (chastised by Galileo for his lack of geometry) as the very embodiment of the new science:

> Among the honourable Assertors of this Liberty, I must reckon *Gilbert*, who having found an admirable Correspondence between his *Terrela*, and the great *Magnet* of the Earth, thought, this Way, to determine this great Question, and spent his studies and Estate upon this Enquiry; by which *obiter,* he found out many admirable magnetical Experiments: This Man would I have adored, not only as the sole Inventor of Magneticks, a new Science to be added to the Bulk of Learning, but as the Father of the new Philosophy.

But if any thinker hovered as a guiding spirit over the group it was the thoroughly empiricist Francis Bacon. Bacon had dreamed of a science that would operate in the way of a collaboration, a 'Fellowship' to take the place of individual geniuses working in isolation; it was all of a piece with his utopian ambitions for the new knowledge, and the members of the Royal Society called themselves 'Fellows' in homage to the Lord Chancellor's vision.

And yet intimations of a union between the 'physico-mathematicall' and 'experimentall' there had no doubt been. It is in the chemist Robert Boyle, the most important scientist among the twelve original Fellows, that we can see the two approaches groping somewhat dazedly toward one another. Boyle was certainly, in many ways, a disciple of Bacon – but not in all ways. He preserved an interest in the practical control of nature through knowledge of cases, which had been such a prominent feature in Francis Bacon, and which both men regarded as closely related to the empirical method; and yet he also had been touched by the Galilean spirit. Though not himself a profound mathematician, Boyle was keenly aware that astronomy and mechanics had outstripped chemistry. He was eager to carry chemistry forward by allying it with an atomistic interpretation of matter, and he recognised that mathematics was integral to the atomistic interpretation of physical phenomena.

But he also contended that chemistry, in its vigorous experimentalism, had something to teach the fields of astronomy and mechanics that had been so transformed by its mathematical reconfiguration. These latter endeavours 'have hitherto presented us rather a mathematical hypothesis of the universe than a physical, having been careful to show us the magnitudes, situations, and motions of the great globes, without being solicitous to declare what simpler bodies, and what compounded ones, the terrestrial globe we inhabit does or may consist in'.[10]

Boyle's suggestion is that the new science, as understood by Galileo et al., is all very well and good, but that, in its overly abstract mathematical demonstrations and idealised formulations, it had travelled too far in the direction of apriorism. Robert Boyle is proposing that chemistry, though lagging behind on the theoretical side, might yet have something to offer the fledgling methodology in the way of getting one's hands stained with the stuff of 'the terrestrial globe we inhabit'. His distinction between mathematical and physical hypotheses is important, and we shall see it again. It reveals Boyle's intuition that there was still something missing in the systems of Galileo and Descartes, no matter how impressive they were.

It is relevant that Boyle was a chemist. The example of the alchemists, though they strayed too near to mysticism and magic for Boyle's taste, was not purely negative, for they had defied the old system's passivity toward nature. (Bacon, too, had praised alchemy as a scientia operativa.)

But though Boyle seemed to have sensed the presence of a unified methodology binding together the activist approaches of the new rationalism and new empiricism, he does not manage to bring it forth, perhaps because he himself lacked mathematical muscle.[11] The best that he can offer is a reconciliation wrought by relativism: if what one is after is knowledge of nature then quantitative deductions on the model of Galileo and Cartesianism will yield satisfaction; but if one's aim is control of nature in

---

10 Robert Boyle, *The Works of The Honourable Robert Boyle*, ed. Thomas Birch (6 vols, London, 1672), vol. I, p. 356.
11 Ironically, it was to be the whole-number arithmetical laws of chemical reactions that would provide, some centuries later, the most direct evidence for the atomic theory of matter.

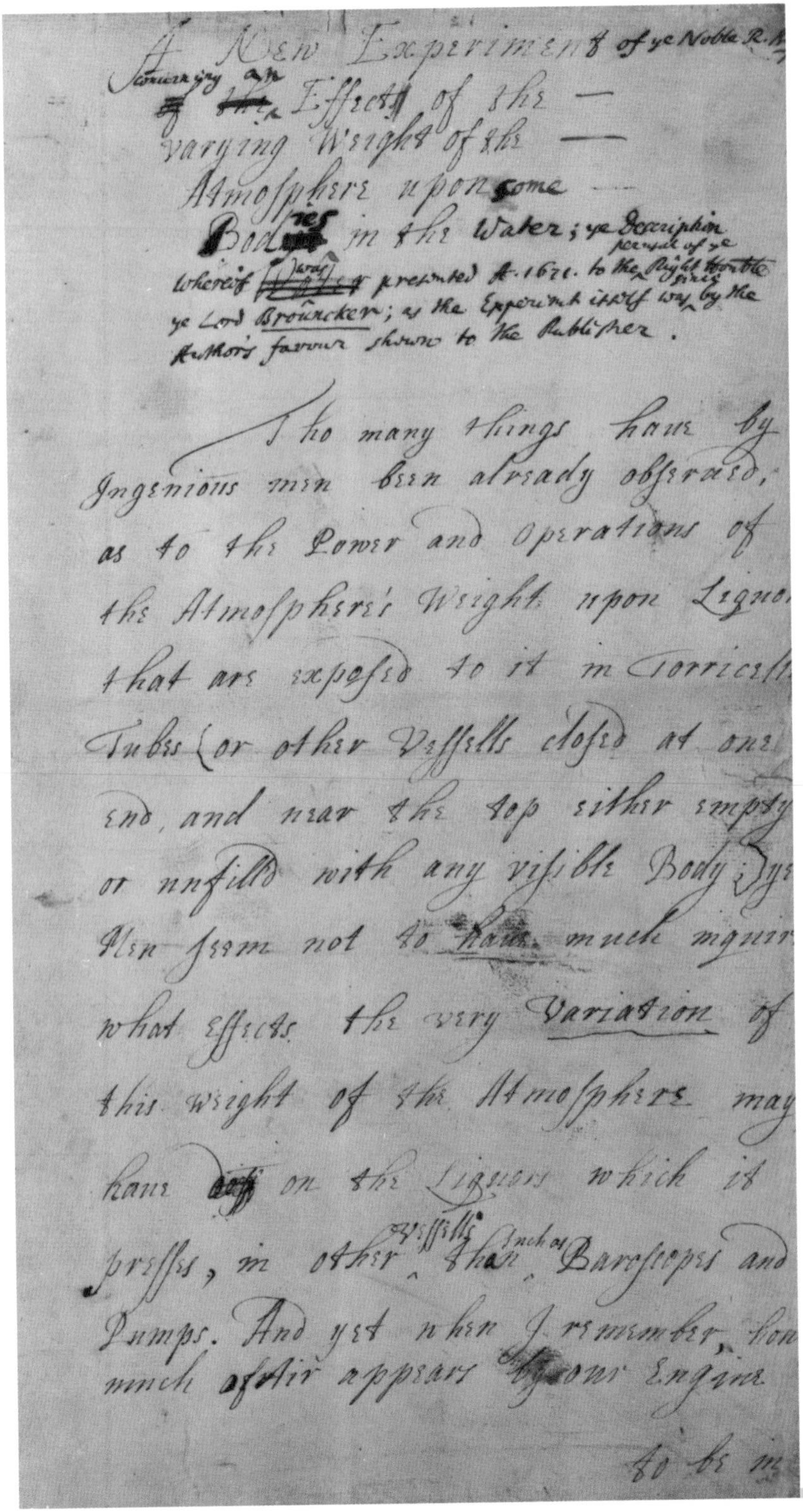

A New Experiment of ye Noble R. b.

concerning an Effects of the —
varying Weight of the —
Atmosphere upon some —
Bodies in the Water; ye Description
whereof was presented A. 1671. to the Right Honble
ye Lord Brouncker; as the Experiment itself was, by the
Author's favour shewn to the Publisher.

Tho many things haue by
Ingenious men been already obserued,
as to the Power and Operations of
the Atmosphere's Weight upon Liquors
that are exposed to it in Torricellian
Tubes (or other Vessells closed at one
end, and near the top either empty
or unfilld with any visible Body) yet
Men seem not to haue much enquird
what effects the very Variation of
this weight of the Atmosphere may
haue on the Liquors which it
presses, in other Vessells than Baroscopes and
Pumps. And yet when J remember how
much of Air appears by our Engine.

to be in

An undated
example of Robert
Boyle's writing on
chemistry –
'Effects of the
varying Weights of
the Atmosphere
upon some bodies
in the Water…'

the interest of particular ends, the necessary relations can often be discovered between qualities immediately experienced or drawn forth from experiments. It all depends on what one wants out of one's science, he writes, although the implication is that true knowledge, if that's what one wants, will require something more deductive than experimental.

The true blending of the two rivals for replacing the teleological understanding of explanation finally arrived in a work whose very title is telling: *Philosophiæ Naturalis Principia Mathematica*, The Mathematical Principles of Natural Philosophy. With Isaac Newton, a scientist who saw mathematics as essential to physical understanding had entered the ranks of the Royal Society. And yet the experimental aspect is also of fundamental importance to his methodology.

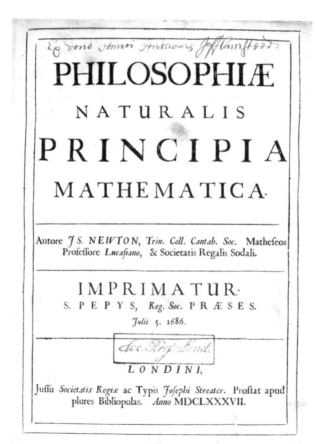

The title page of *Principia*. The inscription is written by John Flamsteed noting that the book is a gift from the author – Newton.

Newton observes in his preface to the *Principia* that 'all the difficulty of philosophy seems to consist in this – from the phenomena of motions to investigate the forces of nature, and then from these forces to demonstrate the other phenomena'. The phrase to 'demonstrate the other phenomena' reiterates the message of the work's title: the fundamental place of mathematics in Newton's method:

> We offer this work as mathematical principles of philosophy. By the propositions mathematically demonstrated in the first book, we then derive from the celestial phenomena the forces of gravity with which bodies tend to the sun and the several planets. Then, from these forces, by other propositions which are also mathematical, we deduce the motions of the planets, the comets, the moon, and the sea.

As it was for Aristotle, so it was for Newton: to investigate nature is to investigate motions. Only, of course, Newton has inherited Galileo's transformed conception of motion, reconfigured by, and restricted to, mathematical expression. The mathematical imagination of Newton, surpassing that of Galileo or Descartes, made possible the mathematical absorption of far vaster reaches of physical phenomena. The language of the Book of Nature is not confined to geometry, as it had been according to Galileo and Descartes; rather it is analysis that becomes the more important means of expressing what is physically relevant. His invention of fluxional calculus afforded him a powerful tool whose operations could not be fully represented geometrically. On the question of mathematical type, Newton is pragmatically flexible, writing in his preface to the *Principia,* 'For you may assume any quantities by the help whereof it is possible to come to equations; only taking this care, that you obtain as many equations from them as you assume quantities really unknown.'

But Newton follows as much in the footsteps of Bacon, Gilbert and Harvey, as in those of Copernicus, Kepler, Galileo and Descartes. This is most sharply brought home by his reiterated denunciation of 'hypotheses'. By hypothesis, Newton means empirically unattached claims about reality, and by his emphatic rejection of 'hypotheses', he is emphasising the necessity of tying scientific statements down to experience. Unlike Galileo or Descartes, Newton distinguishes between mathematical truth and physical truth (echoing the intuition in Boyle's complaint against the rationalists). That the resistance of bodies is in the ratio of the velocity, 'is more a mathematical hypothesis than a physical one', he says in *Principia* II, 9, and makes similar statements in connection with his discussion of fluids (*Principia,* II, 62). A mathematical truth that has not been made manifest in experience has not advanced to a physical truth. And experience must be experimentally manipulated in order for the mathematical truth to be made manifest in it. Galileo and Descartes were right that the mathematical structure that is latent in physical processes provides their explanation; but Bacon, too, had been right that nature requires prodding by way of experimentation in order for the mathematical and the physical to be rendered one.

In fact – and here is where the two anti-Aristotelian strains are finally brought together – it is precisely because ultimate explanation is mathematical, and this mathematical structure is not immediately given up in passively observed nature, that experimentation is necessary. The explanation of the motion is to be found in uncovering the mathematical structure within it; but experience as such does not readily give up the latent mathematical structure. Experiments are necessary to tease out the implicit mathematics, whose consequences can then be mathematically drawn, leading to further mathematical conclusions that must again be tied down to experience by way of experiment.

Newton's work on optics is as instructive as his mechanics, demonstrating both the fundamental place of mathematics and the necessity for experiment. His eagerness to reduce yet another sphere of phenomena to mathematical formulae results in a science of colours. And yet mere observation could not have given Newton the phenomena that would yield to mathematical formulae. His famous interventions – for example, placing two prisms within the path of a light beam, one that would split white light into the spectrum, the other that would reconstruct white light out of the spectrum – were as essential to the science as the resultant mathematical equations. To paraphrase Immanuel Kant (who was three years old when Newton died in 1727): Experimentation without mathematical explanation is blind; mathematical explanation without experimentation is empty.

UNREASONABLE EFFECTIVENESS

Looking back now, there seems something almost accidental about the emergence of both the new rationalism and the new empiricism as coevals, each offering a rival substitute for the disputed teleology of the old system, each appealing to different sorts of intellects, tending toward divergent opinions as regards the ultimate worth and purpose of knowledge. All these centuries later, the methodological amalgamation can still call forth our wonder – most memorably expressed by the late physicist and Nobel laureate Eugene Wigner, in the phrase 'the unreasonable effectiveness of mathematics in the physical sciences'.

It is appropriate to be amazed. Who could have hoped that both the new rationalism and the new empiricism could be joined together in the most successful experiment in human thought to date? Here is a means of exploring nature which, though embedded in the empiricism of experimentation, is also capable of challenging (by way of the theory of relativity) our psychological sense of time, or (by way of quantum mechanics) our notions of causality, two linchpins of common-sense experience.

Who could have hoped? To that question, at least, we have an answer: the men who formed a 'Colledge for the Promoting of Physico-Mathematicall Experimentall Learning'.

...if they meet with your concurrence I think they will add some weight

...view *Virginia* opinion as sent to the Marquis of Rockingham so long

...so in 1784 and with which you have so properly closed your report

...the Board of Ordnance — I am Sir Your Obedient Humble

...gay 24th March 1782 — Servant D Gamble

Fig. 1.

Fig. 2.

SOUTH WEST VIEW of the EAST WING and part of the C...

shewing the junction of the two Conductors that were nearest th...

6

SOCIETAS REGIA LONDINENSIS

NVLLIVS IN VERBA

SIMON SCHAFFER

CHARGED ATMOSPHERES:
PROMETHEAN SCIENCE
AND THE ROYAL SOCIETY

Simon Schaffer is Professor of History of Science at Cambridge University and Trustee of the National Museum of Science and Industry. His books include *Leviathan and the Air Pump: Hobbes, Boyle and the Experimental Life* (with Stephen Shapin) and in 2006 he presented the BBC4 history of science series *Light Fantastic*.

EXPERIMENTS AND MATHEMATICAL DESCRIPTIONS OF THE WORLD SEEM FAMILIAR PARTS OF SCIENCE. A CENTURY AFTER IT GOT GOING, THE ROYAL SOCIETY WAS ALSO DEEPLY COMMITTED TO ANOTHER FAMILIAR OBLIGATION OF NATURAL PHILOSOPHERS: ADVICE TO THE GOVERNMENT. AND AS SIMON SCHAFFER RELATES, IT WAS ALREADY RAISING A VERY MODERN QUESTION – WHEN THE STAKES ARE HIGH, WHOSE EVIDENCE SHOULD BE TRUSTED?

It is not without Reason, that Norwich has been called *the City of Wonders*; if we examine that great Collection of Miracles, the Transactions of the *Royal Society*, we shall find more than ten Times as many strange and wonderful Events dated from this City as from any City of the World. The strangest Things that can be devised are of all others the fittest for the Entertainment of the *Royal Society*.[1]

In search of a key moment in the story of the last 350 years of science and of the Royal Society, I've chosen an eighteenth-century and East Anglian episode of *Promethean science*. I use this term to mean an experimental enterprise that mixes a vaulting ambition to safeguard humanity against a

1 John Hill, *A Review of the Works of the Royal Society*, 2nd edition (London, Lady Hill, 1780), pp. 48–9.

major threat with the troubling hazards of following this science's recipes. The episode grabs attention because we also live in an age when expert disagreement is wrongly treated as a sign of fatal ignorance and when it's hard to make space for all the groups who care about the sciences' direction. The problem lies in the relation between matters of fact, powerful because they seem to escape from human interests, and matters of concern, which count because people find them so interesting. That relation is the theme of this chapter. There's local detail and lots of talk in this tale. The private life of public sciences is where we best see why we should not fear if Fellows fight. This otherwise forgotten moment of fireballs and flooded drains is at least dramatic: 12 June 1781, a dozen miles south-east of Norwich at the Heckingham House of Industry, then a recently built workhouse for the rural poor. Here's what happened, as far as I can tell.

It was a Sunday, the Lord's Day. After a showery Norfolk morning under a harsh south-westerly wind, the couple of hundred residents were given their usual Sunday dinner of meat, dumplings and beer. Between two and three in the afternoon a severe thunderstorm came up, with violent lightning and hail. Rain flooded the front courtyard. Just as the sky was clearing and the wind began to drop, the inmates heard a loud explosion and three of them fainted. A sheet of fire entered their rooms and, so they said, even came up to their waists. A woman at the dining-hall door saw three fireballs fall into the court, others saw them at the corner of the House and towards the east wing. Within a couple of minutes the corner of the south-east roof near the stables was burning. At least seven men worked quickly to save the building by digging a hole in the nearest part of the flooded courtyard to get water to extinguish the flames. The stroke had already smashed windows, raised the lead gutters and broken tiles and bricks. The men removed more bricks and lead to get at the smouldering roof beams. Eventually, the fire was out. Within a few days, local glaziers, carpenters and bricklayers had fixed most of the damage. An ironmonger from nearby

Bungay was paid to repair the sharply pointed iron rods rising high above each of the eight chimneys. He'd installed these lightning rods at the House just four years earlier. Three weeks later the gentry of the management committee voted cash rewards to the men whose efforts had saved its House of Industry after the dreadful lightning strike.

I know all this because of the many reports of the events at Heckingham gathered during the next eight months, including a very detailed account assembled by a couple of Fellows of the Royal Society sent to Norfolk to find out exactly what had happened. Before this inquisitive journey to the House of Industry, the Royal Society Fellowship had to rely on hearsay, with all its typical problems of trust and credibility. 'I cannot hear of any persons seeing it at the instant it happened', reported one of their Norwich correspondents, though he had reason to believe that 'it would soon have destroyed the whole building'.[2] This episode illuminates the fundamental relation in the history of the sciences between what people say and who they are. Much of the best-known science relies on judging others' stories. Three days after the publication of *The Origin of Species* Darwin wrote to Thomas Henry Huxley recalling an informative evening in a South London 'gin-palace amongst a set of pigeon fanciers'. Darwin told Huxley that 'the difficulty is to know what to trust'.[3] Knowing something of the storyteller helps in assessing the worth of the story. In the eighteenth century there were now stylish barometers in the houses of the gentry and some of the middling sort. But in rural society many were expert at reading the sky for signs; most still got their long-range weather forecasts from their pocket almanacs, based on planetary aspects and traditional lore. I can learn a little of the Norfolk weather almost twenty-three decades ago thanks to the work of the modern Climatic Research Unit, now based at the University of East Anglia in Norwich. The unit's long-term data show that on 17 June 1781 a threatening low-pressure region dominated the atmosphere above south-east England and had done so for a fortnight. By these modern scientific standards, nothing meteorologically unfamiliar seems to have taken place at Heckingham that summer.

In other respects that season's wider world seems strangely familiar. The summer was distressingly wet. An increasingly unpopular Westminster government soldiered on with a reduced majority before being thrown out the following spring. Shares were in trouble, unemployment rising and the economy in crisis. British troops overseas were enmeshed in a long-running war against radical insurgents – before surrendering to American and French forces at Yorktown in Virginia in October 1781. The following March, all the bells of Norwich, the second largest town in the country, rang out to mark the prospects of peace. The witty and learned Edward Gibbon published two more volumes of his history of a great empire's decline and fall. The papers were full of celebrity gossip, mainly about disreputable actresses and politicians' mistresses. Shopkeepers touted new gadgets such as fountain pens and automatic clocks. In July 1781 Norwich even hosted an auction of 'every article curious and rare' brought back from the late and glorious Captain James Cook's voyages into the Pacific Ocean: 'shells, cloaks, helmets, capes and necklaces curiously wrought with feathers'.[4]

Public taste for knowledge and novelty, however exotic or dubious, was evident everywhere during those months. In Norwich that summer journals puffed lectures by the notorious therapist Dr James Graham on electric sex. One Norwich onlooker was astonished that this 'impudent empiric' imagined he could restore virility by 'the addition of an atmosphere charged with electrical particles and this proposal was privately defended by many persons of information as perfectly philosophical'.[5] A professional musician, William Herschel, had just announced what some reckoned must be a new planet to be named *George* in honour of His Majesty. We now call it Uranus. In July 1781 the Norfolk newspapers

2 Richard Stevens to Mr Brown, 20 December 1781, British Library MSS Add. 30094, fol. 204.
3 Charles Darwin to Thomas Huxley, 27 November 1859, in F. Burkhardt and S. Smith (eds), *Correspondence of Charles Darwin* (Cambridge, Cambridge University Press, 1985—), 7: 404; J.A. Secord, 'Darwin and the breeders' in David Kohn (ed.), *The Darwinian Heritage* (Princeton, Princeton University Press, 1985), 519–42, on p. 534.
4 *Norfolk Chronicle* (28 July 1781), 3.
5 George Cadogan Morgan, *Lectures on Electricity*, 2 vols (London, J. Johnson, 1794), 2: 248.

A meeting of the Royal Society at Somerset House, the Society's home from 1780 to 1857.

reported this 'new discovery of an orb behind the Sun', but worried that 'at a certain period it will burst'.[6] That summer brought news of the Scottish engineer James Watt in Birmingham who'd developed a new mechanism for getting rotational motion out of a vertical steam engine. In London it was said the experiments of a fabulously wealthy aristocrat, Henry Cavendish, obtained pure water by sparking a mixture of airs. At a coffee house near St Paul's Cathedral, a regular club met during the early summer of 1781 to watch the instrument maker Edward Nairne show off his new electric pistol.

Meanwhile, the Royal Society was settling into its plush if somewhat cramped new quarters at Somerset House on the Thames. Cavendish and Nairne were already Fellows, while Herschel and Watt soon would be. Dr Graham never was. Though the Society's rooms were no longer where experimental inquiry happened, membership certainly added lustre. 'Wherever I come', one travelling lecturer and instrument maker had plaintively written, 'I am constantly asked, if I am a Fellow of the Royal Society? And I as constantly find it no small disadvantage to say, No.'[7] The advantages of Society membership didn't flow from the high status of scientists. The Royal Society contained no scientists, because there was no such thing in 1781. The Society's status depended on late eighteenth-century social order. Ironmongers, bricklayers, glaziers and the women at the workhouse, whose parts in the Heckingham events were so salient, were not generally credited as informants by Royal Society gentlemen. There were no women among its Fellows and wouldn't be until 1945. The Society was a focus of debate and a target of satire. The irascible botanist John Hill, whose marvellous remarks on Norwich provide my epigraph, suggested the Society should be displaced by a more efficient Royal Academy of Sciences. The Royal Society's President, the Lincolnshire landowner, man-about-town and Captain Cook's former botanising travel companion Joseph Banks, had just been honoured with a baronetcy. Candidate Fellows were vetted at one of his weekly breakfasts, then dined at the Society's supper club. A London wit cruelly put words into Banks' mouth: '*untitled* members are mere swine: / I wish for princes on my list to shine. / I'll have a company of stars and strings; / I'll have a proud society of *kings*!'[8] Within eighteen months civil war erupted at Somerset House between the President and those who reckoned he was turning the Society into 'a cabinet of trifling curiosities'.[9] In at least one respect the Society's concerns that summer match ours. Banks' men sought to use their powers to influence the British government with evidence-based public knowledge. Which takes us back to the Norfolk thunderstorm.

6 *Norfolk Chronicle* (14 July 1781), 2.
7 John R. Millburn, *Benjamin Martin: Author, Instrument Maker and 'Country Showman'* (Leiden, Noordhoff, 1976), p. 35.
8 [John Wolcot], *Peter's Prophecy, or the President and the Poet, or, an Important Epistle to Sir J. Banks* (London, G. Kearsley, 1788), p. 12.
9 Russel McCormmach, 'Henry Cavendish on the proper method of rectifying abuses' in Elizabeth Garber (ed.), *Beyond History of Science* (Bethlehem, Pa., Lehigh University Press, 1990), 35–51, on p. 43.

Fig·II.

SOUTH WEST VIEW *of the* EAST WING *and part of the* CENTER RANGE
shewing the junction of the two Conductors that were nearest the Stable Lean-to.

The House of Industry at Heckingham as drawn for Edward Nairne and Charles Blagden in January 1782. The pointed lightning rods high above chimneys D and E passed downwards into a drain at C; they failed to prevent lightning damage to the stable block at g on the far right of the picture. (*Philosophical Transactions of the Royal Society*, vol. 72 (1782), 378, plate IX, figure II.)

It was the Heckingham lightning rods that caused the furore. The rods were supposed to save the House from damage but had failed. They might even have helped cause the strike. There was disagreement about the details of the storm, the strike and the behaviour of the lightning rods. When installed at the House of Industry in 1777 by the Bungay ironmonger, a man with the resonant name of John Bobbitt, these rods embodied state-of-the-art experiments, so were newsworthy and dodgy. But surely it was easy to tell whom to trust about the June 1781 events? Simply check whether a story matched the relevant authorities' reliable knowledge about how lightning behaved and rods worked. But this authority and this knowledge were exactly the matter of dispute. The Fellows of the Royal Society had been involved in two decades of argument about the behaviour of lightning rods. The Heckingham event was seen as 'an *experiment* where a house armed with eight pointed conductors had been set fire to by lightning'.[10] Yet for the strike to be a worthy experiment, Society Fellows already had to know whose story to believe. But to know whom to believe, they had to know how the experiment should run.

To resolve this apparently intractable puzzle, the Fellows had to rely on their deep sense of who should be trusted: gentlemen were judged more reliable than servants, local worthies more credible than the poor and indigent. So they commissioned stories, drawings and three-dimensional models from men they already had reasons to trust. Perhaps these accounts would settle the matter without having to be on the spot. Unlike the names of the workhouse inmates, the Society recorded exactly who these valued correspondents were. They included Samuel Cooper, one of the Heckingham overseers, an eminent doctor of divinity and a wealthy landlord. He'd already sent the Society thunderstorm reports from Norfolk. The Fellows also heard from Dixon Gamble, a merchant and town steward from Bungay; from George Cadogan Morgan, a Welsh radical of sophisticated philosophical interests and fierce politics who'd become a unitarian preacher at Norwich's famous Octagon Chapel; and from that city's principal bookseller Abraham Brook, who marketed electrical and optical instruments in Norfolk. These gentlemen had apparently scoured the building and interviewed the poor inmates, the reliability of whose recollections they barely accepted. During these interviews, they worried about the tale of the spectacular fireballs reported by 'one of the cripples in the House of Industry, a middle-aged woman', then wondered 'if any credit could be given to the testimony of such a person in a matter like this'.[11] According to Morgan, who quizzed Heckingham's residents soon after the strike, 'the contradictory absurdities which they asserted and maintained, are scarcely conceivable'.[12]

By the year's end these confused reports got to London. The effect was almost as explosive as the original strike. If the best technique for preserving buildings against lightning were in question because of some Norfolk oddity, this mattered to the government. The Heckingham stories soon reached the ears of the King, and through him the Board of Ordnance, one of the largest state departments, supplier of military munitions for the American War. Based at the Tower of London, the Board was concerned

10 Benjamin Wilson to Dixon Gamble, 28 January 1782, British Library MSS Add. 30094, fol. 212.
11 'Proceedings relative to the accident by lightning at Heckingham', *Philosophical Transactions of the Royal Society*, 72 (1782), 355–78, on p. 377.
12 Morgan, *Lectures*, 2: 234.

with the protection of its arsenals against fire. Ordnance officers heard about the apparent failure of the Heckingham lightning rods in December 1781: 'the whole Board are much alarmed'.[13] The Royal Society seemed the obvious organisation to contact, because they had a long track record in these matters. Over Christmas the Board's secretary wrote to Joseph Banks. This was 'a matter of the highest importance', but 'no *authentic* account has yet come to the knowledge of the Royal Society'.[14] Since the stories they got from Norfolk were so confused and the details were such a matter of concern, within a few days Banks and his Somerset House colleagues decided to send a pair of Fellows to Norfolk to investigate.

CHARGED ATMOSPHERES, OR HOW TO MAKE A LIGHTNING ROD

The principle of such lengthy and lofty pointed metal rods as a defence against lightning rested on a mix of old and recent thinking. Since the early eighteenth century, experimenters had been able to make electric sparks and shocks using friction machines of glass, leather and metal. These were lucrative items in their shows. 'Lightning is in the hands of nature what electricity is in ours', the London instrument maker George Adams put it, 'the wonders we now exhibit at pleasure are little imitations of the great effects which frighten and alarm us'.[15] The imitation analogised the stormy atmosphere with glass jars and metal rods inside their well-stocked rooms. According to the Royal Society's leading electrical experimenter, the apothecary William Watson, 'we see every day more and more the perfect analogy (to compare great things with small) between the highly electrified glass jar in the experiment and a cloud replete with the matter of thunder'.[16]

In early 1748 Watson read the Society a letter from an ingenious print-er in Philadelphia, second city of the British empire. Quaker networks link-ing London with the City of Brotherly Love helped news of Benjamin Franklin's experiments reach the Society. His demonstrations were

The Copley Medal, which has been awarded since 1731 for outstanding achievements in research in any branch of science, and alternates between the physical sciences and the biological sciences.

supposed to show that electrical fire was an unevenly distributed active fluid gathered in atmospheres round bodies: the fluid would flow so as to restore balance, a satisfying thought for a prudent book-keeper, between excess (or positively charged) and deficient (or negatively charged) regions. Sparks and lightning were such restorative flows, if in dramatic form. As often, the Society initially held that what was right in Franklin's story was already well known and what was wrong must be rejected. Even so, these stories about charged atmospheres were judged prize-winning achievements in electrical philosophy. In 1753 the Society's new President, the Earl of Macclesfield, otherwise preoccupied with persuading a slightly unwilling nation to accept a foreign Gregorian calendar and thus seemingly lose eleven days of its precious time, awarded the Society's prestigious Copley Medal to Franklin. 'True it is', observed the noble Earl, 'that several learned Men, both at home and abroad, do not entirely agree with him in all the Conclusions he draws, and the Opinions which he thinks may be deduced from the Experiments he has made.' However, he remarked,

13 Daines Barrington to Benjamin Wilson, 26 December 1781, British Library MSS 30094, fol. 206.
14 George Cadogan Morgan to Samuel Cooper, 4 January 1782, Norfolk Record Office MSS C/GP 12/12, p. 221; Joseph Banks to Board of Ordnance, 29 December 1781, Public Record Office MSS WO 47/26 series II, fol. 515.
15 George Adams, *Lectures on Natural and Experimental Philosophy*, 5 vols (London, Hindmarsh, 1794), 4: 370.
16 William Watson, 'An account of a Treatise entituled *Letters concerning Electricity* by the Abbé Nollet', *Philosophical Transactions of the Royal Society* 48 (1753), 201–16, on p. 215.

though not yet entirely convincing nor even a Fellow, at least Franklin was 'a Subject of the Crown of Great Britain'.[17] All that changed in the next two decades: following Franklin's move to Europe, his theory would become Society orthodoxy, he won a Fellowship and helped liberate his nation from British rule.

The colonial medallist's new invention was the lightning rod, first announced in his Philadelphia almanac the same year as his Royal Society prize. Since he found in his experiments that sharp needles could quietly withdraw electrical fire from the atmosphere of charged objects some inches away, so on a grander scale pointed metal rods well connected to damp earth should let electrical fire flow silently between the Earth and thunder clouds. He offered hope of disarming lightning, just as the mythical Prometheus had stolen fire from Olympus for humanity's benefit and was thus punished by Zeus. Many Enlightenment sages, including Immanuel Kant, compared Franklin with the fabled Titan. One popular 1770s English writer on farming and weather put it pithily: 'Dr Benj. Franklin's soaring genius has realised the fable of Prometheus' bringing fire down from heaven'.[18] The Secretary of the French Royal Academy of Sciences apologised to Franklin in 1773 that 'I have never had the happiness to meet the *modern Prometheus*'.[19] The poet, philosopher and botanist Erasmus Darwin admired Franklin's heroism, but guessed Prometheus' punishment after stealing heavenly fire was really an allegory for a gin-soaked hangover. There were some more seriously dissident voices. An eminent French experimenter, sceptical of the worth of these fashionable rods, warned of the lethal dangers 'were we to bring into being the Prometheus of the fable'.[20] Within a generation the American with his lightning rods would be celebrated as victor over both tyranny and thunderbolts in a single evocative image of ingenuity and independence.

It seemed to many storytellers that since the rods were obviously rational and effective, any opposition to their use must stem from popular and religious narrow-mindedness. An English traveller in southern Germany was 'told that the people of Bavaria were at least 300 years behind the rest of Europe in philosophy and useful knowledge', so they still riskily rang church bells during thunderstorms to ward off threats.[21] When fierce storms hit not only Norfolk but also lands across the North Sea in 1781, many Dutch and Flemish bell ringers died. From summer 1781 the city of Arras in northern France was racked by a lawsuit because of citizens' opposition to a new lightning rod: the rod's safety was successfully defended by a precise young lawyer with the schoolboy nickname 'The Barometer'. His real name was Maximilien Robespierre, a man soon to be identified with Terror.[22] One East Anglian minister reflected on an old story about members of a congregation marked with the sign of the cross after lightning hit their cathedral and wished 'the Bishop's attention had not been so much absorbed in the wonderful'.[23] When a reckless Russian experimenter tried the electricity of his woefully arranged rod in a thunderstorm, he was killed. In response to this electric martyrdom, London's *Gentleman's Magazine* commented that 'we are come at last to touch the celestial fire, which if we make too free with, as it is fabled Prometheus did of old, like him we may be brought too late to repent of our temerity'.[24]

However fabulous such tales, resistance to these devices was not entirely based on prejudiced ignorance. It is just as wrong to assume that scriptural fundamentalism completely explains why many nineteenth-century commentators challenged Darwin's model of natural selection. Promethean science is debatable and its standing is never explicable by rough-shod appeals to lack of knowledge and to bigotry. There were reasons to wonder about, as well as wonder at, the modern Prometheus. Franklin's account was the best the Society's Fellows knew, but ambiguous

17 George Parker, Earl of Macclesfield, 'Speech awarding the Copley Medal' (30 November 1753), in *Papers of Benjamin Franklin*, ed. L. Labaree et al. (New Haven, Yale University Press, 1959—), 5: 126–33.
18 John Mills, *An Essay on the Weather*, 2nd edition (London, S. Hooper, 1773), p. 19.
19 Condorcet to Benjamin Franklin, 2 December 1773, in *Papers of Benjamin Franklin*, 20: 489.
20 Jean Nollet, *Lettres sur l'électricité* (Paris, Guérin & Delatour, 1753), 1: 19.
21 Charles Burney, *The Present State of Music in Germany, the Netherlands and the United Provinces*, 2 vols (London, Becket, 1775), 1:183.
22 Jessica Riskin, 'The lawyer and the lightning rod', *Science in Context*, 12 (1999), 61–100, on p. 85.
23 Thomas Harmer to John Canton, 11 December 1753, in Royal Society Library MS/598, p. 28.
24 *Gentleman's Magazine* 25 (1755), 312.

and in several ways false. His small-scale experiments suggested to him that rods must be sharply pointed and could silently draw electrical charge from the dangerous atmospheres of thunder clouds. Modern sciences say both claims are untrue. On the vast scale of a lightning strike, the difference between pointed and blunt rods doesn't matter. There's evidence that pointed tips can make lightning rods into bad receptors. These rods cannot quietly discharge a cloud and their presence in an electrically charged region can make a strike more likely. But Franklin never abandoned his claims that rods could prevent a strike and had to be sharply pointed, just like those Mr Bobbitt erected at Heckingham in 1777 and which failed to work in 1781. 'A long *pointed* rod', Franklin told the Royal Society in 1772, 'may *prevent* some strokes as well as conduct others that fall upon it.'[25] Throughout the period these compelling but dubious claims were among the Royal Society's major preoccupations.

When news broke that the Heckingham House of Industry had been equipped with high pointed rods but nevertheless caught fire, one of Franklin's closest allies told him the Ordnance Board and the King were involved because 'these events have a tendency to discredit conductors'.[26] In ways familiar from more recent episodes of public science, such as the fracas surrounding food safety and BSE, the MMR vaccine, or the environmental effects of genetically modified crops, matters of concern seem to demand sure-fire judgments from trusted experts. So authorities called on the Royal Society for unequivocal decision. It is familiar, too, with sensational reports and rival experts in question, that public debate seems very wayward.

In these respects the Heckingham catastrophe was neither unprecedented nor straightforward. For two decades before 1781 the Society faced many episodes when across southern England houses, churches, powder magazines and other buildings guarded by rods had been struck or damaged by lightning. The Board of Ordnance, the clergy of St Paul's Cathedral and the monarch all demanded certainty. The Fellows developed a kind of electrical fieldwork, involving visits to the stricken buildings, interviews with workmen, excavation of the rods' connections and collection of melted metal despatched to the Society. They trusted gentry 'well known to many in the Royal Society'.[27] The Fellows treated these events as so many 'great electrical experiments' then argued that such real-world experiments reinforced Franklin's story about high points.[28] But there were characteristic troubles of interpreting these experiments. If the protection had failed this might be because these rods were wrongly set up, so electrical orthodoxy was safe. But it might be because the orthodoxy was wrong and all such rods fundamentally unsafe. To solve this puzzle, Fellows had to appeal to some prior sense that they alone were masters of the facts.

Yet in the rough and tumble of society gossip and political crisis this trust was hard to win. The Society wasn't on message. Franklin's notions of high pointed rods and silent atmospheric discharge were backed by prestigious Fellows such as Watson, Nairne and Cavendish. But there were vocal critics inside the Society. The newspapers gleefully reported the schism. Opposition was led by the fashionable painter and theatre manager Benjamin Wilson, veteran Royal Society Fellow and pugnacious enemy of Franklin's philosophy and politics, especially of 'the magical point'.[29] Wilson's coterie had good connections. He was employed both by the Board of Ordnance and by the King, and won support from one of the Royal Society's Copley medallists, the able chemist Edward Delaval, from senior military officers, noble courtiers and foreign academicians. Wilson's experiments convinced many others that high pointed conductors were dangerous, for they would invite a lightning stroke and never safely disarm electrical atmospheres. The modern Prometheus was wrong. 'Sharp points are put there only to invite an enemy which otherwise might not have troubled us.'[30] Better, so Wilson urged, to build lower blunted rods much closer

25 Benjamin Franklin, 'Experiments supporting the use of pointed lightning rods', August 1772, in *Papers of Benjamin Franklin*, 19: 251. My emphasis.
26 Richard Price to Benjamin Franklin, 7 January 1782, in *Papers of Benjamin Franklin*, 36: 406.
27 William Henly, 'Experiments concerning the different efficacy of pointed and blunted rods, in securing buildings', *Philosophical Transactions of the Royal Society* 64 (1774), 133–52, on p. 141.
28 William Watson, 'Observations upon the effects of lightning', *Philosophical Transactions of the Royal Society*, 54 (1764), 201–27, on p. 224.
29 Benjamin Wilson, *Further Observations upon Lightning* (London, Lockyer, Davis, 1774), p. 22.
30 Benjamin Wilson, *Observations upon Lightning* (London, Lockyer, Davis, 1773), p. 57.

Wilson's lightning and gunpowder display at the Pantheon.

to threatened roofs and walls. In the midst of these histrionics the Fellows inevitably became the target of vicious satire. There were fraught votes within the Society about whether Wilson's protests should be aired. One of his friends denounced the 'factious illiterati' of the Royal Society.[31] According to an aged earl, 'The Royal Society may if it pleases decide in favour of the pointed conductors, but its decisions cannot oblige me and I hope will not induce any of my friends to adopt them.'[32]

The Society's system of experiment and trust was in trouble. Wilson and Delaval staged their own site visits after spectacular strikes to get different stories from those obtained by Nairne and Watson. Matters got serious in

May 1777. The gunpowder stores run by the Ordnance Board down the Thames at Purfleet were hit by lightning. Rods installed there on the recommendation of a Royal Society committee five years earlier seemed to have failed. In the midst of the American War, British military supplies at Purfleet were no longer safe. Wilson exploited the disaster brilliantly. Supporter of metal points and transatlantic rebels, Franklin was put in the wrong electrically and politically, 'as bad a man as he is a philosopher'.[33] In summer 1777 Wilson set up a vast show to demonstrate the fallacies of his enemies. With royal funds and Ordnance Board gunpowder, he took over the Pantheon, a gorgeous Oxford Street dance hall, and installed a model of the Purfleet arsenal under a huge artificial charged cloud. The theatrical Wilson aimed to prove the dangers of elevated and pointed conductors. Many Londoners, including the royal family, watched the model's spectacular sparks. Nairne and other Fellows tried to heckle Wilson and designed their own models of lightning and gunpowder to show the errors of his ways. Wilson's confidants grumbled about Franklin's 'junto', especially 'setting Nairne to put you in the wrong'. Franklin's allies launched a politically venomous attack on royal policy and the Pantheon displays: 'those butchers sent by our infamous Ministry to exterminate the Americans are no more courageous in their hellish profession than our daring philosopher B. Wilson has been in his drum tricks'.[34]

For many months these tricks were satirised mercilessly in the press.[35] The King reportedly ordered pointed rods replaced by lower blunt ones at Ordnance buildings and royal palaces. Some even said the fight forced the resignation of the Royal Society's President: the Secretary of the French Academy of Sciences certainly thought this is what happened, and the

31 S. Martin to Benjamin Wilson, 30 October 1775, British Library MSS Add. 30094, fol. 161.
32 Lord Harcourt to Benjamin Wilson, 12 August 1777, British Library MSS Add. 30094, fol. 179.
33 Daines Barrington to Benjamin Wilson, 25 September 1777, British Library MSS Add. 30094, fol.188.
34 Daines Barrington to Benjamin Wilson, 25 September 1777, British Library MSS 30094, fol. 188; Jean Hyacinthe de Magellan to Achille Lebègue de Presle (copy), 15 September and 3 October 1777, Library of Congress, Franklin MSS, at fol. 7v.
35 [Review of William Swift, 'Account of some electrical experiments'], *Monthly Review*, 60 (1779), 417; Trent A. Mitchell, 'The politics of experiment in the eighteenth century: the pursuit of audience and the manipulation of consensus in the debate over lightning rods', *Eighteenth-Century Studies*, 31 (1998), 307–31, on p. 324.

resignation was soon followed by Joseph Banks' assumption of the presidency.[36] This ghastly history explains the high tension around the Heckingham story a couple of years later and the pointed political interest in its details. The Royal Society had bad form in its management of lightning strikes where pointed conductors had failed. As soon as he heard from Norfolk, Wilson again mobilised his extensive networks to make the most of the fact. He 'began to apprehend there might be an intention to smother the matter and keep it secret from the public'.[37]

PROMETHEAN SCIENCE, OR HOW TO BE AN EXPERT

If the high and pointed rods had been badly set up, the Royal Society's view would be safe. If, however, they'd been competently designed, that view would be in trouble. If the rods were plunged deep enough into damp soil or their bases covered in flood-water, the official view would have expected them to work: their failure would count as a challenge to Royal Society doctrine. So Wilson gathered stories about floods and the rods' grounding. Royal Society envoys sent to Heckingham would seek to show the rods were not well set up and that this explained their failure. It didn't help that the Society's delegates were Banks' right-hand man the suave physician Charles Blagden and Edward Nairne, Wilson's old enemy. Wilson sent Banks details of Nairne's 'troublesome manner' at the Pantheon show and support for Franklin's doctrine. The President boldly answered that Nairne's 'veracity is preferred by the public and the Royal Society in general'.[38]

So the Fellows' Norfolk fieldwork was initially difficult. Blagden and Nairne did their homework by re-reading reports from Purfleet and recent electrical textbooks. They needed to show the Heckingham lightning rods were badly set up. Mr Bobbitt had allegedly been at fault by letting them reach only a few inches below ground where they led into a

36 Condorcet, 'Éloge de M. Pringle' (delivered 1782), published in *Oeuvres de Condorcet*, ed. A. Condorcet O'Connor and F. Arago, 12 vols (Paris, Firmin Didot, 1847), 2: 513–28, on p. 524; C.R. Weld, *History of the Royal Society with Memoirs of the Presidents*, 2 vols (London, John Parker, 1848), 2: 101–2.
37 Benjamin Wilson to Dixon Gamble, 28 January 1782, British Library MSS Add. 30094, fol. 213.
38 Benjamin Wilson to Joseph Banks, 12 and 17 January 1782, British Library MSS Add. 30094, fols 208–10; Joseph Banks to Benjamin Wilson, January 1782, Royal Society MSS CB/6/104.

Detailed drawings of the connections between the lightning rod and the drain at Heckingham. Nairne and Blagden claimed this drain was never full of water so the rod was not securely earthed, but their critics Wilson and Gamble insisted it was.

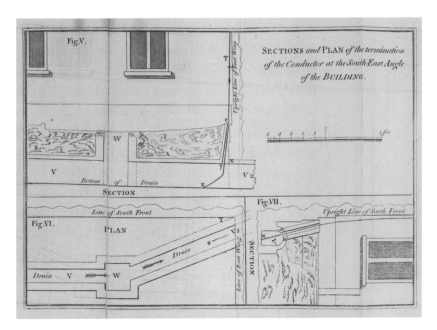

The damage caused by the lightning strike to the lead on the roof of the Heckingham stable block. (Both images from *Philosophical Transactions of the Royal Society*, vol. 72 (1782), 378, plate XI, figure V.)

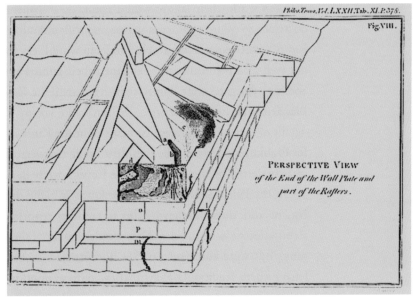

drain 'without being in contact with anything but air'.[39] A broken rusty iron pole whose lower end was in contact with nothing but air wasn't really a lightning rod at all. The strike hit the lead on the stable roof simply because 'the lightning picked out the best and nearest conductors to the moist earth'.[40] The Fellows seized on any story that the drains were dry even during the storm. Blagden and Nairne got the House workmen to put back everything as they recalled it was just before the strike. Three different lightning paths might explain why the rods had not taken the strike, so the Fellows accepted the story of spectacular fireballs, even if the source was a dubious female inmate. Then they toured county gentry for evidence that the electrical defence of the House of Industry was inadequate and their theory of lightning conductors safe. Wilson did the opposite. He contacted Norfolk friends for signs the rods were in a good state, drenched with drain water and well maintained. 'Have you been able to learn from anyone of good judgment how high the black cloud was at the time it hung over the House? And whether any of the flashes of lightning were seen to make towards the pointed conductor?'[41] Wilson got Gamble to build a model of the House like the one of Purfleet, then showed it to the King and the Ordnance Board. He reckoned it showed the high pointed lightning rods had failed. If so, Royal Society doctrine had failed too.

The metropolitan outcome of the Heckingham inquiry was managed by Nairne, Blagden and Banks. The report they sent the Ordnance Board in February 1782 showed the imperfections of the Norfolk lightning rods and strengths of the received theory of their behaviour. It was publicised by the Society and copies sent to foreign papers.[42] With the status of the Fellows and the select group they interviewed, they could secure agreement in the capital. Back in Norfolk things were less sure. In the 1780s 'there was more mind afloat in Norwich than is usually found outside the

Opposite:
In March 1782 Dixon Gamble in Bungay sent Benjamin Wilson in London these diagrams of how pointed lightning rods incite lightning strikes by producing vertical conducting columns in the air. Above, a charged cloud is acted on by a pointed rod causing a strike to the house; below, a charged cloud 'throws its superabundance' on to a lower cloud, which then 'passes a stroke' to the house. These views were designed to counter those of Nairne and Franklin on the effects of points. (British Library mss Add. 30094, fol. 220.)

39 Royal Society MSS CB/6/105, fol. 1.
40 Edward Nairne to Jean Hyacinthe de Magellan, 5 March 1782, American Philosophical Society Library MSS BP 85, vol. 25, fol. 26.
41 Benjamin Wilson to Dixon Gamble, 28 January 1782, British Library MSS Add. 30094, fol. 213.
42 Jean Hyacinthe de Magellan to Benjamin Franklin, 13 April 1782, in *Papers of Benjamin Franklin*, 37: 150.

the Cloud A being previously in a Charged state is acted upon by point-
-ed Conducting rods erected on the House B. — here I have taken the 220
liberty of rendering Visible what we know is going forward in Nature
altho' not sensible to the Eye. — I mean the Column of the Electric Matter
adjoined to the Air, and through which, as his Lordship says a longer stroke
is taken to pass by the Conducting power of the Electricity which is adjoin-
-ed to the Air in the whole of this Column. — and the effect may then
take place as represented in Fig. 2 where A being an highly charged Cloud
comes within the striking distance to B, and throws its superabundance into
it, which ⟨= as suddenly⟩ cause the Cloud B to pass a Stroke through the Column C, upon the House D.
this I apprehend will ever be the Case where one Cloud strikes to Another
which may happen to be within the sphere of Action of pointed Conductors,
and indeed seems to agree with the Course of the Strokes which fell upon
the Purfleet building, and the House at Rockingham. — I earnestly
wish you will consider these thoughts before you send them into the World
and if they meet with your concurrence I think they will add some weight
to your Original opinion as sent to the Marquis of Rockingham so long
ago as 1764 and with which you have so properly closed your report
to the Board of Ordnance — I am Sir, Your Obedient Humble
Bungay 24th March 1782 — Servant D. Gamble

Fig. 1.

Fig. 2.

literary circles of the metropolis'.[43] The Fellows' informants were gentlemen with their own views of electricity and lightning. None lined up in an orderly fashion behind Nairne and Blagden. The Heckingham governor Samuel Cooper insisted his House's rods were well earthed, 'nothing wonderful or even extraordinary' had happened, and complained to Banks that 'some of those who spend their time chiefly in making of experiment are too apt to treat those who do not with a dogmatism bordering upon contempt, would the latter venture to deduce by the legitimate principles of logic a plain and obvious conclusion from the experiments of the former'.[44] While Cooper questioned London experimenters' authority, Gamble had his own story of how electricity worked. Along with his model of the House, he made a diagram of lightning discharge. He insisted against Nairne that the rods were perfectly grounded, 'these pointed rods were the cause of the stroke's taking place in their vicinity', and couldn't accept the Fellows' notion that the House was struck because the rods were surrounded by insulators. 'For God's sake, what should it be connected with so proper to keep the effect of the storm from entering the House!'[45] According to Mr Gamble, the Society's story simply didn't make sense.

Even the Society's best Norwich allies, Morgan and Brook, broke ranks. Brook had major experimental interests in electricity. He'd been the informant who'd insisted there'd been little rain before the strike and that the rods were not grounded at all. He designed his own electric models of thunderstorms and an ingenious electrometer that helped determine the atmospheric charge. He and Morgan showed Blagden and Nairne their own electrical experiments and the lightning rods atop Norwich Cathedral that Wilson designed.[46] Brook joked with Nairne about whether Norwich soil had special electric properties. But Brook rejected the Society's account, insisting that electrical fluid moved always from the soil towards the clouds. Unlike those of the Royal Society his instruments 'speak so as to be understood universally'.[47]

Morgan was more radical about London doctrine. The Unitarian minister admired Franklin's politics and experiments and aided the Society's Heckingham fieldwork. Supporter both of the American and French revolutions, Morgan preached the cause of Promethean liberty: 'In all ages the thunder of heaven has contributed more powerfully to promote the cause of imposture and tyranny. By the science of electricity, however, the future possibility may be exterminated of renewing these frauds. It has enabled the most common artificer to avert every danger attending a thunder-storm. It teaches the vulgar mind to smile at a thousand religious ceremonies.'[48] But like his friend Brook, Morgan doubted Franklin's explanation of this enlightened practice. 'By guarding your house you make it of all objects that which is the most likely to become the circuit of a cloud.' Franklin was wrong to imagine that pointed rods could silently and safely discharge the electrical atmosphere in the skies.[49] Such views became common. The instrument maker George Adams had no doubts that pointed rods were ineffective and unsafe. 'It is evident', Heckingham's events showed, 'that the effect of conductors in general is too inconsiderable either to *lessen fear* or *animate hope*.'[50] Soon Franklin's electrical atmospheres and the Heckingham workhouse would both be under fierce attack. Galvanism and electrodynamics preoccupied experimenters on life and matter. The workhouse was burnt to the ground by Norfolk protesters against the poor laws.

Promethean science claimed it was grounded in experiences available to all, yet it proved hard to organise experiences so all agreed about these

43 C.B. Jewson, *The Jacobin City: A Portrait of Norwich in its Reaction to the French Revolution* (Glasgow, Blackie, 1975), p. 143.
44 Samuel Cooper to Henry Hammond, 17 October 1781, and to Joseph Banks, 13 January 1782, Royal Society Library MSS CB/1/3/82 and /83.
45 Gamble to Wilson, 24 March 1782, British Library MSS Add. 30094, fols. 219–20.
46 Charles Blagden, *Diary 1776–88*, Yale University Library MSS Osborn fc 16, entry for 26 January 1782.
47 Abraham Brook, 'Account of a new electrometer', *Philosophical Transactions of the Royal Society*, 72 (1782), 384–8, on p. 387; Brook, 'On thunder storms', Royal Society Library MSS Letters and Papers, vol. 8 (1789), 129; Brook, *Miscellaneous Experiments and Remarks on Electricity, the Air Pump and the Barometer* (Norwich, Crouse and Stephenson, 1789), 101.
48 Morgan, *Lectures on Electricity*, 1: xxix–xxxii.
49 Morgan, *Lectures on Electricity*, 2: 298; [Obituary of George Cadogan Morgan], *Monthly Magazine*, 6 (December, 1789), 475–80, on p. 476; D.O. Thomas, 'George Cadogan Morgan', *Price-Priestley Newsletter*, 3 (1979), 53–70, on p. 65.
50 Adams, *Lectures*, 4: 381–2.

principles. Only certain places and people could be trusted. Even close allies could waver from Royal Society orthodoxy. The problem was evident in 1780s Norfolk. At the same time as the Heckingham controversy, a lawsuit began about the security of north Norfolk harbours. Leading engineers and Royal Society Fellows were witnesses. This case led to a crucial legal decision on the status of the scientific expert: 'In matters of science', the Lord Chief Justice declared, 'the reasoning of men of science can only be answered by men of science.'[51] The problem was to determine who counted as 'men of science', so how to establish riskily Promethean science. The Titan's theft of fire and subsequent vicious punishment stands for the rights of free inquiry and its penalties. In her brilliant commentary on the French Revolution, the feminist Mary Wollstonecraft wrote in 1794 about the Prometheus story 'on which priests have erected their tremendous structures of imposition'. Rather, she argued, 'we shall find that men will insensibly render each other happier as they grow wiser'.[52] Within a generation, her daughter Mary Shelley composed one of the most important accounts of scientific ambition and its fearful consequences. *Frankenstein*'s subtitle was *The Modern Prometheus*.

Promethean science matters because of the hopes it offers and the demands it places on disputable knowledge and puzzling threats. It still counts. *Promethean Science* is the title of a 2000 World Bank report on the promises of genetic engineering and biotechnology for global food crises. The authors apparently chose this striking phrase because it has come to mean '*daringly original and creative*'.[53] However, that's not all it means. Promethean science has a long and troubled history involving the many groups who claim the right to describe and intervene in the world. The same year as the World Bank report, the then head of Monsanto, Hendrik Verfaillie, spoke in Washington DC about the crisis surrounding genetically modified crops: 'when we tried to explain the benefits, the science and the safety, we did not understand that our tone – our very approach – was seen as arrogant. We were still in the "trust me" mode when the expectation was "show me". And so, instead of happily ever after, this new technology became the focal point of public conflict, the benefits we saw were jeopardised, and Monsanto became a *lightning rod*.'[54] This is an appropriately highly charged image of the troubles of public trust in science.

51 Tal Golan, *Laws of Men and Laws of Nature: The History of Scientific Expert Testimony in England and America* (Cambridge, MA., Harvard University Press, 2004), p. 24.

52 Mary Wollstonecraft, *An Historical and Moral View of the Origin and Progress of the French Revolution; and the Effect it Has Produced in Europe* (London, J. Johnson, 1794), 17; Jane Goodall, 'Electrical Romanticism', in Jane Goodall and Christa Knellwolf (eds), *Frankenstein's Science: Experimentation and Discovery in Romantic Culture 1780–1830* (Aldershot, Ashgate, 2008), 117–32 on p. 125.

53 Ismail Seragildin and G.J. Persley, *Promethean Science: Agricultural Biotechnology, the Environment and the Poor* (Washington DC, Consultative Group on International Agricultural Science, 2000), v. Stress in the original.

54 Hendrik A. Verfaillie, 'A new pledge for a new company', Farm Journal Conference, Washington DC, 27 November 2000, online at www.monsanto.com/monsanto/media/speeches/new_pledge_speech.html.

7

RICHARD HOLMES

A New Age of Flight:
Joseph Banks Goes Ballooning

Richard Holmes, biographer and travel writer, is a Fellow of the British Academy and author of celebrated works on Shelley, Coleridge and young Dr Johnson. His latest book, *The Age of Wonder*, is an examination of the life and work of the scientists of the Romantic age who laid the foundations of modern science. It was shortlisted for the Samuel Johnson Prize, and won the 2009 Royal Society Prize for Science Books.

Newton's theories impressed the intellectuals, but the mass appeal of ballooning really spread the idea that a new age was dawning. The Royal Society maintained a proper scientific scepticism. But as Richard Holmes reveals, its President was a good deal more intrigued than he let on in public.

BALLOMANIA

On 6 November 1783, the recently elected President of the Royal Society, the botanist Joseph Banks, called a special meeting of the Fellows at their splendid new premises in Somerset House. The subject up for discussion was a controversial one: the extraordinary phenomenon of the French 'aerostatique Machines'.

Banks had received two long and confidential 'papers' from Benjamin Franklin, the American Ambassador in Paris, describing the experiments of the Montgolfier brothers with hot-air balloons; and of Dr Alexander Charles with hydrogen balloons. Franklin prophesied – correctly – that the first manned flight in history was about to occur. A balloon would

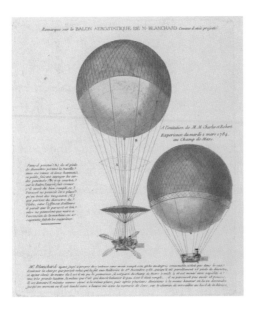

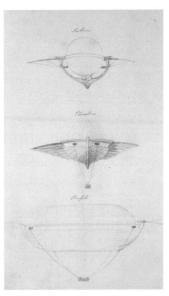

Right: Jean-Pierre Blanchard's design for the 'Vasseau Volant' hydrogen balloon, which he flew from the Champ de Mars in Paris to Billancourt on the Haute-de-Seine on 2 March 1784.

Far right: These drawings, sent to the Royal Society on 4 March 1784, were found in correspondence from George Cumberland to Joseph Banks and included sketches by Henry Smeathman. It is believed that they are of Blanchard's hydrogen balloon 'Vaisseau Volant' which flew two days earlier and shows extendable wings around the canopy, perhaps as a suggestion for modification.

inevitably 'carry up a Man'. Pilâtre de Rozier and the Marquis d'Arlandes duly took to the air on 19 November 1783. So what, Franklin wondered mildly, did the British intend to do about it all?

After the meeting, Banks wrote back thanking Franklin for his 'Philosophical amusements', but playing down any notion of Anglo-French competition in balloon technology. Instead he sounded a note of ironic caution. 'I think I see an inclination in the more respectable parts of the Royal Society to guard against the *Ballomania* which has prevailed, and not to patronise Balloons merely on account of their rising in the Atmosphere, till some Experiment likely to prove beneficial either to Society or Science, is proposed to be annext to them.' Banks's witty coinage – 'ballomania' – was destined to float quite as far as the balloons themselves.

It is usually said that the Royal Society subsequently – and wisely – made little attempt to sponsor, fund or even foster rival British balloon experiments. Its Fellows were gently discouraged by Banks, who continued to dismiss 'ballomania' as a typically French craze for novelty and display. It was a passing fashion that could have no scientific outcome. Like the exactly contemporary French craze for Mesmerism (also reported by

Franklin), it would soon dissipate and be utterly forgotten.

Certainly, all the early balloon ascents made in England in the following months, unlike those in France, were privately funded through commercial exhibitions or subscriptions. There was no official sponsorship from the Society or the Crown, or from any university or public institution – unless one counts the glamorous Georgiana, Duchess of Devonshire as a public institution. Moreover nearly all the successful British ascents were in fact made by foreign aeronauts and showmen, such as the young Neapolitan Vincenzo Lunardi, the Italian Count Francesco Zambeccari, the Frenchman Jean-Pierre Blanchard and the American Dr John Jeffries.

Banks' views appeared to express a mixture of sensible scientific scepticism, combined with a somewhat aloof disapproval of French excitability. Patriotically, he always insisted that the science of ballooning had been originated by the British, in the 'inflammable air' experiments of Henry Cavendish, Joseph Black and Joseph Priestley. Only the French, he joked, would have turned Cavendish's elegant soap bubbles of hydrogen into the seventy-foot monster of 'Montgolfier's flying Medusa' (appropriately powered by hot or 'rarefied' French air).

The ballomania which ensued over the next two years is often remembered in terms of the sudden rage for balloon fashion accessories which seized Paris (and to some extent London). This might now be termed Montfgolfier merchandising. Both the Musée de l'Air at Le Bourget and the Blythe House section of the Science Museum, London, are crammed with a wild selection of these astonishing, and sometimes rather beautiful, artefacts. They include popular prints, paintings, satirical cartoons, fans, snuffboxes, teapots, chinaware, lampshades, tobacco pipes, ladies' garters, milk jugs, hair clips, coat buttons, desk handles, parasols, pen-holders, and even (at Le Bourget) a ceramic toilet bowl with 'Bon Voyage' glazed on the interior.

But the element that Banks truly distrusted in ballomania was its demagogic potential. His secretary, Dr Charles Blagden FRS, a chemist who also worked for Cavendish and travelled frequently in France, perhaps encouraged these misgivings. So in August 1783 he informed Banks: 'all Paris is in an uproar about the flying machines'. In October he noted: 'It appears that the enthusiasm, I almost said madness, which prevailed in Paris on the subject of balloons, has taken a turn more characteristic of the [French] nation, and is converted into a most violent party spirit. Ridicule and invective, verse and prose, are employed without mercy on this occasion.'

Blagden enjoyed passing on comic or frankly scabrous material. He obtained a French satirical pamphlet purporting to recount 'the supposed conversations between the three animals which went up in [Montgolfier's] globe' at Versailles. The cockerel (symbol of France) seemed somewhat subdued on its return to earth, and 'all the animals' complained about the novel experience of air-sickness.

Blagden also gleefully reported the open war in Paris between supporters of hot air and those of hydrogen, quoting an unacademic phrase of Dr Alexander Charles: '*La belle cacade que Faujas et Montgolfier ont fait.*' He then added primly: 'I know no decent English translation of this term [*cacade*].' Banks (a product of Eton and Oxford) knew of course that *cacade* meant *a heap of shit*. Blagden concluded sententiously: 'Every thing that occurs relative to this business makes me rejoice that during all the Heat & Enthusiasm of our Neighbours we retained in this country a true Philosophical Tranquillity.'

A year later, in September 1784, he was happy to pass on the opinion of his friend, the distinguished French chemist Claude Berthollet. 'Aerostatic globes and Animal Magnetism have, during the whole of this past year, so filled people's heads in this country that *useful research* has been utterly neglected.' Blagden added pointedly that this now expressed the view of 'the soberer part' of the French Academy.

French ballooning certainly generated the most powerful outpouring of popular feeling. It also assembled enormous crowds in Paris, full of dangerous utopian dreams and heady aspirations. The kind of eyewitness account of such balloon launches which would have alarmed Banks is well illustrated by *Le Tableau de Paris*, 1 December 1783:

Extrait du Journal du 21 Novembre 1783

The swarm of people was itself an incomparable sight, so varied was it, so vast and so changing. Two hundred thousand men, lifting their hands in wonder, admiring, glad, astonished; some in tears for the intrepid philosophers should they come to harm; some on their knees overcome with emotion; but all following the aeronauts in spirit, while these latter, unmoved, saluted, dipping their flags above our heads. What with the novelty, the dignity of the experiment; the unclouded sun, welcoming as it were the travellers to his own element; the attitude of the two men themselves sailing into the blue, while below their fellow-citizens prayed and feared for their safety; and lastly the balloon itself, superb in the sunlight, soaring aloft like a planet, or the chariot of some weather-god! It was a moment which can never be repeated, the most astounding achievement the science of physics has yet given to the world.

Yet such wild enthusiasm could strongly appeal to a British physician and inventor like Dr Erasmus Darwin. Though Darwin was a Fellow of the Royal Society (elected in 1761), he was also part of a radical and non-conformist network of provincial philosophers, and a leading light of the Lunar Society based in Birmingham and Derby. Moreover he was a poet.

Darwin saw the Montgolfiers as the pioneers of a new age, and was untroubled by Banks' scientific reservations or patriotic anxieties. He celebrated the Montgolfiers' early flights in a long, ecstatic passage from his poem *The Loves of the Plants* (eventually published in 1789). A botanical description of a flying thistle-seed from Canto 2, 'Air', was suddenly transformed into an image of an airborne French balloon:

> ... *So on the shoreless air the intrepid Gaul*
> *Launch'd the vast concave of his buoyant ball,*
> *Journeying on high, the silken castle glides*
> *Bright as a meteor through the azure tides;*

Erasmus Darwin –
an enthusiast of the
early balloon flights.

O'er towns, and towers, and temples wins its way,
Or mounts sublime, and gilds the vault of day …

Darwin presents the Montgolfier balloon's ascent first as seen from *below,* as the watching crowd gaze upwards, torn between rapture and terror:

Silent with upturn'd eye unbreathing crowds
Pursue the floating wonder to the clouds;
And flush'd with transport, or benumb'd with fear
Watch, as it rises, the diminish'd sphere.
Now less and less – and now a speck is seen –
And now the fleeting rack obtrudes between!

Then Darwin ascends to the imagined view from *above,* looking out from the aeronaut's basket, at the very edge of the stratosphere. The aeronaut gazes down upon the Earth, but also upwards at the stars revealed above him. He observes the clouds and the weather systems. He even sees (like the future astronauts) the curvature of the planet itself, in the blue horizon-line of the ocean:

The calm Philosopher in ether sails,
Views broader stars and breathes in purer gales;
Sees like a map in many a waving line,
Round earth's blue plains her lucid waters shine;
Sees at his feet the forky lightning glow
And hears innocuous thunder roar below ...

Finally, in a quite extraordinary passage, Darwin sends the balloon on a fantasy voyage right through the solar system. Here science is frankly abandoned for science fiction. 'Rise, great Montgolfier! Urge thy venturous flight / High o'er the moon's pale ice-reflected light.' He sails past 'the red eye' of Mars, floats round Jupiter, and surges beyond Saturn with its 'crystal rings' and Herschel's newly discovered 'Georgian' planet of Uranus. He elevates magnificently through the Milky Way, and the glittering constellations of the zodiac. He becomes a new North star, 'to blaze eternal round the wondering pole', a beckoning light to all future space 'mariners'. (*The Loves of the Plants,* 1789, Canto 2, lines 27–66.)

British journalists, though more sceptical than poets, were not entirely immune to such fantasy journeys either. They also saw the immense possibilities of balloon flight, and responded to the first reports of the French experiments. A long article on ballooning in the *Monthly Review* at the end of 1783 concluded:

We found our imaginations warmed by the gigantic idea of our penetrating some day into the wildest and most inhospitable regions of Africa, Arabia, and America, of our crossing chains of mountains hitherto impervious, and ascending their loftiest summits, of our reaching either of the two poles and in short, of our extending our dominion over the creation beyond any thing which we now have conception.

REPORTS

It was exactly this kind of ballomania and unscientific speculation that Joseph Banks is reputed to have dismissed out of hand. Yet the recent publication of Banks' *Scientific Correspondence*,[1] and a re-examination of the Royal Society archives suggests a more complex and intriguing state of affairs.

First, it turns out that no fewer than fifty letters on the subject of ballooning were exchanged between Banks and his scientific correspondents between 1783 and 1786. Not only Banks himself, but several other Fellows of the Royal Society (besides Darwin) were evidently fascinated by ballooning, and became far more closely involved than has been previously assumed.

It is also clear that the latest balloon news, including extensive cuttings from the French newspapers, and many 'a parcel of pamphlets and journals', was regularly supplied to Banks not only by Charles Blagden, but also by Banks' personal assistant and librarian at Soho Square, the Swedish botanist Jonas Dryander.

This began in September 1783 when Dryander excitedly passed on a package from Paris: '*Journal de Paris* from August to September 17 is just come. I have only had time to turn over some of the last numbers to hunt for information about the great aerostatique experiment. I'll copy here the description of the machine ...'

Banks continued to receive such detailed reports of all the French balloon ascents throughout the rest of 1783 and 1784. He was informed that the French Academy of Sciences, under his opposite number the Marquis de Condorcet, had appointed an official commission to investigate 'aerostation', and were funding further ascents by Pilâtre de Rozier. He also received various communications from Barthélemy Faujas de Saint-Fond, a geologist and official from the Jardin du Roi, who had set himself up as a commercial promoter of ballooning in France. Saint-Fond published one of the earliest books on the subject, *Descriptions des Experiences des Machines Aerostatiques de MM Montgolfier;* Banks had obtained his own copy by the end of November 1783.

1 See the wonderful new edition, *The Scientific Correspondence of Joseph Banks 1765–1820*, edited by Neil Chambers, 6 vols (London, Pickering & Chatto Ltd, 2007). Further sources are given in my bibliography on page 486.

A letter dated 20 October 1784 from Sir Joseph Banks to Charles Blagden. In it he explains how on this particular flight the Duke of Chartres and his companions quickly experienced problems and were 'frightened beyond the power of assisting thence'. However, one companion 'kept his presence of mind' and directed the balloon away from 'a deep pond'.

Apart from Blagden, Dryander and Saint-Fond, Banks' most important source of balloon information was Benjamin Franklin, by then in his seventies and wise in the ways of both men and machines. It has not been appreciated how significant this contact was. Franklin's 'two papers', submitted to the Society in November 1783, were in reality just part of an extensive exchange of confidential letters and ballooning documents between him and Banks, amounting to no fewer than sixteen items, which continued virtually unbroken from July 1783 to April 1785.

Banks' fascination with ballooning is expressed much more openly to Franklin than to anyone else. In September 1783 he wrote: 'Most agreeable are the hopes you give me … I consider the present day which has opened a Road in the Air, as an Epoch … the more immediate Effect it will have upon the concerns of mankind, [is] greater than anything since the invention of Shipping …'

Portraits of Benjamin Franklin (left) by Joseph Wright, 1782, and Joseph Banks (right) by Thomas Phillips, 1815.

It was to Franklin, rather than to Blagden, that Banks wrote so warmly on 28 November 1783, immediately after the Montgolfiers' first manned flight.

The Experiment now becomes interesting in no small degree. I laughed when Balloons of scarce more importance than soap bubbles occupied the attention of France. But when men can with safety pass, and do pass, more than 5 miles in the first Experiment I begin to fancy that I espy the hand of the Master in the education of the Infant Knowledge, who so speedily attains such a degree of maturity …

If not a 'ballomaniac', Franklin was certainly in favour of balloons. He had interviewed Joseph Montgolfier and the Marquis d'Arlandes at the American Embassy, the evening after that first manned flight. He had also witnessed Dr Alexander's first ascent by hydrogen balloon from the Tuilleries on 1 December, and sent Banks a most eloquent account. 'All Paris was out, either about the Tuilleries, on the quays and bridges, in the fields, the streets, at the windows, or on the tops of houses …'

It was after this flight that Franklin was reported to have made his famous remark, when asked what was the use of a balloon: 'I replied – *what's the use of a newborn baby?*' Perhaps he was inspired by Banks' earlier reference to 'the Infant Knowledge'.

Banks again wrote enthusiastically to Franklin on 9 December:

The new Art of Flying … makes such rapid advances in the country you now inhabit … Charles's Experiment seems decisive, and must be performed here in its full extent. I have hitherto been of the Opinion that it is unwise to struggle for the honour of an invention that is about to be Effected. Practical flying we must allow to our rivals, Theoretical flying we claim ourselves … When our Friends on your side of the water are cooled a little … they shall see that we will visit the repositories of the Stars and Meteors.

The question now became, what were the realistic applications of 'Practical flying', as opposed to theories and fantasies?

PRACTICAL FLYING

Indeed it was not at first clear, either to the Royal Society or the French Academy of Sciences, what the true purpose or possibilities of ballooning really were. In fact 'flight' was itself a novel and surprisingly unexplored concept, despite an extensive literary tradition from Icarus and Pegasus onwards. What, in practice, could balloons actually do for mankind, except provide a hazardous journey interspersed with fine aerial 'Prospects'?[2]

According to Saint-Fond they might, for example, provide observation platforms: for military reconnaissance, for sailors at sea, for chemists analysing the Earth's upper atmosphere, or for astronomers with their telescopes. It is notable that most of these applications were based on the notion of a *tethered* balloon. In fact many of the Montgolfiers' early experiments were made with tethered aerostats, held to the ground by various ingenious forms of harness, guy ropes or winches.

Despite his poetical effusions, Erasmus Darwin's first practical idea of balloon-power was paradoxically that of shifting payloads along the ground. He suggested to his friend Richard Edgeworth that a small hydrogen balloon might be tethered to an adapted garden wheel-barrow, and used for transporting heavy loads of garden manure up the steep hills of his Irish estate. This convenient aerial skip would allow one man to shift ten times his normal weight. Indeed it might revolutionise manual labour.

Similarly, Banks himself had the initial idea that balloons could increase the effectiveness of earth-bound transport, by adding to its conventional

2 The idea that the 'Prospect' itself – the free ascent, the magnificent views, the whole 'aerial experience' – was the real point of ballooning, only truly arrived with the sporting, propane-powered hot-air balloons of the late twentieth century. However, one early pioneer of this existential attitude was Thomas Baldwin, whose remarkable *Aeropaedia* (1786) was an entire book dedicated to a single flight, made from Chester on 8 September 1785. It contained the first ever paintings of the view from a balloon-basket; an analytic diagram of the corkscrew flight path projected over a land map; and a whole chapter simply given up to describing the astonishing colours and structures of cloud-formations. One typical observation reads: 'The river Dee appeared of a red colour; the city [Chester] very diminutive; and the town [Warrington] entirely blue. The whole appeared a perfect plane, the highest buildings having no apparent height, but reduced all to the same level, and the whole terrestrial prospect appeared like a coloured map.' [p. 204].

horsepower. He saw the balloon as 'a counterpoise to Absolute Gravity': that is, as a flotation device to be attached to traditional forms of coach or cart, making them easier to move over the ground. So 'a broad-wheeled wagon' normally requiring eight horses to pull it, might only need two horses with a Montgolfier attached. This aptly suggests how difficult it was, even for a trained scientific mind like Banks', to imagine the true possibilities of flight in these early days.

Franklin, 'the old fox' as Blagden called him, was quick to suggest various menacing military applications, perhaps deliberately intended to fix Banks' attention. 'Five thousand balloons capable of raising two men each' could easily transport an effective invasion army of ten thousand marines across the Channel, in the course of a single morning. The only question was, Franklin implied, which direction would the wind be blowing from?

His other speculations were more light-hearted. What about a 'running Footman'? Such a man might be suspended under a small hydrogen balloon, so his body weight was reduced to 'perhaps 8 or 10 Pounds', and thus made capable of running in a straight line in leaps and bounds 'across Countries as fast as the Wind, and over Hedges, Ditches & even Water…' Or there was the balloon 'Elbow Chair', placed in a beauty spot, and winching the picturesque spectator 'a Mile high for a Guinea' to see the view. Then there was Franklin's patent balloon icebox. 'People will keep such Globes anchored in the Air, to which by Pullies they may draw up Game to be preserved in the Cool, & Water to be frozen when Ice is wanted.' This contraption would surely have appealed to that twentieth-century illustrator Heath Robinson.

Many other ingenious suggestions were made, including the use of balloons as buoyancy tanks for ships, as aerial river-ferries, and for air mail between towns. The latter merely required that the recipients were always precisely downwind of the sender. Indeed, Erasmus Darwin attempted to pioneer balloon-post by sending a Christmas letter in December 1783, attached to a small hydrogen balloon. It was meant to fly northwards carrying seasonal greetings from the Philosophical Society in Derby to Matthew Boulton's garden in Birmingham. In the event it overshot by fifteen miles when 'the wicked wind carried it to Sir Edward Littleton's'.[3]

Thomas Martyn, a Professor of Botany at Cambridge, published an illustrated pamphlet appealing directly to the Royal Society, *Hints of Important Uses for Aerostatic Globes*, 1784. Martyn's big idea was high-speed visual communications by tethered balloon. He urged the use of balloons as signal platforms, invaluable for directing armies on land or fleets at sea. A day-time system of flag semaphore could be replaced by fireworks at night – a rather more problematic suggestion. 'These Experiments … might be beyond measure enlarged and extended under the direction of a public body, such as our Royal Society.'

Finally even Professor Martyn succumbed to aerostatic fantasy, by fixing an astonishing frontispiece to his pamphlet. It showed a huge, beautiful dream-balloon soaring magnificently amidst the clouds, carrying beneath it a solid, wooden ocean-going 'air-ship', with square-rigged sails, large sea-going rudder and elegant anchor on a chain, evidently ready to circumnavigate the entire globe.

NAVIGATION

The great emerging scientific question became this: could an aerostat be navigated? Was it truly an 'air-ship'? Could a balloon be steered against the prevailing air current, to a previously chosen destination? Could it ever, quite simply, provide a sure method of getting from A to B? Throughout 1784 Banks closely followed the British balloon flights of Lunardi and Blanchard with this navigation question in mind. Several distinguished Fellows of the Royal Society were sent to observe them. Blagden and Cavendish, together with the astronomers Herschel and Aubert, stationed themselves at various

3 The supremely impractical suggestion of balloon mail was to be strangely vindicated by the French some ninety years later. During the Prussian siege of Paris in 1870–71, no fewer than sixty-six hydrogen balloons, each carrying 125 kilos of domestic mail and government despatches, sailed successfully over the Prussian lines, landing as far afield as unoccupied Brittany, whereupon the mail was rapidly distributed by horse across the nation. The first balloon, the *Neptune*, carried a letter from the photographer Felix Nadar to *The Times*. Subsequent balloons, with that touch of French genius, teased the Prussians by having patriotic names emblazoned on their canopies in huge letters – the *Victor Hugo*, the *George Sand*, the *Armand Barbès*.

rooftop vantage points in London, equipped with telescopes and quadrants. They carefully sent back their data to Banks, and made a special point of observing the effects of wings, oars and rudders on the balloon's horizontal flight-path. Could it be diverted against or across the wind, however marginally? Lunardi favoured simple wooden oars for this task, while Blanchard proclaimed his faith in silken wings, cotton rudders and a complex propeller-type device known as a *moulinette* ('a sort of ventilator that could be turned by means of a handle'). Despite their repeated claims, none of this equipment produced the least observable effect.

These negative observations were significant, because aeronauts in France had been claiming that they could produce a slightly diverted flight-path across the line of the wind, using sails and rudders. During an impressive 150-mile flight made from Paris to Artois on 19 September 1784, the Roberts brothers, who had helped design Dr Charles' original balloon, stated with pseudo-scientific precision that they had achieved a 'deflection of 22 degrees', and 'might have obtained 80 degrees'. This, they argued, was almost as efficient as a close-hauled sailing ship moving through the comparable medium of water. Banks now had reason to believe that they were deluded.

The one scientific instrument which proved effective in balloon navigation was the mercury barometer. It was already established that air pressure dropped with an increase in altitude. In some sense, not entirely understood, the air got 'thinner' the higher one went. So as a balloon rose, an onboard barometer would give a steadily lower reading; and conversely, as the balloon descended, the barometric reading would rise. So an appropriately calibrated barometer (with an adjustable scale set at zero immediately before launching) could act as an altimeter, indicating a balloon's changing height above the ground.

Banks was therefore particularly scathing when he learned that Lunardi had forgotten to take a barometer on his first historic ascent in September 1784, and had pretended to calculate his maximum altitude from the length of the icicles formed on the lower edge of the balloon canopy. He concluded that the pilot was a brilliant charlatan. Banks feared that Lunardi, having entranced the fashionable and susceptible Duchess of Devonshire, would go on to ensnare the gullible Prince of Wales, and even King George III (already rather less than stable) with his 'balloon madness'.

But there was an alternative to Lunardi: the Frenchman Jean-Pierre Blanchard. In the autumn of 1784, two Fellows of the Royal Society decided to purchase private passages aboard Blanchard's hydrogen balloon, making proper observations and taking appropriate equipment with them. The first was John Sheldon, Professor of Anatomy at the Royal Academy, who flew from Chelsea in October 1784.

Despite much anticipation, Sheldon's flight was largely abortive from a scientific point of view. 'The balloon was so loaded at first,' recorded Blagden dryly, 'that it fell down in a neighbour's garden.' Alarmed by the whole experience, Sheldon broke his barometer shortly before take-off, while Blanchard threw overboard the rest of his equipment immediately after. Blanchard mercifully off-loaded the terrified Sheldon at Sunbury, in Middlesex. He then claimed that he had successfully navigated with his wings and rudder some seventy-five miles into Hampshire.

But the first half-hour of the ascent was observed by Blagden and Cavendish from the roof of a house at Putney Heath 'with instruments', triangulating their observations with another observer from a house in Earls Court. Their meticulous calculations showed that the balloon 'floated along with the wind uniformly and regularly, seeming to pay no regard to the operation of the machinery they had taken up'. There was still no indication that a balloon could be navigated.

Blagden estimated that Sheldon had spent £500 on the ascent, and concluded that he had 'made himself so ridiculous in this business, as to reflect little credit on the Royal Society'. Banks noted, with perhaps pardonable ambiguity, that 'Mr Sheldon and Mr Blanchard have probably fallen out, as I have not heard a word from them for some time.'

The next philosopher to purchase a flight with Blanchard was the American physician Dr John Jeffries, in November 1784, ascending from Grosvenor Square. In fact, Jeffries was not yet a Fellow of the Royal Society, but hoped to be elected on the strength of his ballooning experiments. Accordingly, he carefully prepared a suite of scientific instruments to take with him: a mercury barometer, a thermometer, a hygrometer and an electrometer, to measure the much-feared electrical charges in clouds. In addition he packed maps, a compass and special note-making equipment. He also strapped aboard special air flasks, to sample the upper atmosphere at different altitudes, which he promised to give to Cavendish for analysis.

Jeffries drew up a memorandum for the Royal Society before they left, stating the main scientific objectives of the ascents, to be achieved by 'a variety of experiments' and 'not for mere amusement'. He was quite precise:

Four points need to be more clearly determined. First, the power of ascending or descending at pleasure, while suspended or floating in the air. Secondly, the effect which oars or wings might be made to produce towards this purpose, and in directing the course of the Balloon. Thirdly, the state and temperature of the atmosphere at different heights above the earth. And fourthly, by observing the varying course of the currents of air, or winds, at certain elevations, to throw some new light on the theory of winds in general.

On this trip, going across the Thames into Kent, Jeffries made the first truly scientific record of a balloon ascent. He meticulously recorded a mass of data – height, direction, air temperature, electrical charges, appearance of clouds, horizon line – at regular time intervals. One of the details which emerged was a 'profile' of the characteristic flight-path of a hydrogen balloon: not a single smooth parabola, as had been supposed, but a series of looping ascents and descents, as the balloon moved above and below its 'equilibrium point'. It was also clear to Jeffries that wind directions often changed at different altitudes. But on the crucial question of navigation, Jeffries could observe no controlled alteration of flight-path, for all Blanchard's 'heroic' rowing and flapping and spinning.

Jeffries went on to take part in the most significant of all the early balloon ascents in Britain, the first crossing of the English Channel with Blanchard on 5 January 1785. He wrote an outstanding account, which exists in at least three versions. The first was sent as a private letter to Banks from Paris shortly after the flight on 13 January 1785, the second as a formal paper published by the Royal Society in the *Philosophical Transactions* for January 1786, and the third as a retrospective diary.

Despite its apparent triumph, both sporting and diplomatic, the main scientific significance of this flight was that it proved conclusively that a balloon was not navigable, either over land or sea. As Jeffries expressed it privately in his diary, he could only 'thank God' and a favourable wind for his survival. He never flew again.

By the end of 1785, Banks too was rapidly losing interest in ballooning. His correspondence with Franklin tailed off into a courteous exchange of medals and compliments. His doubts could be summed up succinctly: balloons were not navigable, and – as he had originally thought – they should be left to the French. Yet at the last Banks may have encouraged a book by a younger Fellow of the Royal Society that would inspire a new generation of aeronauts.

RETROSPECTIVE

In 1785 Tiberius Cavallo FRS published *A Treatise on the History and Practice of Aerostation.* Cavallo was a brilliant Italian physicist who had moved to London at the age of twenty-two, and had already written extensively on magnetism and electrical phenomena. Elected a Fellow in 1779, he quickly turned his attention to ballooning. He had some claims to be one of the first to inflate soap bubbles with hydrogen as early as

Opposite:
Tiberius Cavallo
by an unknown
artist, circa 1785.

1782. Although a handsome portrait is held by the National Portrait Gallery in London, he is now largely and unjustly forgotten. Yet his study emerges as the most authoritative early treatise on the subject, either in English or French. The copy of Cavallo's book held by the British Library is personally inscribed 'To Sir Joseph Banks from the Author' – in firm, black, racy ink.

Cavallo adopted a considered and even sceptical tone, well calculated to appeal to Banks. Of his fellow-countryman Lunardi's historic flight he noted:

> Besides the Romantic observations which might be naturally suggested by the Prospect seen from that elevated situation, and by the agreeable calm he felt after the fatigue, the anxiety, and the accomplishment of his Experiment, Mr Lunardi seems to have made no particular philosophical observation, or such as may either tend to improve the subject of aerostation, or to throw light on any operation in Nature.

He analysed and dismissed most claims to navigate balloons, except by the use of different air currents at different altitudes. He emphasised the aeronaut's vulnerability to unpredictable atmospheric phenomena, such as downdraughts, lightning strikes and ice formation. He deliberately included the alarming account of those who survived when a French balloon was caught in a thunderstorm, during an ascent from St Cloud in July 1784, and dragged helplessly *upwards* by a thermal:

> Three minutes after ascending, the balloon was lost in the clouds, and the aerial voyagers lost sight of the earth, being involved in dense vapour. Here an unusual agitation of the air, somewhat like a whirlwind, in a moment turned the machine three times from the right to the left. The violent shocks, which they suffered prevented their using any of the means proposed for the direction of the

balloon, and they even tore away the silk stuff of which the helm was made. Never, said they, a more dreadful scene presented itself to any eye, than that in which they were involved. A unbounded ocean of shapeless clouds rolled one upon another beneath, and seemed to forbid their return to earth, which was still invisible. The agitation of the balloon became greater every moment …

Yet for all this, Cavallo was a passionate balloon enthusiast. He recorded and analysed all the significant flights, both French and English, made from the Montgolfiers' first balloon at Annonay in June 1783 to Blanchard and Jeffries' crossing of the Channel in January 1785. He distinguished carefully between hot-air and hydrogen balloons, and their quite different flight characteristics. He looked in detail at methods of preparing hydrogen gas, noting that Priestley had come up with one that used steam rather than sulphuric acid. He also examined the different ways of constructing balloon canopies from rubber ('cauchouc'), waxed silk, varnished linen and taffeta.

In a longer perspective, he stressed the astonishing speed of aerial travel over the ground – 'often between 40 and 50 miles per hour' – combined with its incredible 'stillness and tranquillity' in most normal conditions. This he thought must eventually revolutionise transport and communications, even if the moment had not yet arrived. He pointed out that in achieving altitudes of over two miles, balloons opened a whole new dimension to mankind's observations of the Earth beneath. Man's growing impact on the surface of the planet became visible from the air for the first time, as did the vast tracts of the Earth – mountains, forests, deserts – yet to be traversed. Above all he stressed that the full potential of flight had not yet been remotely explored.

Cavallo considered the whole range of possible balloon applications. But he finally and presciently championed its relevance to the infant science of meteorology:

The philosophical uses to which these machines may be subservient are numerous indeed; and it may be sufficient to say, that hardly anything of what passes in the atmosphere is known with precision, and that principally for want of a method of ascending into the atmosphere. The formation of rain, of thunder-storms, of vapours, hail, snow and meteors in general, require to be attentively examined and ascertained.

The action of the barometer, the refraction and temperature of air in various regions, the descent of bodies, the propagation of sound etc are subjects which all require a long series of observations and experiments, the performance of which could never have been properly expected, before the discovery of these machines. We may therefore conclude with a wish that the learned, and the encouragers of useful knowledge, may unanimously concur in endeavouring to promote the subject of aerostation, and to render it useful as possible to mankind.

It was largely due to Cavallo's book that, a decade later, ballooning received a signal acknowledgment and consecration. The third edition of the hugely influential *Encyclopaedia Britannica*, published in 1797, for the first time recognised the existence of 'Aerostation'. It described it with all due formality as 'a science newly introduced to the Encyclopaedia', and gave it a comprehensive article of fourteen pages. This included two full spreads of diagrammatic illustrations, showing every known kind of aerostats that would actually fly. Almost all the material was drawn, unacknowledged, from Tiberius Cavallo.

The editors of the great *Encyclopaedia* made one symbolic gesture. They placed as the frontispiece to the opening volume of the new edition a prophetic engraving. It showed a traditional gathering of 'natural philosophers' in a Roman forum, arrayed in classical togas and surrounded by pillared Doric temples. (Could they have intended a sly reference to the

Royal Society?) They then introduced one striking anachronism. High overhead, a hydrogen balloon (complete with wings) sails imperiously into some unknown future.

Such prophetic dreams would soon be taken up by a new generation of British aeronauts, such as James Sadler and Charles Green. But as for Sir Joseph Banks PRS, now perhaps made more earth-bound by his knighthood, aerostation virtually disappears from his letters after 1790. When in January 1800 he received a charming inquiry from Ireland suggesting a scheme to build a balloon railway beneath a 'mile-long covered gallery' at Greenwich, he replied with barely a sigh: 'The Royal Society have no Funds destined for the Execution of Projects so Expensive as yours must be; nor indeed have they in any one instance *interfered* in the business of Aerostation.'

8

Cacao
Ray.Hist. 1679
The Cacao tree

EXPLANATION.

RICHARD FORTEY

ARCHIVES OF LIFE:
SCIENCE AND COLLECTIONS

Richard Fortey FRS is a geologist and palaeontologist and spent his career in research at London's Natural History Museum from where he retired in 2006. His widely acclaimed books include *The Hidden Landscape, Life: An Unauthorised Biography, Trilobite!: Eyewitness to Evolution, Fossils: The Key to the Past* and *The Earth: An Intimate History.* His latest book, *Dry Store Room No.1*, is a portrait of the Natural History Museum.

OBSERVATION WAS A CRUCIAL FOUNDATION FOR THE NEW SCIENCE. IN BIOLOGY, THAT MEANT THE CLOSEST EXAMINATION OF SPECIMENS. KEEPING THEM, SO OTHERS COULD REFINE THE OBSERVATIONS YEARS, DECADES, OR EVEN CENTURIES LATER, PROVED TO BE JUST AS IMPORTANT, AS RICHARD FORTEY EXPLAINS.

Safely stored behind the scenes at the Natural History Museum in South Kensington is a slightly twisted vertebrate skeleton preserved on a slab of creamy white limestone. This particular specimen was discovered in quarries near Solnhofen in southern Germany in 1861. The fine limestones of Solnhofen are ideally suited to making lithographic stones, and in the nineteenth century lithographs provided one of the most important means of book illustration – indeed lithographic stones of this quality are still in demand by artists today. Vast quantities of this lithographic limestone of Jurassic age – about 150 million years old – have been taken out of opencast workings, where the rocks can be split into convenient slabs a centimetre or two thick; the German word *plattenkalk* appropriately describes their lithological character. On many of these flat-surfaced pieces of rock, fossils are laid out like gifts on a salver.

Some Solnhofen fossils are rather common, such as those of delicate little sea lilies. Others are both rare and more spectacular. There are a great variety of fish species known nowhere else, for example. The fossil horseshoe crab *Mesolimulus* provides evidence that its living relatives breeding each year along the Atlantic coast of America have changed little over tens of millions of years. Delicate flying reptiles – half a dozen species or so of pterodactyl – testify by contrast to creatures that have vanished from the Earth for ever. A few species of dinosaur are known, of the most delicate sort (*Compsognathus*), and quite unlike the monsters of popular imagination. Insects include dragonflies (*Aeschnogomphus*) whose every wing-vein is visible as delicate tracery. All these creatures are preserved in rocks which originated as tacky muds flooring a lagoon that lay offshore from a richly biodiverse habitat. Such special circumstances sampled and preserved a much wider variety of organisms than the usual fossil locality, and the wide range of fossils provides a rare window into an entire habitat from a very different world. Yet if the remains were not kept carefully in museums all this evidence of past life would perish, and new generations of children and scholars could not interrogate the past. Local museums at Eichstätt and Solnhofen fulfil that function for those who would come to Bavaria and marvel at its geological treasures. But some of the specimens from the Solnhofen limestone have a relevance that extends far beyond the reconstruction of the late Jurassic scene, and these specimens are treasures in the collections of museums around the world. None more so than that specimen – a mere 35 cm at its longest – safely curated in the Natural History Museum in London.

For this is the first example ever discovered of the early bird *Archaeopteryx*. It remains one of the most important specimens in the British national collections. The next complete fossil bird of the same species – the so-called Berlin specimen – was found sixteen years later. It would be difficult to overstate the importance of this London specimen of *Archaeopteryx* in the history of biology.

First, the date of its discovery is only two years after the publication of *The Origin of Species,* the sesquicentenary of which we celebrated in 2009. Charles Darwin famously described what he called 'difficulties on theory' in that work, where he anticipated a number of criticisms that he expected his great idea to encounter. Prime among these was 'the rarity or absence of intermediate forms' in the fossil record. Second, the detailed scientific description of *Archaeopteryx* was an accomplishment of Richard Owen in 1863; he was later to become first director of the Natural History Museum. Owen was no Darwinian, but he was an able anatomist. It must have proved anathema to him when *Archaeopteryx* was recruited as probably the best example of an 'intermediate form' and one that had turned up with the impeccable timing usually associated with a good piece of theatre. Its amalgam of reptilian and bird features (feathers and wishbone among them) was a striking vindication of the notion of descent with modification, and a rebuttal to those who might wonder how it was possible for animals to make the transition from earth to the skies.

In this sense *Archaeopteryx* became a kind of talisman for evolution. Owen was enough of a 'Museum man' to ensure that this fossil was safely curated, and part of any museum's function is just that – to protect material regardless of the current explanations of its importance. The old bird has now been joined by half a dozen or so subsequent examples worldwide, but its importance has not diminished over the years. Periodically, it has been taken out from storage and re-evaluated. Sir Gavin de Beer described it in great detail in 1954. Twenty years later more bits of it were manually prepared, and new details revealed, and in the last few years the brain case of the early bird has been CAT-scanned and its endocast reconstructed. All these endeavours have served to confirm the transitional nature of *Archaeopteryx* – but have also confirmed that in most important functional respects it is closer to the birds than to the dinosaurs. This in turn has contributed to the debate about whether birds descended from one particular group of dinosaurs: most palaeontologists nowadays concur that they did. One might say that the *meaning* of *Archaeopteryx* has changed, while

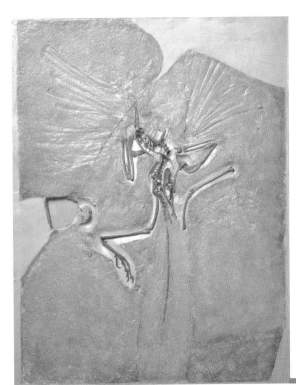

The London specimen of *Archaeopteryx*, on display at the Natural History Museum.

the information that has been extracted from this specimen (and other new discoveries) has increased fitfully as scientific hypotheses have shifted.

I begin with the London specimen of *Archaeopteryx* because it is an emblem for the importance of collections in science. Collections provide the ground truth on which hypotheses are built. Physics has laboratories; systematic biology has collections. It would be misleading to claim that the millions of specimens stored in cabinets and bottles in the galleries of national natural history museums are all, individually, as important as the type specimen of *Archaeopteryx*. But well localised, properly documented natural history archives have been, and continue to be, central to under-standing many kinds of scientific questions: the course of evolution; the relationships between animals and plants (the 'tree of life'); biogeography and biodiversity; how climate change has affected the biota. Human memories are short and inaccurate. Our shifting perceptions need to be tested against archives which are – as near as possible – permanent records of the fauna and flora.

This concept of collections developed or evolved rather like those organisms kept in drawers or herbaria. There is evidence that humankind made collections from the earliest times, if claims about pierced snails and tusk shells from Africa are to be believed. These first collections were assuredly made for ornament, but humans evidently had a taxonomic eye from the outset, by picking out matching individuals belonging to a single species. Development of a 'working taxonomy' – distinguishing edible from poisonous plants, for example – is clearly of adaptive value. Collections made for cultural purposes accompanied early civilisations, and Adrienne Mayor has argued that fossil mammal collections made from the Cenozoic rocks of the Mediterranean region were displayed in Classical times as concrete evidence of the battles between races of giants and men: evidence of a kind, but mostly spectacle.

The growth of scientific collections in a more modern sense frequently also had a comparable connection with display. The major figures in the early intellectual history of collections made what were essentially personal acquisitions, and a genuine love of scholarship happily mixed with a certain showmanship. They wanted to elicit admiration from their peers as well as understanding. John Evelyn (1620–1706) was both active in the Royal Society at its inception and was one of the outstanding virtuosi of his age; he corresponded with Boyle and Wren and other scientifically minded Fellows. It would be incorrect to categorise Evelyn as a scientist (after all, the term itself did not exist) though rationalist he assuredly was. His garden at Sayes Court, Deptford, was in a sense a research laboratory, a living catalogue of plants, and Evelyn was a pioneer in recognising what would now be regarded as the balanced diet and the importance of nutrition. He was justly proud of his garden and liked to show it off to his influential friends. But the idea of a living collection of plants was a natural extension from the medicinal gardens of the herbalists, and only a step away from the botanical gardens of today. The 'system' of specimen arrangement might change from one of curative 'virtues' to one of botanical classification, but on the ground that is only a matter of moving plants from one bed to another.

As in so many other fields, Joseph Banks (1743–1820) contributed to the evolution of collections for scientific ends. When the young Banks embarked on the *Endeavour* under the captaincy of James Cook he was intellectually omnivorous, for all his official label as the expedition botanist. The expedition arrived in Tahiti on 13 April 1769 and stayed for three months. It is clear from the *Endeavour Journal* that Banks had a remarkably open attitude towards the manners and customs of the Tahitians; his observations cover the sexual mores, tattooing procedures, food and cooking, and organisational hierarchies of the native peoples, and are engagingly frank, without any sense of patronisation. The latter was to change, particularly in Victorian times, but Banks' non-partisan approach speaks highly of the feisty aristocrat, and it was an attitude that he maintained despite several assaults that would have daunted a lesser man. One could also argue that his methods anticipated those of social anthropologists more than a century later. He even learned something of the language of the Tahitians, which is now regarded as the first thing any aspiring anthropologist must do.

Banks' ethnological and natural history collections were displayed to a wondering public at an apartment in New Burlington Street in 1772. They caused something of a sensation. In three rooms he exhibited different collections of the objects acquired on the famous voyage: militaria and sailing paraphernalia in one room; in a second, cooking utensils, dresses, jewellery and the like, together with 1,300 new species of plants; while a third room displayed a range of natural history specimens – reptiles, amphibians, birds, insects and many more, most new to science. The exhibition was more than just showmanship and display. It established the veracity of what Banks and his colleagues had seen on Cook's voyage. The specimens became vouchers for the truth, and as such acquired permanent value. To be sure, his written observations of native peoples do constitute another kind of 'collection', but Banks was also assiduously developing the routines of making scientific and permanent collections of the natural world, in the company of his faithful friend, the botanist Daniel Solander. His *Journal* abundantly attests to a *routine*, and such steady behaviour always seems to characterise the scientist – as opposed to the poet, perhaps. Because of the perishable nature of living organisms it was also necessary to preserve the animal or plant in an image, and during the voyage Sydney Parkinson was on hand (until his untimely death*) to sketch and then colour the new finds with exquisite delicacy. Parkinson provided what has been termed a 'virtual museum' – a testimony to biological reality that could eventually be distributed among savants throughout Europe. Australia and New Zealand's botanical wonders could be experienced on paper. The herbarium specimens were permanent, but pallid.

Curiously, though, Banks never fully published Parkinson's splendid drawings of the flora encountered on the *Endeavour*'s voyage. This is all the odder because Banks had spent £7,000 between 1771 and 1784 from his

<hr />

* Sydney Parkinson tragically died of dysentry on the way to Cape Town, 17 January 1771.

personal fortune to have copper plates of superb quality engraved from the watercolour drawings. Two centuries passed before the botanical engravings were finally published in their full glory; this happened between 1980 and 1990 as *Banks' Florilegium,* produced in several parts to the highest standards by Alecto Historical Editions and the Natural History Museum. Banks eventually bequeathed both his plates and his specimens to what was then the British Museum, where they remain to this day. The reasons for Banks' reluctance to publish are not clear; doubtless perfectionism was part of the problem. Then he was always busy with his duties as seemingly perpetual President of the Royal Society. The death of his friend Solander in 1782 did not help either – nor did the drop in share prices in the years leading up to the Napoleonic Wars.

However, an interesting idea is suggested by Banks' removal of the collections to a permanent house in his London address in Soho Square. Here they were freely available to visiting scholars, including those from abroad. They became proper reference specimens, like the London *Archaeopteryx* with which this chapter began. Although not described in so many words, Banks had created a museum with pretensions for the public good. When he was taken on board the *Endeavour* the emphasis might have been on the discovery of commercially significant plants, or, in the words of the Council of the Royal Society 'for the advancement of useful knowledge'. Although Banks had a good eye for business possibilities this was not the *raison d'être* of Soho Square, which was directed equally towards the scholar and naturalist. Maybe urgency of publication for Banks was diluted by the ready availability of his collections to those who desired to see the spoils of exploration, or make comparisons with some other plant to hand. The 'virtual museum' could wait.

Banks also had a central role in the promulgation of *living* collections. He was closely involved with what eventually became the Royal Botanical Gardens at Kew, and by 1773 was *de facto* director. He planted eight hundred trees and shrubs originating from North America. In the Thames-side soil west of London 'useful knowledge' of plants could indeed be turned to potential gain. The fashion for hot houses full of exotics was in turn taken up by many members of the aristocracy – often for reasons of conspicuous display as much as botanical enthusiasm. The organised collection of living plants at Kew Gardens continues splendidly to this day, and the important role of these collections in conservation of rare species is something of which Banks would doubtless have approved. However, rather like *Archaeopteryx*, new scientific interrogations are constantly being made of the plant collections: molecular and genetic studies are currently most fashionable, but new areas of research will continue to open as science advances into the twenty-first century.

Collections need to have a system for their arrangement; otherwise, how can an individual example be retrieved? The larger the collection, the greater is the problem of organisation. How should the plants be arranged as Kew expanded? The eighteenth century was a time when collections grew from a few cabinets to whole galleries, and gardens occupied many acres: retrieval of information became a logistical necessity. The publication of *Systema Naturæ* by the Swede Carolus Linnaeus (Karl von Linné) in 1735 provided the key – for once in a rather literal sense.

It has become a popular cliché to summarise Linnaeus' achievement as providing the binomial name for organisms – the familiar form of *Albus dumbledorus*. It is certainly true that the provision of a unique name for a species did provide a labelling system that has proved indispensable for more than 250 years. Linnaeus' methods have rubbed awkwardly up against twentieth-century phylogenetics – but that is another matter. From the point of view of collections what Linnaeus provided was a system – a hierarchy – that fed into practical arrangement. The higher levels of the Linnaean system – genera, families and so on – became an effective way of organising the mass of material that was being provided by 1770 from the fruits of global exploration: herbaria, museum galleries and gardens alike. Translation of Linnaeus' works into English in the 1760s, together with popular accounts like William Withering's *Botanical Arrangement … &c* of 1776, ensured that his ideas penetrated far into educated circles. Linnaeus' contribution was far more than that little label stuck in the flower bed beside a strange herb – he was the intellectual designer of the garden as a whole.

Linnaeus' system was not without its conceptual antecedents. For example, no English writer should fail to acknowledge the contribution of John Ray (1627–1703) whose emphasis on morphology in plant classification in *Historia Plantarum* of 1686 anticipated Linnaeus in several respects. However, Linnaeus provided the impression of comprehensiveness, the authority that seemed to be able to embrace the *whole* of nature into a manageable hierarchy. Museum cases could now be labelled with the names of taxa that could be understood by all savants of the age in a similar way. As the Reverend Gilbert White wrote to his friend Daines Barrington FRS on 2 June 1778: 'without system the field of nature would be a pathless wilderness'. Linnaeus provided both 'system' and a basis for systematics. His higher classification of flowering plants according to the sexual parts of the flowers has not survived unscathed, but probably no scientist other than Max Planck has had so many scientific institutions named for him. The Linnean Society of London holds many of his original papers and specimens. Linnaeus himself did not adequately characterise some of his plants in relation to a particular type specimen. This job has just been completed in 2007, with the publication of *Order Out of Chaos* by the Linnean Society. Once more those old herbarium specimens have been revisited, like so many floral *Archaeopteryx*, to live again in a new scientific context.

Linnaeus was far from being an 'ivory tower academic'. He knew how to put on a show. Perhaps the most spectacular example of his talent for display was a floral clock that he designed in Uppsala – a flower bed calibrated with species that opened hour by hour together at the appropriate time of day. He knew how to turn erudition into entertainment, and this did his patronage no harm at all. More seriously, his systematic plantings – his book written on to the earth, as it were – became a standard aid for

Lychnis Sylvestris rubello flore. Var. variat flore albo.
Wild Campion

Aloe Africana caulescens folio crasso depuré viridi, ورith ad latera & in dorso armato. Tom. fol. n. 121.

Capsicum fructu rugoso maximo plerumque nutante.

teaching when imitated around the world. Bed after flower bed is typically planted with examples from particular families. This may sound a little mechanical, but in due season does have a certain aesthetic appeal, somewhat akin to listening to a theme and variations. There is a leisurely version of his systematic garden near the River Seine in the heart of Paris in the Jardin Botanique, and the University of Uppsala has maintained Linnaeus' original. Of course, Kew Gardens has a fine example within its walls.

Banks was a convinced Linnaean systematist, so the disposition of plants in Kew Gardens followed the appropriate arrangements. Even in the arboretum the system ruled by generally ensuring planting of species belonging to a single genus or family together in close proximity. Although most critics agree that Linnaeus himself believed in the fixity of species, it seems to me that the juxtaposition of sets of morphologically similar species is almost a precondition to setting a curious mind thinking about how one plant might relate to another (and the same will apply to a drawer full of congeneric butterflies or beetles). The origin of species is embodied in the arrangement of species. One observer might see discrete categories, created individually, another observer might start drawing in his mind 'dotted lines' between species of greater similarity. If a garden were planted out randomly, or according to some traditional system of medicinal virtue, such similarities – the fundamental ones – would scarcely be apparent. But collections systematically arranged became potential maps of relationships. Erasmus Darwin's (1731–1802) famous assertion that animal life may have arisen from 'one living filament' (*Zoonomia* 1794–96) could be envisaged as a path, somewhat as in Gilbert White's metaphor, connecting one organism to another in the garden of life.

Gilbert White was under no illusion that 'system' was the whole story. In the same letter to Daines Barrington he objected to botany as being seen as something 'that amuses the fancy and exercises the memory, without improving the mind or advancing any real knowledge; and where the science is carried no further than a mere classification the charge is but too true'. This sounds a little like Ernest Rutherford's famous fulminations against 'stamp collectors' (this being everyone except physicists). No, the interesting questions were what White termed 'philosophical' – which would broadly mean 'testable hypotheses' in present terminology. Prime among these would prove to be the mechanism for the generation of the diversity of all those species planted out in the systematic beds or gracing the hot houses of the wealthy, or shells and fossils in their 'cabinets of curiosity'. At a time when international travel was expensive, arduous, and almost impossible to remote areas, collections provided the only access for many observers to a true picture of biological diversity.

In a book concerned with giants, some of them unacknowledged, it would be wrong to reinforce an impression that Linnaeus was a kind of lonely systematising hero, even if he himself might have fostered such a view. He did not cover the whole of biodiversity, although it sometimes seems as if his fellow countrymen conspired to do so. Erik Acharius (1757–1819) tackled lichens, for example, and Elias Fries (1794–1878) made astonishing advances with the fungi somewhat later, both aided in part by advances in microscopy. Nor was Linnaeus greatly concerned with fossils, the scientific understanding of which was advancing hugely in the eighteenth and early nineteenth century, as Martin Rudwick has described so well in *Bursting the Limits of Time* (2005). Georges (later Baron) Cuvier (1769–1832) developed comparative vertebrate anatomy in Paris, as did William Buckland in Oxford, while stratigraphic understanding advanced throughout Europe, most famously perhaps through William Smith's (1815) geological map of Britain. Smith regarded his fossil collection as an essential validation of his map, a solid demonstration almost as important as the printed work. This reference collection now resides a floor or two above *Archaeopteryx* in London. *All* of these different systematic endeavours generated important collections. The permanent storage of reference specimens to found public museums is possibly the most important dowry in the marriage of science and collections.

There has always been something of a tension between the private and public ownership of collections. In the seventeenth century the growing interest in antiquarianism led to many individuals of wealth acquiring collections of Classical antiquities – and, somewhat later, of artefacts from Pharaonic Egypt. Interest in more domestic European archaeology merged naturally enough with a growing awareness of prehistory, and many *dilettanti* also began to write up their observations in a burgeoning number of journals. Natural history 'cabinets' often featured conchological collections, of variable scientific value, but fossils also began to become popular objects of interest. Whether or not such collections were retained was often at the whim of the son and heir: many were not. Probably the first example of a public exhibition open to paying customers was 'The Ark' in Lambeth, a miscellany mostly of antiquarian import collected by John Tradescant (d. 1638) and elaborated by his son (also John, 1608–62). Unlikely though it may seem in what is now a very urban part of London, the Tradescants also ran a nursery for exotic plants, particularly from North America where the younger Tradescant visited, and they were equally known for fruit trees – they supplied 'Cherryes' to the royal household. So the conflation of collections of more-or-less scientific importance with 'Botanical Gardens' had a long pedigree. But these collections were definitely part of private enterprise. Elias Ashmole FRS (1617–92) acquired the Tradescants' collection and added much of his own. When the doors of Ashmole's Museum opened in Oxford on 24 May 1683 the concept of an accessible collection was something of a novelty – a 'Publick Place for the Resort of Learned Men' as it was described in a contemporary lexicon. The notion that qualified people and members of the public might learn from objects without expecting a fee was novel, and, even though Ashmole's (and the Tradescants') collections were to suffer a subsequent chequered history, the Ashmolean Museum broke new ground. Robert Plot FRS (1640–96), the first Professor of Chemistry in Oxford, was also first keeper of the Ashmolean collections (and provided early descriptions of fossils). As we have seen, this tradition of public access was followed by Joseph Banks in his house in Soho Square, and, at least within the upper classes, was commonly held among the savant classes of Europe. If not exactly sponsoring a democracy of learning, there was a growing sense that diffusion of knowledge was desirable in general, rather than its protection by an esoteric elite. Collections provided evidence, and should be carefully preserved.

Ashmole's collections were dwarfed by those made by Sir Hans Sloane (1660–1753). He could outspend most of his rivals, and outlived all of them. Sloane was, moreover, a forerunner of Banks in exotic travel. Between 1687–89 he was physician to the Governor of Jamaica, and acquired there his lifelong enthusiasm for botany – and began his own collections and herbaria. These still survive in good condition in the Botany Department of the Natural History Museum. He also established his reputation as a savant with the publication of such weighty works as *A Voyage to the Islands Madera, Barbados, Nieves, S. Christophers and Jamaica with the Natural History of the Herbs and Trees, four footed beasts, Fishes, Birds, Insects, Reptiles &c of the last of those islands* (2 vols 1707–25). Titles have become crisper since the eighteenth century, but at least the reader knew exactly what he was getting. Sloane also encountered cacao in Jamaica, and made a tidy sum from mixing it with milk and providing it as a wholesome chocolate recipe. Sloane's advancement through the social hierarchy depended on his great reputation as a physician, and he eventually became President of the Royal Society.

Sloane continued to recognise the close connection between living and inert collections, leasing extra land to the Chelsea Physic Garden at a nominal rent; this Thames-side garden was originally founded in 1673 to teach young apothecaries their herbal trade, and it played an important part in establishing exotic plants and in exchanging seeds internationally. Sloane had eventually moved to Chelsea when his collections outgrew his Bloomsbury address; his statue remains in the Physic Garden. By then he had a library of more than 48,000 volumes and had added Egyptian mummies and Greek and Roman antiquities to his colossal natural history collections. Many of his plants were the type specimens of species recently recognised.

Sloane had determined to keep his collection together years before he died; he regarded it as his life's work. He offered the collection to the King for the use of the nation for the sum of £20,000 to be distributed between his daughters – undoubtedly a bargain for the nation. He evidently fretted about the fate of the collection – to the extent of having no less than forty Fellows of the Royal Society as Trustees. The 1753 British Museum Act by which Sloane's collections were changed into a public facility includes the instruction that the collections should be 'preserved and maintained not only for the Inspection and Entertainment of the learned and the curious, but for the general use and benefit of the Publick'. The collections were moved back to Montague House in Bloomsbury, and there the Museum officially came into existence in 1756. The head of the permanent staff was known as the Principal Librarian. With the appointment of Banks' old friend and Linnaeus' student Daniel Solander to the staff in 1773 the connections explored in this chapter reached consummation. The classification of the collections on scientific grounds was assured, with all the subsequent implications for discovery of the natural causes that underpinned that arrangement. The permanence of the collections in the public domain was guaranteed, and the modern notion of a scientific museum was established in Britain.

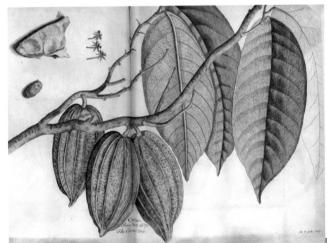

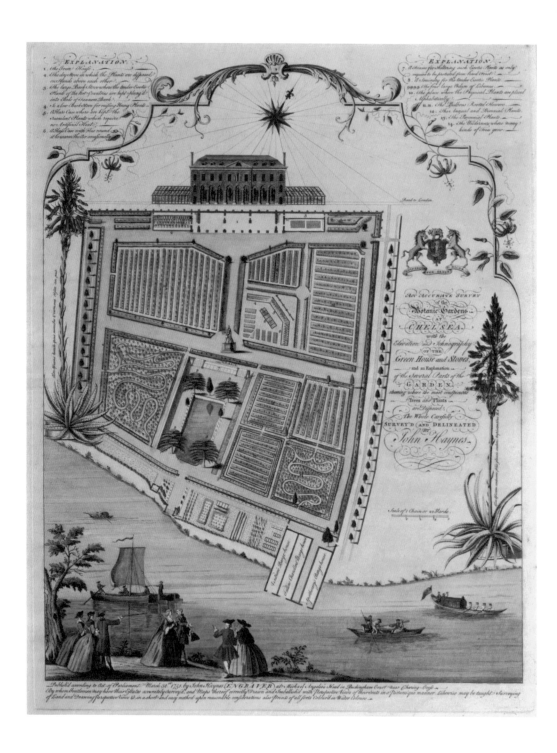

More than a century would pass before the natural history collections parted company with the antiquarian collections and found their own place in Alfred Waterhouse's extraordinary building in South Kensington. The collecting fruits of Empire, and the gradual increase in staff, not to mention the scientific pretensions of the collections, all acted together to ensure better funding. Richard Owen was appointed in May 1856 as Superintendent of the Natural History Departments in Bloomsbury, and worked tirelessly to get separate accommodation for the scientific collections. His contacts with the Royal Family assuredly did no harm: indeed, the progressive spirit of Prince Albert still inhabits all that elegant part of London south of Kensington Gardens. At last, when important specimens were discovered money could be found to acquire them for the nation, not merely to embellish the reputation of the wealthy *aficionado*.

So it was with *Archaeopteryx,* with which this chapter began. The specimen was acquired as part of a collection put together by a Bavarian doctor, Karl Häberlein. His large collection included 23 reptiles, 294 fishes, 194 plants and more than a thousand invertebrate fossils. The price paid was £700 – which historians are always obliged to qualify with the phrase 'a considerable amount of money in those days'. However, no price is relevant when the prize is priceless.

Collections achieved their scientific importance from three innovations: scientific purpose (including collections made on dedicated expeditions); appearance of a rational system for curation; and the museum as a permanent repository for the public good. All this happened before Charles Darwin's birth; but even Darwin began as a collector, and only later became a 'machine for generating hypotheses'. He spent a decade immersed in barnacles, and it is plausible that his ideas on evolution matured during those 'forgotten' years. Science always advances with new techniques and new ideas, but these are frequently applied to collections held for future study. Scientific collections don't die; they are constantly re-invented.

9

RICHARD DAWKINS

DARWIN'S FIVE BRIDGES:
THE WAY TO NATURAL SELECTION

Richard Dawkins FRS is an evolutionary biologist and popular science author. His books include *The Selfish Gene*, *The Extended Phenotype*, *The Blind Watchmaker*, *Climbing Mount Improbable*, *Unweaving the Rainbow*, *The Ancestor's Tale* and *The God Delusion*. His latest book is *The Greatest Show on Earth: The Evidence for Evolution*. He was formerly the first Simonyi Professor of Public Understanding of Science at Oxford University and is a fellow of New College, Oxford.

E VOLUTION WAS IN THE AIR IN THE MID-NINETEENTH CENTURY – A THRILLINGLY RADICAL NOTION WHICH OFFERED A WAY TO MAKE SENSE OF A HUGE ARRAY OF FACTS. WHAT, THEN, WAS DARWIN'S UNIQUE CONTRIBUTION? AS RICHARD DAWKINS TEASES OUT, IT WAS THE COMBINATION OF SEEING THE TRUE POWER OF NATURAL SELECTION, AND EXPLAINING HOW IT WORKED THROUGHOUT THE LIVING WORLD.

Was Darwin the most revolutionary scientist ever? If, by revolutionary, we mean the scientist whose discovery initiated the most seismic overturning of pre-existing science, the honour would at least be contested by Newton, Einstein, and the architects of quantum theory. Those same physicists might have outclassed Darwin in sheer intellectual firepower. But Darwin probably did revolutionise the worldview of people outside science more comprehensively than any other scientist. He may be only one plausible candidate for the most important or most revolutionary scientist ever, but Darwin has a strong claim to be the most *seditious*.

Before Darwin, it took a philosopher of the calibre of David Hume to rumble the illogic of 'if a thing looks designed it must have *been* designed'.

And even Hume, though he could see that the argument to design was a bad argument, couldn't think of a good alternative. Darwin provided the alternative. How Hume would have relished the 'I told you so' moment that Darwin handed him.

The argument to design was familiar to Darwin, for whose cohort of Cambridge undergraduates the Reverend William Paley was compulsory reading. If it looks designed, it was designed. And the more designed it looks, the stronger the argument. 'Looks designed' means something along the lines of 'statistically improbable in a previously specified functional direction'. Paley's watch,[1] and the vertebrate eye, are both statistically improbable in that, if you take their parts and scramble them into random combinations a million times, not once will you hit upon a combination that tells the time to the nearest second, or that sees, in full colour, stereoscopically and with instantaneous light-metering and autofocus.

We must add 'in a previously specified direction' because, with hindsight, every random combination can be made to seem as improbable as any other. How astounding that, of all the blades of grass on the golf course, the ball landed on this particular blade, and no other! The reason a hole-in-one is so rare is that the hole is specified in advance as the target. If you specified any particular blade of grass in advance, and the ball landed on it, it would be as remarkable as a hole-in-one (actually more so, because the hole is larger than a blade of grass).

Watches and eyes have their functions – telling the time and seeing, respectively – specified in advance, and both are functions that are difficult to achieve. Therefore a random scrambling of parts is exceedingly unlikely to perform either function with any efficiency. The fact that a watch does tell the time accurately, and with (at least) two hands to accommodate two conveniently related time-scales, correctly indicates to any reasonable person that it is not the product of random chance. Before Darwin, the only known alternative to random chance was design. Everybody could see the force of the argument that Paley generalised from watch to eye – and to

1 W. Paley, *Natural Theology* (Oxford, Vincent, 1802); R. Dawkins, *The Blind Watchmaker* (London, Longman, 1986).

every other part of every living body. There must have been a designer. And yet intuition was wrong. It is the unholy juxtaposition of 'commonsensically true' with 'now known to be false' that singles out Darwin's great idea as seditious. Darwin discovered the alternative to chance and design that had eluded everybody, even Hume. The answer is cumulative natural selection. Provided that a smoothly cumulative gradient of improvement exists – not a difficult condition to realise – natural selection is likely to find it, and will propel evolution up the slopes of 'Mount Improbable'[2] to apparently limitless heights of perfection, which – if you overlooked the smooth, cumulative gradients – you would think were too improbable to countenance.

Darwin's dangerous idea[3] was seditious, revolutionary, deeply surprising. And yet, having eluded Hume in the eighteenth century, and every great philosopher and scientist before him, it was an idea that came, independently, into the prepared minds of at least two naturalists in the nineteenth century: Charles Darwin and Alfred Russel Wallace. I'm not talking about evolution itself, for that idea had occurred to many, including Lamarck and Darwin's grandfather Erasmus. Nor am I talking about natural selection itself, for that too, as we shall see, had crossed other minds than Darwin's and Wallace's. I am talking about the idea that natural selection is *powerful* enough to drive evolution in such a way as to explain everything about life, including that illusion of design that, in Hume's own words, 'ravishes into admiration all men who have ever contemplated [them]'.[4]

I singled out Darwin and Wallace as the two nineteenth-century naturalists who independently solved the riddle of life. But claims of priority have been made on behalf of at least two other nineteenth-century writers, Patrick Matthew and Edward Blyth. If those claims are upheld, it should be a matter of some national pride that all four independent discoverers of natural selection were British. But should they be upheld?

Edward Blyth (1810–73) was Darwin's near contemporary. Like Darwin and Wallace, he was a naturalist and collector of specimens in the

2 R. Dawkins, *Climbing Mount Improbable* (London, Viking, 1996).
3 D. Dennett, *Darwin's Dangerous Idea: Evolution and the meaning of life* (New York, Simon & Schuster, 1995).
4 D. Hume, *Dialogues Concerning Natural Religion* (1779).

tropics, in his case India. He really did hit upon the idea of natural selection, publishing it in 1835. But his version is only what we would today call *stabilising* selection, that is, natural selection preserving the original type, *not* natural selection driving evolutionary change to ever new types. No wonder he was a staunch creationist. He thought of natural selection as preserving God's original creations in their pristine, archetypal state. He was, indeed, the very opposite of an evolutionist. Natural selection, in his formulation, would amount to a force of resistance against evolutionary change.

Patrick Matthew (1790–1874) used his experience of growing apple and pear trees in his Scottish orchard to write a book, in 1831, on *Naval Timber and Arboriculture.* In an appendix to this work, Matthew recognised that the principles of artificial selection, which he advocated for growing good quality timber for the navy, could be generalised to natural selection. Unlike Blyth, Matthew didn't see natural selection purely as a stabilising force, preserving the original form of the species. He even went so far as to speculate that:

> ... the progeny of the same parents, under great differences of circumstance, might, in several generations, even become distinct species, incapable of co-reproduction.[5]

When *The Origin of Species* was first published, Matthew protested at Darwin's failure to cite him, and Darwin punctiliously did so in the third (1861) and subsequent editions of his book. The passage that immediately follows the above-quoted sentence seems to bear out Darwin's acknowledgment that Matthew 'clearly saw the full force of the principle of natural selection':

> The self-regulating adaptive disposition of organised life may, in part, be traced to the extreme fecundity of Nature, who, as before stated, has, in all the varieties of her offspring, a prolific power much beyond (in many cases a thousandfold) what is necessary to fill up

5 Patrick Matthew, *Naval Timber and Arboriculture* (Edinburgh, 1831).

the vacancies caused by senile decay. As the field of existence is limited and preoccupied, it is only the hardier, more robust, better suited to circumstance individuals, who are able to struggle forward to maturity, these inhabiting only the situations to which they have superior adaptation and greater power of occupancy than any other kind; the weaker, less circumstance-suited, being prematurely destroyed. This principle is in constant action, it regulates the colour, the figure, the capacities, and instincts; those individuals of each species, whose colour and covering are best suited to concealment or protection from enemies, or defence from vicissitude and inclemencies of climate, whose figure is best accommodated to health, strength, defence, and support; whose capacities and instincts can best regulate the physical energies to self-advantage according to circumstances – in such immense waste of primary and youthful life, those only come forward to maturity from the strict ordeal by which Nature tests their adaptation to her standard of perfection and fitness to continue their kind by reproduction.

Like Blyth (indeed, Darwin seems to have been indebted to Blyth's observations on the subject), Matthew saw the importance of overproduction and the consequent struggle for existence, and he clearly went further than Blyth.

But I am left wondering. Did Matthew really grasp the immense power of the discovery that he had made? Did he appreciate that natural selection is the answer to the great riddle of existence? Did he see it as the explanation for all of life, the destroyer of the argument from design? If he had, wouldn't he have published it in a more prominent place than the appendix to a manual on silviculture? Wouldn't he have trumpeted it from the rooftops, as arguably the most important idea anyone ever had? On the contrary, Matthew seems to have found the idea so obvious – almost trivial – as to need no discovery! In a letter to the *Gardeners' Chronicle* of 12 May 1860, he wrote:

To me, the conception of this law of Nature came intuitively as a self-evident fact, almost without an effort of concentrated thought. Mr Darwin here seems to have more merit in the discovery than I have had – to me it did not appear a discovery. He seems to have worked it out by inductive reason, slowly and with due caution to have made his way synthetically from fact to fact onwards; while with me it was by a general glance at the scheme of Nature that I estimated this select production of species as an a priori recognisable fact – an axiom, requiring only to be pointed out to be admitted by unprejudiced minds of sufficient grasp.

With hindsight, we may be tempted to sympathise. But where Huxley, on closing *The Origin,* movingly sighed, 'How extremely stupid of me not to have thought of that', Matthew's response would seem to have been the Victorian equivalent of 'Big deal. So what else is new?' Is this the response of a man who, seven years before Darwin and twenty-seven before Wallace, found himself in possession of the central, unifying idea that dominates all biology and explains almost everything about life?

As a fair parallel, imagine that a seventeenth-century ancestor of Patrick Matthew saw an apple fall (perhaps in the very same orchard, for the Matthews had been farming in the Carse of Gowrie since the sixteenth century). Our earlier Matthew, I imagine to have been a physicist and, as he watched his apple fall, he conjectured that the Earth exerted an attractive force on apples, pulling them towards it. If this hypothetical horticulturalist had later written to Isaac Newton and indignantly claimed priority for the theory of gravitation, Newton (a less generous man than Darwin) would rightly have given him short shrift. The physicist Matthew, let's suppose, confined his theory to apples, or at best to objects falling towards the Earth. He lacked Newton's grand vision of the same force acting throughout the universe, responsible for the elliptical orbits of the planets, for the stars in their courses, ultimately for the very structure of the universe itself.

I agree with W.J. Dempster, Patrick Matthew's modern champion, that Matthew has been unkindly treated by history.[6] 'But, unlike Dempster, I hesitate to assign full priority to him. Partly, it is because he wrote in a much more obscure style than either Darwin or Wallace, which makes it hard to know in some places what he was trying to say (Darwin himself noted this). But mostly it is because he seems to have underestimated the idea, to an extent where we have to doubt whether he really understood how important it was. The same could be said, even more strongly (which is why I have not treated his case in the same detail as Matthew's), of W.C. Wells, whom Darwin also scrupulously acknowledged (in the fourth and subsequent editions of *The Origin*). Wells made the leap to generalise from artificial to natural selection, but he applied it only to humans, and he thought of it as choosing among *races* of people rather than individuals as Darwin and Wallace did. Wells therefore seems to have arrived at a form of 'group selection' rather than true, Darwinian natural selection as Matthew did, which selects individual organisms for their reproductive success. Darwin also lists other partial predecessors, who had shadowy inklings of natural selection. Like Patrick Matthew, none of them seems to have grasped the earth-shattering *significance* of the idea they had lit upon, and I shall use Matthew's name to represent them all. I am increasingly inclined to agree with Matthew that natural selection itself scarcely needed discovering. What needed discovering was the significance of natural selection for the evolution of all life.

Alfred Russel Wallace (1823–1913) was different. Although he discovered natural selection after Matthew (and after Darwin's unpublished manuscripts) he has a genuine claim to be up there with Darwin and Newton, among the immortals.[7] When Wallace hit upon natural selection, he was in no doubt of its immense importance for the whole history of life. The very title of his paper – the one he sent to Darwin, and which set the cat among Darwin's pigeons – says it all: *On the Tendency of Varieties to Depart Indefinitely from the Original Type*. 'Depart indefinitely', that was the key

6 W.J. Dempster, *Evolutionary Concepts in the Nineteenth Century* (Edinburgh, Pentland Press, 1996).
7 Unlike Patrick Matthew or Edward Blyth, Wallace was a Fellow of the Royal Society, although elected rather late – about thirty-five years after his landmark paper on evolution by natural selection. Darwin was elected in 1839, when still not yet thirty. Both Wallace and Darwin were honoured with the Society's Royal Medal and Copley Medal.

Portrait of Alfred Russel Wallace, the Welsh naturalist, explorer, geographer, anthropologist and biologist.

phrase. If they depart indefinitely from the original type, they can branch and eventually spawn all of life. And Wallace made that explicit in his paper.

The drama of how Wallace's letter arrived at Down House on 17 June 1858, casting Darwin into an agony of indecision and worry, is too well known for me to retell it. In my view the whole episode is one of the more creditable and agreeable in the history of scientific priority disputes – precisely because it wasn't a dispute – although it so easily could have become one. It was resolved amicably, and with heartwarming generosity on both sides, especially Wallace's. As Darwin later wrote:

> Early in 1856 Lyell advised me to write out my views pretty fully,
> and I began at once to do so on a scale three or four times as extensive

as that which was afterwards followed in my *Origin of Species*; yet it was only an abstract of the materials which I had collected, and I got through about half the work on this scale. But my plans were overthrown, for early in the summer of 1858 Mr Wallace, who was then in the Malay archipelago, sent me an essay 'On the Tendency of Varieties to depart indefinitely from the Original Type'; and this essay contained exactly the same theory as mine. Mr Wallace expressed the wish that if I thought well of his essay, I should send it to Lyell for perusal.

The circumstances under which I consented at the request of Lyell and Hooker to allow of an extract from my MS., together with a letter to Asa Gray, dated September 5, 1857, to be published at the same time with Wallace's Essay, are given in the *Journal of the Proceedings of the Linnean Society*, 1858, p. 45. I was at first very unwilling to consent, as I thought Mr Wallace might consider my doing so unjustifiable, for I did not then know how generous and noble was his disposition. The extract from my MS. and the letter to Asa Gray … had neither been intended for publication, and were badly written. Mr Wallace's essay, on the other hand, was admirably expressed and quite clear. Nevertheless our joint productions excited very little attention, and the only published notice of them which I can remember was by Professor Haughton of Dublin, whose verdict was that all that was new in them was false, and what was true was old. This shows how necessary it is that any new view should be explained at considerable length in order to arouse public attention.

Darwin was over-modest about his own two papers. Both are models of the explainer's art. Wallace's paper is also very clearly argued. His ideas were, indeed, remarkably similar to Darwin's, and there is no doubt that Wallace arrived at them independently. In my opinion the Wallace paper needs to be read in conjunction with his earlier paper in the *Annals and Magazine of*

Natural History. Darwin read this paper when it came out in 1855. Indeed, it led to Wallace joining his large circle of correspondents, and to his engaging Wallace's services as a collector. But, oddly, Darwin did not see in the 1855 paper any warning that Wallace was by then a convinced evolutionist of a very Darwinian stamp. I mean as opposed to the Lamarckian view of evolution, which saw modern species as all on a ladder, changing into one another as they moved up the ladder. By contrast Wallace, in 1855, had a clear view of evolution as a branching tree, exactly like Darwin's famous diagram, which became the only illustration in *The Origin of Species.* The 1855 paper, however, makes no mention of natural selection or the struggle for existence.

That was left to Wallace's 1858 paper, the one that hit Darwin like a lightning bolt. Here, Wallace even used the phrase 'Struggle for Existence'. Wallace devoted considerable attention to the exponential increase in numbers (another key Darwinian point). Wallace wrote:

> The greater or less fecundity of an animal is often considered to be one of the chief causes of its abundance or scarcity; but a consideration of the facts will show us that it really has little or nothing to do with the matter. Even the least prolific of animals would increase rapidly if unchecked, whereas it is evident that the animal population of the globe must be stationary, or perhaps … decreasing.

Wallace deduced from this that 'The numbers that die annually must be immense; and as the individual existence of each animal depends upon itself, those that die must be the weakest …' Wallace's peroration could have been Darwin himself writing:

> The powerful retractile talons of the falcon – and the cat – tribes have not been produced or increased by the volition of those animals; but among the different varieties which occurred in the

earlier and less highly organised forms of these groups, those always survived longest which had the greatest facilities for seizing their prey. Neither did the giraffe acquire its long neck by desiring to reach the foliage of the more lofty shrubs, and constantly stretching its neck for the purpose, but because any varieties which occurred among its antitypes with a longer neck than usual at once secured a fresh range of pasture over the same ground as their shorter-necked companions, and on the first scarcity of food were thereby enabled to outlive them. Even the peculiar colours of many animals, especially insects, so closely resembling the soil or the leaves or the trunks on which they habitually reside, are explained on the same principle; for though in the course of ages varieties of many tints may have occurred, yet those races having colours best adapted to concealment from their enemies would inevitably survive the longest. We have also here an acting cause to account for that balance so often observed in nature, – a deficiency in one set of organs always being compensated by an increased development of some others – powerful wings accompanying weak feet, or great velocity making up for the absence of defensive weapons; for it has been shown that all varieties in which an unbalanced deficiency occurred could not long continue their existence. The action of this principle is exactly like that of the centrifugal governor of the steam engine, which checks and corrects any irregularities almost before they become evident.

The image of the steam governor is a powerful one which, I can't help feeling, Darwin might have envied.

Historians of science have raised the suggestion that Wallace's version of natural selection was not quite so Darwinian as Darwin himself believed. Wallace persistently used the word 'variety' as the level of entity at which natural selection acts. There was an example in the long passage

I have just quoted, and also an example of Wallace's usage of the word 'race' in a similar sense. Some have suggested that Wallace, unlike Darwin, who clearly saw selection as choosing among *individuals,* was proposing what nearly all modern theorists rightly denigrate as 'group selection'. This would be true if, by 'varieties' or 'races', Wallace meant geographically separated groups of individuals, or indeed races in the more usual sense of the word. At first I wondered myself whether Wallace meant that. But a careful reading of his paper rules it out. By 'variety' and 'race' Wallace meant what we would nowadays call 'genetic type', even what a modern population geneticist might mean by an allele. To Wallace in this paper, variety meant not a local race of eagles, for example, but 'that set of individual eagles whose talons were hereditarily sharper than usual'.

If I am right, it is a similar misunderstanding to the one suffered by Darwin, whose use of the word 'race' in the subtitle of *The Origin of Species* is sometimes misread as supporting group selection[8] or even racialism. That subtitle, or alternative title rather, is *The Preservation of Favoured Races in the Struggle for Life.* Once again, Darwin was using 'race' to mean 'that set of individuals who share a particular hereditary characteristic', such as sharp talons, *not* a geographically distinct race such as the Hooded Crow. If he had meant that, Darwin too would have been guilty of the group selection confusion. I believe that neither Darwin nor Wallace was.[9] And, by the same token, I do not believe that Wallace's conception of natural selection was different from Darwin's.

As for the calumny that Darwin plagiarised Wallace, that is rubbish. The evidence is very clear that Darwin did think of natural selection before Wallace, although he did not publish it. We have his abstract of 1842 and his longer essay of 1844, both of which establish his priority clearly, as did his letter to Asa Gray of 1857, which was read at the Linnean Society in 1858.

Why Darwin delayed so long before publishing is one of the great mysteries in the history of science. Some historians have suggested that he

8 The distinguished physicist Freeman Dyson has read it in exactly this sense, to buttress his own partiality for group selection.
9 The one exception – a rare exception in Darwin's thinking – is his treatment of the evolution of human cooperation and kindness through a kind of group selection among rival tribes.

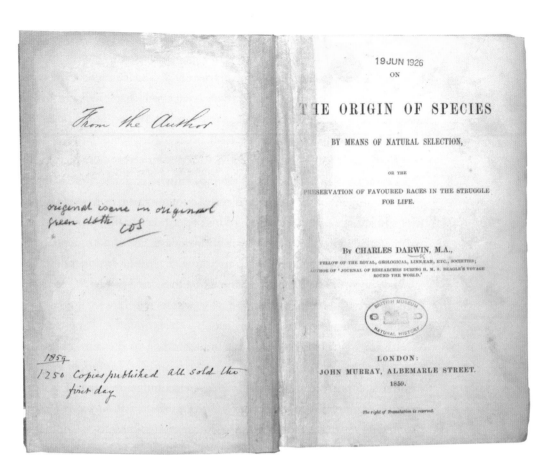

From the Author

original issue in original green cloth. cos

1859
1250 Copies published all sold the first day

19 JUN 1926

ON

THE ORIGIN OF SPECIES

BY MEANS OF NATURAL SELECTION,

OR THE

PRESERVATION OF FAVOURED RACES IN THE STRUGGLE
FOR LIFE.

By CHARLES DARWIN, M.A.,

FELLOW OF THE ROYAL, GEOLOGICAL, LINNÆAN, ETC., SOCIETIES;
AUTHOR OF 'JOURNAL OF RESEARCHES DURING H. M. S. BEAGLE'S VOYAGE
ROUND THE WORLD.'

LONDON:
JOHN MURRAY, ALBEMARLE STREET.
1859.

The right of Translation is reserved.

First edition of *On the Origin of Species by Means of Natural Selection: or the Preservation of Favoured Races in the Struggle for Life* by Charles Darwin, 1859. Inscribed by the publisher.

was afraid of the religious implications, others the political ones. Perhaps he was afraid of upsetting his devout wife. Maybe he was just a perfectionist, keen to have all his evidence lined up and in place before going public. Or did he just get distracted by barnacles?

When Wallace's letter arrived, Darwin was more surprised than we moderns might think he had any right to be. He wrote to Lyell:

> I never saw a more striking coincidence; if Wallace had had my manuscript sketch, written out in 1842, he could not have made a better short abstract of it. Even his terms now stand as Heads of my Chapters.

The coincidence extended to both Darwin and Wallace being inspired by Robert Malthus on population. Darwin, by his own account, was immediately inspired by Malthus' emphasis on overpopulation and competition. He wrote in his autobiography:

> In October, 1838, that is, fifteen months after I had begun my systematic inquiry, I happened to read for amusement Malthus on population, and being well prepared to appreciate the struggle for existence which everywhere goes on from long continuous observation of the habits of animals and plants, it at once struck me that under these circumstances favourable variations would tend to be preserved and unfavourable ones to be destroyed. The result of this would be the formation of new species. Here, then, I had at last got a theory by which to work.

Wallace's epiphany after reading Malthus took longer to happen, but was more dramatic when it came … to his overheated brain in the midst of a malarial fever, on the island of Ternate in the Moluccas archipelago:

> I was suffering from a sharp attack of intermittent fever, and every day during the cold and succeeding hot fits had to lie down for several hours, during which time I had nothing to do but to think over any subjects then particularly interesting me …
>
> One day something brought to my recollection Malthus' 'Principles of Population.' I thought of his clear exposition of 'the positive checks to increase' – disease, accidents, war, and famine – which keep down the population of savage races to so much lower an average than that of more civilised peoples. It then occurred to me …

And Wallace proceeds to his own admirably clear exposition of natural selection, as the guiding principle of all evolution.

I want to recognise four 'bridges to evolutionary understanding', and I can conveniently illustrate them with our four claimants to independent discovery of natural selection. Blyth crossed the first of Darwin's four bridges, Matthew the first two, Wallace the first three and Darwin all four. Bridge One is to natural selection as a force for weeding out the unfit. I have used Blyth as my example of a nineteenth-century writer who crossed this bridge, but really the only reason to single him out is that he has been championed by Loren Eiseley as a predecessor, and even a possible source, of Darwin's ideas. As Stephen Jay Gould has argued, however, the idea of natural selection as a weeder-out, a purely negative force, was already widespread:

> Yes, Blyth had discussed natural selection, but Eiseley didn't realise
> – thus committing the usual and fateful error in this common line of
> argument – that all good biologists did so in the generations before
> Darwin. Natural selection ranked as a standard item in biological
> discourse – but with a crucial difference from Darwin's version: the
> usual interpretation invoked natural selection as part of a larger
> argument for created permanency. Natural selection, in this nega-
> tive formulation, acted only to preserve the type, constant and invi-
> olate, by eliminating extreme variants and unfit individuals who
> threatened to degrade the essence of created form.[10]

Gould even quotes William Paley himself as setting out this purely negative version of natural selection. As I remarked above, it is almost an anti-evolution argument, for it uses natural selection to explain the fixity of species rather than their changing into other species.

Bridge Two is the recognition that natural selection can drive evolutionary change. In modern jargon, it amounts to the difference between Stabilising Selection and Directional Selection. Matthew, Wallace and Darwin all crossed this second bridge.

Bridge Three leads to the imaginative grasp of the importance of natural

10 S.J. Gould, *The Structure of Evolutionary Theory* (Cambridge, Mass., Harvard University Press, 2002).

Ser. 5th edn 1869 28?

to bodily ... organs. Changes of instinct may
sometimes be facilitated by ... same species having,
... different instincts at
different periods of life, or ... in the ... or in which ...
placed under different ... it ... might be
one instinct ... the other ... preserved by
natural selection: and such cases & diversity of instinct
in the same species can be shewn to occur in nature.

Again, as in the case of corporeal structure, & conformably
with our theory, the instinct of each species is good
for itself, but ... never, as far as we can
judge, been produced for the exclusive good of others.
... each species tries to take advantage of
... in ... cases ... some instincts of
others ... the instincts of others ...
cannot be considered as absolutely perfect; but as details
on these heads can ... not indispensable ...
they shall not be here given.

As some degree of variation in instincts under a state
of nature, ... & its inheritance of such variations,
in the indispensable foundation for natural selection
... would be highly advisable here ...
to act on; it ... necessary to
give ... instances of variation; but want of space
prevents me. I can only say that variations
certainly do occur; ... in the migratory instincts of

selection in explaining all of life, in all its speciose richness, and especially
to dispel the illusion of design. Wallace and Darwin certainly crossed it.
Maybe Matthew did too, but I have given reasons for doubting that he
developed the full imaginative vision of the constructive power of
'Darwinism' (as Wallace, in a generous gesture, was later to dub it).

Bridge Four is the bridge to public understanding and appreciation.
Darwin crossed it alone, in 1859, by writing *The Origin of Species*. It is a striking fact, remarked by Darwin himself, that when the Darwin/Wallace papers
were read to the Linnean Society in 1858, nobody took a blind bit of notice,
even among the professional biologists of that august body. The end-of-year
clanger of the hapless President of the Linnean, Thomas Bell, has become
notorious and will ring on down the ages. In his review of the Society's transactions during 1858, he said that the year had 'not been marked by any of
those striking discoveries which at once revolutionise, so to speak, the department of science on which they bear'. The end of 1859 would have to be
reviewed very differently. *The Origin of Species* struck the Victorian solar
plexus like a steam hammer. The world of the mind would never be the same
again, neither science, nor anthropology, psychology, sociology, even – and
here we come close to the dark side – politics. This book, which Darwin
always described as the 'abstract' of the great book that he intended to write
but never completed, achieved what the 1858 papers did not.

It isn't that *The Origin* explained the theory more clearly than Darwin's
and indeed Wallace's brief offerings of 1858. The difference was that a
book-length treatment was required to muster all the evidence and lay it
out for all to see: 'one long argument' as Darwin himself called it. And I
quoted above Darwin's own recognition, when the joint papers of 1858 fell
flat, that 'This shows how necessary it is that any new view should be
explained at considerable length in order to arouse public attention.'

And is there a fifth bridge, which Darwin himself never crossed?
Inevitably, 150 years later, there are several, but the one I shall single out is
the bridge to the so-called 'neo-Darwinism' of the 'Modern Synthesis'.

Neo-Darwinism is a union of Darwinian evolution with Mendelian genetics, but the trouble is that what is *neo* changes all the time. What comes after '*nouvelle vague*'? We don't want to get into a sort of 'infinite progress', in the way that 'modernism' gives way to 'post-modernism' and then neo-post-modernism' and then … what? I shall rename neo-Darwinism 'digital Darwinism'. There may be other things more 'neo' than the neo-Darwinian 'modern' synthesis of the 1930s, but digital Darwinism is here to stay. The essence of Mendelian genetics is that it is digital. Mendelian genes are all-or-none, and they don't blend. Genes are things you can *count* in a population's gene 'pool'. Evolution consists of changing *frequencies* of discrete, digital, countable entities, not changing *quantities* of substances, or changing measurements of dimensions. Changing quantities and measurements apply at the organism level, but not at the gene level. What happens in natural selection is that successful genes become more frequent in the gene pool, and unsuccessful genes become less frequent. Frequent, as in *counted*.

Darwin never crossed the digital bridge. If he had, he would have had a ready answer to Fleming (pronounced Fleming) Jenkin, the Scottish engineer who – independently of his colleague Lord Kelvin (with whom he collaborated on the trans-Atlantic cable) – gave Darwin a hard time over matters of theory.[11] Jenkin pointed out that, on the current non-digital, *blending* view of heredity, variation would be swamped by successive sexual crossings, and after a few generations would disappear. There'd be no hereditary variation for natural selection to work on. Blending inheritance would be like mixing black and white paint: you get grey, and no amount of subsequent mixing of grey with grey will give you back the original black and white.

As a matter of fact, any fool could have seen that Jenkin's premise must be wrong. Variation does not dissolve away as the generations go by. We are not more uniform than our grandparents were, and our grandchildren will retain the same level of variation as we possess. Jenkin thought he was doubting Darwin. Actually he was doubting observable facts. Nevertheless, his criticism worried Darwin.

Enlightened by Mendel's nineteenth-century peas and building on Hardy and Weinberg's elementary algebra, the twentieth-century founders of population genetics, R.A. Fisher, J.B.S. Haldane and Sewall Wright, buried Fleeming Jenkin. If genes are countable, digital entities that don't blend, their frequencies have no inherent tendency to change. If they do change, that is evolution, and it happens for a reason. The most interesting reason is non-random selection, but random drift also occurs – to an extent disputed among the founding fathers but now widely admitted among molecular geneticists. Even those three founding fathers never knew quite how digital genetics really is. In the light of the Watson/Crick revolution, we now see the very genes themselves as digitally coded messages, digital in exactly the same sense – and in the same way to an astonishing level of detail – as computer information is digital.

Of the three founding fathers of population genetics, it was Fisher who, in his great book of 1930, *The Genetical Theory of Natural Selection,* most clearly expressed the evolutionary significance of blending inheritance and its Mendelian antithesis.[12] If genes did indeed blend, the variance available for selection would be *halved* in every generation. It's the grey paint over again, but Fisher proved it mathematically. Mutation rates would have to be colossal – utterly unrealistic – to maintain the variation. Fisher quotes a letter from Darwin to Huxley, tentatively dated to 1857, before *The Origin,* which shows how tantalisingly close Darwin himself came to Mendelism:

> … I have lately been inclined to speculate, very crudely and indistinctly, that propagation by true fertilisation will turn out to be a sort of mixture, and not true fusion, of two distinct individuals, or rather of innumerable individuals, as each parent has its parents and

11 Kelvin's attack centred on his (entirely erroneous) 'demonstration' that the Sun and Earth were too young to allow enough time for evolution. His calculations were based on the assumption that the Sun's energy came from some kind of combustion. Pleasingly, it fell to Sir George Darwin FRS, Charles' second son, to redo the calculations on the assumption that the Sun was a nuclear furnace and thereby vindicate his father.

12 I like to think that Ronald Fisher, arguably Charles Darwin's greatest intellectual descendant, was also his intellectual grandson through his mentor, Major Leonard Darwin, the dedicatee of Fisher's great book. Leonard, Charles' fourth son, lived into my own lifetime and died on my second birthday, 26 March 1943.

ancestors. I can understand on no other view the way in which crossed forms go back to so large an extent to ancestral forms. But all this, of course, is infinitely crude.

Even Fisher didn't know how breathtakingly near Darwin really was to discovering Mendelian genetics, even working on sweetpeas! In 1867, he wrote a letter to Wallace that began as follows:

My Dear Wallace
I do not think you understand what I mean by the non-blending of certain varieties. It does not refer to fertility, an instance will explain. I crossed the painted lady and purple sweetpeas which are very different coloured varieties, and got, even out of the same pod, both varieties, perfect but non-intermediate. Something of this kind I should think, must occur with your butterflies ... Though these cases are in appearance so wonderful, I do not know that they are really more so than every female in the world producing distinct male and female offspring.

That last sentence is a beautiful example of the power of reason, and the importance of seeing through the obvious. When a male mates with a female, you do not get a hermaphrodite. You get either a male or a female, with approximately equal probability. In a way, Mendel never needed to go into his monastery garden. All he had to do was take the inheritance of sex itself, and generalise it to all other cases of inheritance. Digital heredity was staring us in the face, in the most obvious way you could imagine. The trouble was, it was *too* obvious to be noticed. Darwin noticed it, and he came close to making the connection. But, just as Patrick Matthew didn't quite cross the bridge that Darwin and Wallace crossed, so Darwin didn't quite manage to cross the Mendel/Fisher Bridge – at least not decisively enough to answer Fleeming Jenkin.

Opposite:
Francis Crick's original sketch of the structure of DNA, made in 1953.

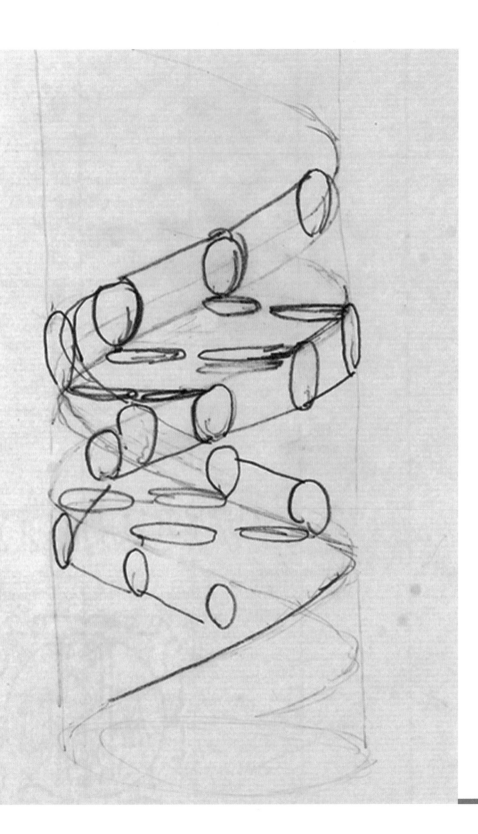

I distinguished Bridge One from Bridge Two as 'stabilising selection' versus 'directional selection'. But there's more to it than that – or perhaps the distinction I am about to make really separates Matthew's Bridge Two from Darwin and Wallace's Bridge Three. I am talking about the distinction between selection as a negative force and selection as a positive, constructive force that puts together complex new 'designs'. My own preferred way – the 'selfish gene' way – of explaining this is again to deploy 'digital genes', so perhaps we really have to cross Bridge Five in order to paint the full picture.

In modern genetic terms, not Darwin's own, natural selection may be defined as the non-random survival of randomly varying coded instructions for how to survive. We see – and admire – the products, the *phenotypes,* of the successful instructions. The instructions are DNA and their most visible products are bodies that survive by doing something impressive such as flying, swimming, running, digging or climbing – all in the service of reproduction, which means they also tend to be good at attracting a mate and warding off rivals. An important part of the environment that each gene must exploit, if it is to ensure its survival in the form of copies of itself, is the other *genes* it encounters in the *genomes* of a succession of bodies – which, because of sexual recombination, means the other genes in the *gene pool* of the species. As a result of this, cartels of mutually supportive genes cooperate to build bodies that specialise in some particular method of surviving, such as grazing or hunting. Different cartels are the gene pools of different species, bound together by the remarkable phenomenon of sexual recombination – and separated from all other cartels, for it is part of the definition of species that they can't interbreed. Occasionally, often through accidents of geography, gene pools find themselves subdivided for long enough to become sexually incompatible, and the subdivisions are then free to go their separate evolutionary ways as distinct species. Eventually, 'separate ways' can mean 'very separate indeed', for animals as different as vertebrates and molluscs originally split apart as members of the same species. Successive branchings of this kind have given rise to hundreds of millions of species, over thousands of millions of years.

At least in sexually reproducing species, evolution consists of changes in gene frequencies in gene pools. I stipulate sexual reproduction, because without it we have no clear idea what 'gene pool' even means. Where there is sexual reproduction, the gene pool is the set of available alleles from which the individual members of a species draw their genomes – 'draw' as in a lottery, the lottery of sex. Each individual genome is like a shuffled pack of cards. The available cards to be shuffled are sampled from the gene pool. The statistical frequencies of these available cards change as the generations go by, and that is evolution. We can monitor evolution by measuring a sample of the phenotypes – the anatomy and physiology of typical members of the population. As the average phenotype changes – as legs get shorter, horns longer, coats shaggier, or whatever happens to be evolving at the time – it is tempting to see natural selection as a sculptor's chisel, carving the bones and flesh of the animals themselves.

But if we want to talk chisels, a sharper representation of evolution sees them as working not on the bodies of animals but on the statistical structure of gene pools. As crests get longer, or eyes rounder, or tails gaudier, what is really being carved by natural selection is the gene pool. As mutation and sexual recombination enrich the gene pool, the chisels of natural selection carve it into shape. We observe the results in the form of changes in the average phenotype, and it is phenotypes that serve as the proxies for genes. As the external and visible manifestations of genes, they determine whether those genes are eliminated, or whether they persist in the gene pool.

Natural selection carves and whittles gene pools into shape, working away through geological time. It is an image that might have seemed strange to Darwin. But I think he would have come to love it.

HENRY PETROSKI

IMAGES OF PROGRESS: CONFERENCES OF ENGINEERS

Henry Petroski, the Aleksandar S. Vesic Professor of Civil Engineering and a Professor of History at Duke University, is the author of more than a dozen books on engineering and design, including *To Engineer Is Human: The Role of Failure in Successful Design* and *Engineers of Dreams: Great Bridge Builders and the Spanning of America*. His newest book is *The Essential Engineer: Why Science Alone Will Not Solve Our Global Problems*. He is a Distinguished Member of the American Society of Civil Engineers; a Fellow of the American Society of Mechanical Engineers, the Institution of Engineers of Ireland, and the American Academy of Arts and Sciences; and is a member of the American Philosophical Society and the US National Academy of Engineering.

A S CELEBRATED IN THEIR DAY AS THE STATESMEN OF SCIENCE WERE THE GREAT ENGINEERS OF THE NINE-TEENTH CENTURY. HENRY PETROSKI EXPLAINS HOW THEY BUILT THESE AWESOME STRUCTURES AND WHY THEY ATTRACTED SUCH ACCLAIM.

One of the great engineering achievements of the nineteenth century was the expansion of the railways into an ever-widening network. Extending the right of way across major bodies of water naturally presented especially difficult problems for engineers, and so early railways often relied upon ferries at these locations. But this solution was not in keeping with the developing image of a fast and uninterrupted journey in a string of carriages pulled by a steam locomotive, and so bridges were built whenever possible. The most daring of these bridges, symbolic of the creativity, resolve, and integrity of the engineers that designed and built them, proved to be great engineering achievements in their own right, especially when the body of water to be crossed presented unique challenges, as it did at the Menai Strait.

This strategic strait, which separates the isle of Anglesey from the mainland of north-west Wales, was controlled by the Royal Navy, and so the Admiralty required that any bridge that was to cross it had to provide a clearance of at least 400 feet horizontally and 100 feet vertically so that tall-masted sailing ships of the day could pass between its piers and beneath its roadway without hindrance. Furthermore, because of the importance of the strait, temporary supports were not allowed in the water during construction. This virtually ruled out the choice of an arch bridge, which traditionally required the use of an elaborate system of falsework upon which the arch was assembled until it was self supporting.

Thomas Telford had already been presented with this problem when he was charged with completing the highway that connected London and Dublin and thereby providing a reliable route for the delivery of, among other things, the royal mail. The Irish Sea could only be crossed by ferry. The ideal location for a terminal was at Holyhead, which is on the west side of the island of Anglesey.

To carry the road from London to Holyhead meant bridging the Menai Strait. Telford initially wanted a cast-iron arch, which in 1811 he proposed to support by cables from above and thereby not obstruct ship traffic during construction. This untried method would have worked, as would be proven a half-century later, but it was not to be tried first at Menai. Instead, Telford designed the only other then-known bridge type that could span the distance and provide enough headroom: a suspension bridge.

The Menai Strait Suspension Bridge was completed in 1826 and remains an aesthetic paragon of what can be achieved with the form. Telford's early experience as a mason enabled him to design graceful viaducts and towers bracketing the main span, which was a record-shattering 580 feet. He employed wrought-iron chains that were tested before installation, and the completed bridge was a structural marvel of its time. Unfortunately, the wooden roadway of the bridge proved not to be as substantial as its stone towers and viaducts and iron chains. When the wind was especially unfavourable, the roadway was susceptible to being tossed about, and on occasion it was destroyed.

When the Chester & Holyhead Railway was being laid out, routing its

The Menai
suspension bridge.

tracks across the Menai Bridge seemed the natural thing to do. However, as the wind had demonstrated, the structure's roadway was light and flexible, and this would not serve the purpose of the contemporary railway. As well as the possibility of the road being destroyed in a storm, there was also the problem of a heavy steam locomotive causing the roadway of the bridge's main span to deflect so much that the engine would have had to climb out of a valley of its own creation. The engineer George Stephenson suggested decoupling the train of carriages from its locomotive and using horses to pull the carriages to the other side of the bridge, where they could be coupled to another locomotive for the continuation of the journey. This was not what engineers would call an elegant solution.

Stephenson's son, Robert, had a different idea. It involved designing a bridge that relied on neither the arch nor the suspension principle. Stephenson identified a site about a mile south of Telford's Menai Suspension Bridge, where a large rock formation divided the strait into two wide navigation channels. Since this natural formation, known as Britannia Rock, was a recognised and accepted obstacle to shipping, there could be no reasonable objection to constructing a tall stone tower upon it. Similarly tall towers could also be erected outside the navigation channels on either side of the rock. Massive wrought-iron girders could then be installed at a sufficient height between these towers so that the vertical clearance was equal to that beneath the suspension bridge.

Robert Stephenson's scheme was acceptable to both the railway company and the government, and so the detailed design and construction of the bridge was begun in the mid-1840s. Since no such structure had ever been designed, let alone built, it fell to Stephenson to organise what would today be called a research-and-development project. In order to keep the weight of deep girders exceeding 450 feet in length within acceptable bounds, it was decided early on that they should be hollow. At the time there existed no structural theory sufficiently advanced whereby the design of such girders could proceed by calculation alone. An experimental programme was thus embarked upon.

The experimentalist-engineer William Fairbairn, who had established a shipyard and had tested cast-iron beams years earlier, was responsible for conducting scale-model strength tests to establish the preferred shape and detailed design of the wrought-iron tubes. He began with small-scale models to compare the relative strengths of different shapes and arrived at the conclusion that a rectangular cross-section was the best. The model tubes were tested by hanging from their centre weights that represented the load of a heavy locomotive. Weights were added until the tube failed,

William Fairbairn
by Benjamin
Faulkner, 1872.

which revealed the weakness of the structure and thereby provided guidance for how to modify it in the next model. By progressively increasing the scale of his models, Fairbairn was able to establish trends of behaviour, and from the experimental data the theorist Eaton Hodgkinson established an empirical formula by means of which he could extrapolate to the requirements for the full-size tube.

To build a full-scale model and test it to destruction would have been essentially to build the bridge itself. So, as is typical in the engineering of large structures to this day, there comes a point when judgment dictates that the model testing must end and the real thing begin. In order to keep the navigation channels of the Menai Strait unobstructed, the longest tubes were fabricated along the banks. When completed, the tubular beams were floated into position between the towers and lifted into place by means of hydraulic jacks. This critical stage in the construction sequence was accomplished in a relatively short period, during which ships used the channel on the other side of the rock.

Although there were some anxious moments in the floating and hoisting process, the tubular girders were finally in place by 1850. However, since they had not been tested at full scale, there remained legitimate questions as to how they would perform. Such heavy girders might deflect so much under their own weight that they would be noticeably bowed and so present to a steam locomotive little better a roadway than the flexible deck of the suspension bridge. In anticipation of this possibility, the towers had been deliberately designed to be tall enough to accept iron chains from which the weight of the tubes could be partially supported. If this were necessary, then the bridge would effectively be a suspension bridge with a very heavy roadway. However, the tubes proved to be sufficiently stiff so that no supplementary support was necessary. Thus, the height of the towers in the finished bridge appeared to serve no structural purpose, a condition that some structural critics have seen as a flaw of its form.

The ultimate structural test would, of course, be when the first trains crossed the bridge. In anticipation of that, it was customary to conduct a 'proof test'. In the case of the Britannia Tubular Bridge, as the structure came to be known, as many heavy steam locomotives as could be assembled were driven end to end upon it. The girders barely moved under the unreasonably heavy load, and so the design was 'proven' to be sound. (There was hardly a thought given to the structure's ability to resist the wind, for it was so heavy that the strongest winds could no more move the tubes from their piers than a breath of air could a brick sitting on a table.) The structural performance of the bridge established it to be everything that Robert Stephenson claimed it would be. He, Fairbairn and Telford were all elected to the Royal Society within a year of the completion of the respective Menai crossing on which they worked.

The period during which the Britannia Bridge was under construction also saw rapid progress in the development of the new technology of photography. An engineering construction project was a perfect subject for the new art because it provided a static scene that was ideal for the long exposure times required. Indeed, photographs of the building of the Britannia Bridge are among the first of the genre. It was not practical to photograph groups of project engineers, however, because it was unlikely that they could all stay still long enough to capture a sharp image. Thus, the traditional art of painting was more likely to be employed to capture an occasional assembly of engineers.

The famous group portrait, *Conference of Engineers at Britannia Bridge*, was produced by the artist John Lucas shortly after the structure was completed, though it purports to show the partially completed bridge in the background. In any case, it conveys a sense of how such an ambitious engineering project advances in stages and that it takes a team to bring it to fruition. The completed Britannia Bridge may be attributed to its conceptualiser and chief engineer, Robert Stephenson, but like any other great structural achievement it owes its realisation to a host of other engineers advising and working on various details of the design and construction. To the engineers must also be added the often anonymous foremen and workers. These were represented in the painting by the two men kneeling on the floor and leaning against a wall, but clearly paying attention to the goings on.

Some key figures of the Britannia Bridge project are missing from *Conference of Engineers*. Neither Fairbairn nor Hodgkinson, without whose physical experiments and empirical formulas a successful full-scale tube design might never have been achieved, is depicted. This suggests that Lucas' intent was not to capture a scene where all of the responsible parties are assembled, but rather to depict an example of what was probably a not infrequent occurrence at the construction site. On the occasion that Lucas visited, the final tube of the bridge was being floated into place to complete the bridge shown in the background of his painting.

The progress of a project like the Britannia Bridge was followed by engineers and contractors around the world. It was not only the design that interested them but also the manner in which it was executed and the erection of the parts accomplished. Anyone with such an interest who would

A group portrait of the *Conference of Engineers at Britannia Bridge* by John Lucas.

have been travelling in the vicinity of the Menai Strait would likely have wanted to visit the construction site and experience for himself its scale and the energy and atmosphere surrounding it. Conferences of engineers and others, including everyone from members of the railway's board of directors to foremen responsible for key operations, would take place regularly.

It was evidently at a board meeting that took place near the Britannia Bridge site that Joseph Paxton daydreamed and sketched on a blotter before him the rudiments of his design for a building to house the Great Exhibition of 1851 – which became known as the Crystal Palace. Though not an engineer himself, Paxton had been responsible for the design of the Great Stove and the Lily House at Chatsworth and believed that structural principles embodied in those achievements could be applied to making an iron-and-glass building to accommodate the Great Exhibition. The following week, with the help of the railroad engineer and Royal Society Fellow Peter Barlow, the design was fleshed out. When Paxton shared his scheme with Robert Stephenson, he declared the concept sound and encouraged Paxton to proceed. Such one-on-one conferences between engineers, architects and interested parties occurred too frequently and privately to be captured by Lucas or, apparently, by anyone else.

Engineers with no direct connection to a project would also visit it, much as those on a busman's holiday do today. In the conference at Menai that Lucas did recreate in oils, the engineer Isambard Kingdom Brunel is depicted sitting to the extreme right. Brunel and Stephenson, among the most prolific of the great Victorian engineers, had had different views on railway gauges, with Brunel favouring the broad and Stephenson what came to be known as the 'standard' gauge. After the initial spate of building independent railways throughout the land, the lack of a common gauge among them made interconnecting them problematic. Brunel eventually lost the battle of the gauges, but he was to best Stephenson in designing a bridge to carry railway trains.

As much of a structural success that the Britannia Tubular Bridge was, it was an economic and environmental failure. The enormous amount of material and labour entailed in riveting relatively small sheets of wrought iron together to form massive tubular girders made the bridge very costly. In addition, since the trains ran through rather than atop the tubes, the ride could be a very hot and sooty experience. When Brunel was faced with designing a bridge to carry trains across the River Tamar near Plymouth, he had to achieve structurally essentially what Stephenson did at Menai, while at the same time doing it more economically and in a more environmentally acceptable way. His solution was to exploit in combination both arch and suspension principles to produce a significantly lighter bridge that was also open to the atmosphere and so presented a more pleasant ride. Brunel's Saltash Bridge – officially known as the Royal Albert Bridge and bearing the inscription 'I.K. Brunel, Engineer' above its portals – as well as the wind-resistant suspension bridge of the German-American engineer John Roebling, proved that Stephenson's solution to carrying trains over great spans was, in his own words, 'a magnificent blunder'. Only about a half-dozen tubular bridges would be built throughout the world.

However, just as the design and construction of the bridge itself remains significant as a case study of how an overwhelming problem was solved and an epochal building project accomplished, the symbolism embodied in Lucas' group portrait is timeless.

Confluences of engineers and the physical embodiments of the designs from their mind's eye have been recorded with conventional optical cameras on many occasions, especially when failures occurred. After the high girders of the Tay Bridge collapsed in 1879, a photographer from Dundee was hired by the investigative body to record on film the remains in place. The set of systematic photographs was generally forgotten for over a century, until Peter Lewis of the Open University came across them in the Dundee City Library. Employing high-resolution and digitally enhanced scans of the old photos, he found on the piers distinct signs of brittle fracture of many of the cast-iron lugs. This led him to his revisionist explanation of the cause of the failure: the repeated movement of the

bridge under passing trains and wind caused fatigue cracks to grow, which eventually led to the fractures. This made the cross bracing dependent on the lugs ineffective and the bridge consequently became more flexible. On the late December night in 1879, the combination of a fast-moving train, a howling storm, and a weakened structure proved to be fatal.

Bridge failures have been dramatic both structurally and photographically. At the turn of the twentieth century, the Firth of Forth rail bridge, the world's first significant all-steel bridge, had the longest spans (1,710 feet) of any bridge in the world. In response to the collapse of the girders of the Tay Bridge, the Forth Rail Bridge had been designed as a robust cantilever structure, an old form that had recently been revived and popularised in Britain by engineers William Fowler and Benjamin Baker. The heavy look of the completed Forth Rail Bridge led some engineers to believe that it was grossly over-designed, and they sought to produce cantilevers lighter in form and fact. In 1907, a cantilever bridge under construction over the St Lawrence River near Quebec was on its way to achieving a record 1,800-foot span. Photographs of the incomplete bridge show it to have been of a very much lighter and lacier design than the

Forth. Indeed, the Quebec Bridge proved to be overly slender and unable to support even its own weight. The bridge collapsed before it could be completed, claiming the lives of seventy-five construction workers. Photographs show it to have dropped into a tangled pile of steel.

A commission appointed to look into the causes of the collapse found that the weight of the bridge had been underestimated by the design engineer, who also made errors in his calculations of the stresses in the structure. The principal consulting engineer, Theodore Cooper, who was also the de facto chief engineer, had been remiss in not overseeing the work closely enough. After the causes of the failure were understood, the bridge was redesigned as a heavier cantilever structure and one whose geometry was much more amenable to analysis.

The failure of the first bridge brought uncommon attention to the rebuilding project. In one case, the board of engineers charged with redesigning the structure – the American and Canadian team of Ralph Modjeski, C.C. Schneider and chairman C.N. Monsarrat – were caught by the camera standing in the individual chambers of one of the key compression members (a redesign of the inadequate component that had initiated the collapse) awaiting assembly into the new bridge. In another photo, Monsarrat and Modjeski, along with the engineer of construction G.F. Porter and the chief engineer of the bridge company, G.H. Duggan, are sitting in a line on one of the thirty-inch-diameter pins – as if it were a beast of burden – that were awaiting installation.

When the central section of the redesigned bridge was being lifted into place to complete the structure, a fracture in one of the hoisting devices caused the entire section to fall into the river. A photograph of the impact of the steel on the water, complete with the accompanying dramatic splash, provided a rare example at the time of a failure caught on film. In spite of its troubled construction history, the Quebec Bridge was finally finished in 1917 and has stood for almost a century as the longest cantilever span in the world, a testament to the consequences of a failure. For longer spans, engineers looked to suspension bridges, which thanks to John Roebling and his successful approach to designing wind-resistant structures, were no longer looked upon as the frail descendants of the Menai Strait Suspension Bridge. Indeed, it was Roebling's Niagara Gorge Suspension Bridge, completed in the mid-1850s, that had become the first suspension bridge to carry railway trains. The principles on which it was based – weight, stiffness and stay cables – also guided the design of his masterpiece, the Brooklyn Bridge. Through the last part of the nineteenth and the first couple of decades of the twentieth century, engineers designed suspension bridges with longer and longer spans, almost always stiffened by a truss.

Among the most watched suspension bridge projects of the 1930s was the Golden Gate Bridge across the entrance to San Francisco Bay. The structure's 4,200-foot-long main span was to remain the longest in the world for over a quarter of a century. San Francisco had long wanted a bridge to connect it with Marin County across the strait – known as the Golden Gate – and thus with other northern California counties, but engineering proposals came with a prohibitive price tag. When Joseph Strauss, whose bridge company had specialised in movable bridges of modest span and appearance, proposed a hybrid cantilever-suspension bridge that he promised to deliver for a very attractive price, local movers and shakers paid

attention. Not only did he assure them that he could design the bridge but
also that he could help promote the bond issue needed to pay for it. When
he was made chief engineer of the project, he invited engineers who did
have direct experience with suspension bridge design to serve as consultants. At the first conference of the engineering advisory board, held in
Sausalito in August 1929, the participants posed for a photo on the steps of
the Alta Mira Hotel.

The President of the Board of the Golden Gate Bridge and Highway
District, William P. Filmer, is naturally front and centre. Close to him, on
his right, is chief engineer Strauss, hands on hips, elbows out, in a defiant
stance that at the same time signals keeping others at bay. Directly behind
Filmer is an army officer; as was the case at the Menai Strait, the approval
of the military was essential in allowing any bridge to be built across the
strategic Golden Gate.

Charles A. Ellis, the designing engineer, is standing on the same step as
Strauss and Filmer, but away from them, a placement that may have been
directed by the photographer to keep the tall Ellis from appearing to tower
over everyone else. Still, his height emphasises Strauss' small physical
stature – something about which he was reportedly sensitive. Though the
difference in their heights is ameliorated somewhat by Ellis' standing

almost off by himself, it is very likely that Strauss' stance was prompted by this placement of Ellis on an equal footing. The tension between Strauss and Ellis suggested in this group portrait presaged that which would grow as the designing of the bridge progressed.

No chief engineer can be as fully informed about design details as those who are working directly on the calculations. Strauss had never carried to completion the design of a suspension bridge, let alone one that would break the world record for span length. The detailed design work fell to Ellis, working under the consultants, and specifically under the supervision of Leon S. Moisseiff. At one public presentation of progress on the project, questions of substance about the design could only be answered by Ellis, making it clear to all who did not already know it that Strauss was uninformed about critical details of his own bridge. Not one to like being found in such a position, Strauss effectively exiled Ellis from San Francisco by sending him back to the Chicago office to continue the design work out of the public eye.

With little staff help, Ellis worked away on the bridge's design, but did not work fast enough to suit Strauss. After a confrontation over the design of the towers, Strauss ordered Ellis to take a vacation, from which he was never welcomed back. Ellis was replaced by Clifford E. Paine, who was identified as principal assistant engineer when construction on the bridge was completed in 1937. The engineering team listed on the dedicatory plaque located on the bridge tower did not include Ellis. This omission went generally unremarked upon for almost five decades, until the story of Ellis' involvement was told by John van der Zee in his 1986 book, *The Gate: The True Story of the Design and Construction of the Golden Gate Bridge.*

While the Golden Gate Bridge was under construction, an even larger and arguably more ambitious project was underway to connect San Francisco with its neighbours across the bay to the east. Comprising two suspension bridges in tandem, a large-bore tunnel, a 1,200-foot cantilever span and a long viaduct, the San Francisco–Oakland Bay Bridge was the most expensive publicly funded highway project undertaken to that time. Since no state highway department possessed within its ranks all the expertise needed to undertake such an ambitious project, California enlisted expert consultants to help with the job. At its completion, which occurred about six months prior to the completion of the Golden Gate Bridge, the team of engineers making the final inspection of the Bay Bridge posed for a photo against the backdrop of one of its large suspension cables. Among the engineers were specialists in foundations, superstructure and traffic, emphasising the multiple disciplines needed to carry out a work of such magnitude and complexity.

With the Golden Gate and Bay bridges finished, there were few large metropolitan areas left in America that needed – and could afford – such spectacular bridges. But there remained the need for more modest suspension bridges in special locations for special purposes. New York City was preparing to host the 1939 World's Fair, and it wished a new highway link in the vicinity. Elsewhere in the US, remote areas with the political clout and will also felt the need for suspension bridges. These were designed according to a new aesthetic, which dictated that a bridge's deck should be as slender-looking as possible. One way of achieving this look was to eliminate the trusswork that had become a hallmark of American suspension bridge design.

The first significant departure from the use of a stiffening truss had occurred in the design and construction of the George Washington Bridge, which opened in 1931 and crosses the Hudson River between New York and New Jersey. The exceptional width of this bridge's roadway and the consequent weight did make a truss unnecessary in this case, but in regions where light traffic meant that only two lanes were required, a narrow and shallow bridge deck meant also a much lighter and more flexible structure. Suspension bridge designers sought to fit their structures with ever more slender decks. By the end of the 1930s, this trend produced bridges whose

The collapse of the Tacoma Narrows Bridge in 1940.

roadways moved a suprising amount in the wind. There was no satisfactory theoretical explanation for this behaviour, but engineers felt confident that their bridges were in no danger of collapse.

They were disabused of that in 1940, when the Tacoma Narrows Bridge, whose deck had been undulating in the wind for months, began to twist and soon collapsed. Since the undulations had been occurring for some time, the bridge was the object of an ongoing study. Its misbehaviour was being investigated experimentally through a scale model, and the real bridge was being filmed. On 7 November, when the vertical undulations changed over to torsional oscillations, a film crew was despatched to capture the new behaviour. The twisting lasted for hours, and the final writhing of the steel structure caught on film made the bridge infamous. Indeed, before the collapse of the New York World Trade Center twin towers, the failure of the Tacoma Narrows Bridge was the most widely viewed structural collapse in engineering history.

Today, bridges of unprecedented scale and unchallenged beauty continue to be designed and built worldwide, and they require no less of a team than did their predecessors. The seemingly unrelated aims of functional strength and aesthetic appeal had been not only successfully integrated in many of the classic suspension bridges of the past two centuries but also commonly achieved by engineers alone or leading teams. Thomas Telford was in fact both engineer and architect of his Menai Suspension Bridge, and John Roebling was both engineer and architect of his Brooklyn Bridge. That these engineering structures especially have come to be regarded as architectural icons demonstrates the aesthetic heights that an engineer can achieve.

Engineers less artistically confident than Telford and Roebling have engaged consulting architects to advise them on the design of everything from the façades placed on massive anchorages and skyscraper-high towers to the finishing details like deck railings and lampposts. Othmar Ammann, the chief engineer of the George Washington and many other New York City bridges, often sought the help of famous architects. When the George Washington was but an idea on paper, Ammann engaged Cass Gilbert, the architect of the Woolworth Building and other landmarks, to depict how the towers might be finished in stone. Since money was tight when the bridge was being completed, however, the steel-framed towers were left bare – a look that the Swiss architect Le Corbusier found extremely appealing – and bare steel became the new aesthetic standard for monumental bridge towers. For his Bronx–Whitestone Bridge, Ammann engaged the 'architect to the elite' Aymar Embury II in designing the structure's anchorages. It was Embury's suggestion that they express the force that they exert against the pull of the suspension cables and show its trajectory into the monolithic bookends of the bridge proper.

But relationships between architects and engineers were generally strained in America in the 1920s and 1930s. There had been continuing tensions over which of these professionals should control bridge projects. The architects argued that they were better prepared to choose the form and site for a bridge, leaving it to engineers working under them to figure out how to build the structure. But, unlike large buildings, long-span bridges had traditionally been sited, designed and constructed under the

direction of a chief engineer. The increasing structural challenges presented by long-span bridges kept the engineers in control.

In a series of articles in *Civil Engineering* magazine, the architect Embury
described his working relationship with the engineering team for the
Bronx–Whitestone and made it clear that the chief engineer always had the
final decision. According to Embury, in a bridge project engineers and architects alike were 'instruments' of the one chief engineer and 'were guided by his
desires as to the lines along which we should proceed'. But neither was
Embury uncritical of his colleagues in either camp. He did not approve of
engineers pursuing 'design by drawing instruments', by which he meant that
they tended to use certain angles in their structures because they were the ones
of the drafting instruments close at hand. He was also critical of his fellow
architects, who he felt too often followed 'the easiest way'. In an attempt to
promote a meeting of the minds, Embury believed that 'engineers should be
good architects, and architects good engineers'. Who could argue with that?

In more recent years commissioning organisations have tried to force
engineers and architects to be equal partners in bridge design. The design
competition guidelines for the Gateshead Millennium Bridge, the strikingly
original arch-and-cable 'blinking-eye' movable structure that carries pedestrians over the River Tyne between Gateshead and Newcastle, made it clear
that multidisciplinary teams were expected to produce entries 'of sufficiently
high technical and aesthetic merit'.

The design competition for the London Millennium Bridge, the low-
slung suspension bridge for pedestrians that spans the Thames to tie
together St Paul's Cathedral and the Tate Modern museum, went further
than the Gateshead one. For the London crossing, it was required that
design teams comprise not only engineers and architects but also artists.
The winning entry was a collaboration among the engineering firm of Ove
Arup, the architectural firm of Norman Foster and the sculptor Anthony
Caro. The resulting long, slender-decked bridge has been described as a
'blade of light', which it resembles when viewed from a distance up or

down the river. As was the case with the Tacoma Narrows Bridge four decades earlier, aesthetics dominated structure, and the unconventional design of the Thames crossing allowed its deck to move sideways excessively under the crowds of pedestrians that flocked to its opening in June 2000. After three days of movement that was deemed potentially dangerous for people, if not the bridge itself, the structure was closed. Much of the public blame for the fiasco fell on the engineers, who were sent back to the drawing board. After being retrofitted with dampening devices, some of which may be said to compromise its aesthetics, the bridge was reopened and has become a popular tourist attraction.

Artistic designs like the Gateshead and London Millennium bridges may not be suitable for large-span bridges that carry vehicles as well as pedestrians. But that is not to say that such large-scale bridges cannot also have a strikingly innovative aesthetic component. The relatively new bridge form that has become a favourite for achieving striking profiles and dramatic effects is the cable-stayed bridge. In contrast to the suspension bridge, from whose two or four main cables a roadway is hung, the cable-stayed bridge employs multitudes of cables that stretch directly from towers to deck. The

great number of cables allows for a wide variety of arrangements, and so each cable-stayed bridge can have a distinctive look. This characteristic has led to the design of unique bridges known as 'signature bridges'.

Among the most widely admired new bridges of this type is the Millau Viaduct, which carries a very high roadway across the Tarn Valley, formerly a traffic bottleneck on the road between Paris and Barcelona. The Millau is a breathtakingly striking design that is commonly attributed to the architect Norman Foster, and it certainly is an architectural achievement in its sculptural form and the way it harmonises with its dramatic natural setting. However, the structural design and construction of such a towering bridge are not architectural but engineering achievements. Unfortunately, the French bridge engineer Michel Virlogeux, who was responsible for the structural design, is largely forgotten when the bridge is marvelled at.

Architects may be more extroverted and therefore the more visible members of a bridge design team today, but they are not always the most essential. Perhaps we ought to revive the grand tradition embodied in John Lucas' *Conference of Engineers* to remind us of what was obvious in the nineteenth century, but may now be forgotten.

11

GEORGINA FERRY

X-RAY VISIONS:
STRUCTURAL BIOLOGISTS AND SOCIAL
ACTION IN THE TWENTIETH CENTURY

Georgina Ferry is a science writer, broadcaster and biographer. She is the author of *Dorothy Hodgkin: A Life*, *The Common Thread: A Story of Science, Politics, Ethics and the Human Genome* (with John Sulston), *A Computer Called LEO: Lyons Teashops and the World's First Office Computer*, and most recently, *Max Perutz and the Secret of Life*.

S CIENTISTS NEED TENACITY: NONE MORE SO THAN THOSE WHOSE PAINSTAKING, DRAWN-OUT WORK PICTURED THE THREE-DIMENSIONAL STRUCTURES OF THE VAST MOLECULES BUILT BY LIVING CELLS. AS GEORGINA FERRY RELATES, THE STRENGTH OF CHARACTER DEMANDED BY CRYSTALLOGRAPHY OFTEN WENT ALONG WITH STRONG CONVICTIONS ABOUT THE ROLE OF SCIENCE IN SOCIETY.

Anyone crossing the courtyard of Burlington House in Piccadilly on a certain day in 1945 would have seen a contrasting couple sitting on the steps of the East Wing, then home to the Royal Society. She was slight, girlish, a fair-haired woman in her thirties with penetrating blue eyes. He was a decade older, shock-headed, fleshy-faced and physically imposing. Waiting for a colleague, they were discussing her latest scientific result. Delightedly she confided that after two years of wartime work, still officially a secret, she and her colleagues had solved the structure of penicillin. 'You'll get the Nobel Prize for this,' he said. She countered that she would far rather be elected one of the Fellows on whose doorstep she was sitting. Without irony, he told her 'That's more difficult.'[1]

Wrapped up in this anecdote about Dorothy Crowfoot Hodgkin and

John Desmond Bernal is a whole chapter of interlocking stories: about collegiality, about scientific workers of both sexes, about the impact of war on research, but above all about the conviction that knowing how biological molecules were constructed from atoms in three dimensions would fundamentally alter our understanding of life. Hodgkin (FRS 1947) and Bernal (FRS 1937), her former PhD supervisor and lifelong mentor, were among the founders of a project that at first seemed hopeless, even quixotic in its ambition: to use physical techniques to reveal the structure of life in atomic detail. Today their inheritors are at work every day, using essentially the same techniques to build a catalogue of the shapes of every molecule in the living body, and applying that information to understand health and disease and to design new drugs.

The legacy of the structure pioneers, however, is richer than the sum of their scientific achievements. Each in his or her own way, they gave their considerable energies to causes such as scientific education, the organisation of research, international understanding, gender equality, human rights, prison reform and world peace. Partly because the subject crossed so many disciplinary boundaries, but also because of the personalities involved, they also developed a way of doing science that valued collaboration over competition, and fostered egalitarianism in relation to rank, gender and class.

FATHER AND SON

Everyone in this story can trace a scientific lineage back to William Henry Bragg (FRS 1907) or his son William Lawrence Bragg (FRS 1921).[2]

Most would also credit the Braggs with establishing the egalitarian outlook that the early structural biologists shared. William H. Bragg was born in the UK and studied at Cambridge, but in 1885, at the age of twenty-three, he was appointed to the professorship in physics at the University of Adelaide. In 1909 he returned from Australia to take up the chair in physics at Leeds. His nineteen-year-old son Willie immediately went to

1 Hilary Rose, *Love, Power and Knowledge: Towards a Feminist Transformation of the Sciences* (Cambridge, Polity, 1994), p. 117.
2 For more on the Braggs, see Graeme Hunter, *Light is a Messenger: The Life and Science of William Lawrence Bragg* (Oxford, Oxford University Press, 2004).

Cambridge to study natural sciences.

In 1912 the Munich-based physicist Max von Laue and his junior colleagues reported that a zinc sulphide crystal could diffract a beam of X-rays, producing a characteristic pattern of spots on a photographic plate and demonstrating the wave-like nature of X-radiation. Bragg *père*, who at that time inclined to the view that X-rays consisted of particles, was tipped off about the paper by a colleague who was working in Germany. When Willie came home for the long vacation they pored over the problem, and in subsequent months began their own experiments. Willie confirmed that X-rays formed diffraction patterns on passing through crystals (in the same way that light does on passing through narrow slits), and therefore behaved like waves. He went on to demonstrate that a simple mathematical formula (now known as Bragg's Law) could relate the positions and intensities of the spots in the pattern to the positions of the parallel layers of atoms in the crystal from which the X-rays were reflected. The formula required a figure for the wavelength of the X-rays, and the Braggs were able to measure this using an X-ray spectrometer of Bragg senior's invention. Applying the formula to X-ray photographs of simple compounds such as sodium chloride, Willie Bragg was able to draw a picture of the sodium and chlorine atoms neatly alternating throughout the cubic lattice, like the simplest of wallpaper patterns but in three dimensions.

William Henry Bragg *(left)* and William Lawrence Bragg.

The Braggs had turned X-ray diffraction from an intriguing observation into a tool for exploring what matter is made of in the range that was too small to be seen with a microscope, and too large for chemical analysis. They shared the 1915 Nobel Prize in physics for their discovery. The announcement came when the younger Bragg, aged only twenty-five, was in France developing sound-ranging techniques to help the allies in the war against Germany to fix the coordinates of enemy artillery batteries. He remains the youngest person ever to win a Nobel. Years later Max Perutz summed up the range of discoveries that subsequently flowed from the Braggs' achievement:

> Why water boils at 100°[C] and methane at -161°, why blood is red and grass is green, why diamond is hard and wax is soft, why graphite writes on paper and silk is strong, why glaciers flow and iron gets hard when you hammer it, how muscles contract, how sunlight makes plants grow and how living organisms have been able to evolve into ever more complex forms ... The answers to all these problems have come from structural analysis.[3]

Knighted in 1920, Sir William Bragg moved to London as Professor of Physics at University College (UCL), and then Director of the Davy-Faraday Laboratory at the Royal Institution (RI), a post that he held from 1923 until his death in 1942. A central figure in British science, he was also President of the Royal Society from 1935 until 1940. Long before 'public understanding of science' became a topic of debate, Bragg retained the nineteenth-century assumption that new discoveries in science should be part of public discourse, and was an enthusiastic writer and speaker. In 1919 he gave the Christmas Lectures for children at the Royal Institution on the subject 'Concerning the Nature of Things'; six years later he again fascinated his young audience with his series 'Old Trades and New Knowledge'. Both series were published as books, and contained some of the first public descriptions of the capacity of X-ray crystallography to open up new perspectives:

3 Letter from Max Perutz to Gerald Holton, 9 July 1966, private papers, quoted in Georgina Ferry, *Max Perutz and the Secret of Life* (London, Chatto & Windus, 2007), p. 28.

The discovery of X-rays has increased the keenness of our vision …
a thousand times, and we can now 'see' the individual atoms and
molecules.[4]

From the early 1920s Bragg began to use X-ray crystallography to investigate organic molecules (those containing carbon, which include all the molecules that make up living things) rather than the simple, inorganic salts that his son continued to work on as a very young professor at Manchester. Now well into his sixties and with heavy administrative responsibilities at the RI, Sir William recruited young men and women to join his endeavour in the laboratories where Humphry Davy and Michael Faraday had conducted their chemical and electrical experiments a century before.

Bill Astbury (FRS 1940), the son of a potter from Stoke on Trent, had gone to Cambridge on a scholarship and graduated with a First in Natural Sciences. Joining Bragg as a postgraduate at UCL and the RI, he began to use X-ray diffraction to study the structure of natural fibres such as wool that are made of large, complex protein molecules. In 1928 he moved to the University of Leeds, an important centre of the textile industry, where he continued to develop the technique of fibre diffraction. During the 1930s he was the first to take X-ray photographs of DNA fibres (long before anyone had established its significance as the molecule of heredity). Although he was not able to obtain definitive structures, his insights into the 'coiled' nature of these molecules were fundamental to later discoveries by Linus Pauling (the alpha helix of proteins) and Maurice Wilkins, Rosalind Franklin, James Watson and Francis Crick (the DNA double helix).

Kathleen Yardley (later Lonsdale, FRS 1945) was the tenth child of an Irish postmaster who had a drink problem. Her mother brought the family to England for a better life, and in 1922 Yardley graduated from Bedford College (a women's college of London University) with the highest mark in physics that anyone in the university had achieved for ten years. Bragg immediately wrote to recruit her as his research assistant. When she married fellow researcher Thomas Lonsdale and had three children, Bragg kept her supplied with work she could do at home, then found her a grant to pay for domestic help so that she could come back to the lab. This concern to create conditions in which a married woman could pursue a scientific career was wholly exceptional at the time, as was Thomas Lonsdale's willingness to share domestic chores and support his wife in her career. Kathleen Lonsdale clarified the structure of a number of small organic molecules, notably confirming that benzene was a flat ring of six carbon atoms, each with a hydrogen attached.

Tiny, courageous and independent, Lonsdale dealt with glass ceilings by refusing to see them, and achieved a series of notable firsts for British women in science. In 1945 she and Marjorie Stephenson, the Cambridge biochemist, signed the Register of Fellows of the Royal Society, the first women to do so since its foundation in 1660. Their election followed delicate political manoeuvring largely on the part of the then President, Sir Henry Dale (who became Lonsdale's boss at the RI after William Bragg's death in 1942) to overturn what prejudice remained among the Fellowship after legal obstacles were removed in 1919. She also benefited from the energetic advocacy of her erstwhile fellow graduate student, Bill Astbury. It was his presentation of a correctly drawn up certificate of her candidacy that prompted Dale to win the majority of the Fellowship over to this revolutionary move.[5]

In 1949 Lonsdale became the first woman to be appointed to a professorship at University College London, and in 1968 the first to become President of the British Association for the Advancement of Science. A Quaker and conscientious objector, in 1943 she refused to pay the fine of £2 for non-registration for civil defence work, an action that earned her a month in Holloway prison. Appalled at the monotony of prison life, she became an active supporter of prison reform after her release. At the height of the Cold War she wrote a book, *Is Peace Possible?*,[6] giving a personal response to her 'sense of corporate guilt and responsibility that scientific knowledge should have been so misused' as to develop atomic weapons. Her example was an inspiration to the generations of women crystallographers who followed.

4 W.H. Bragg, *Concerning the Nature of Things* (London, G. Bell & Sons, 1925).
5 Rose, *Love, Power and Knowledge*, pp. 115–35.
6 Kathleen Lonsdale, *Is Peace Possible?* (London, Penguin, 1957).

Kathleen
Londsdale and
John Desmond
Bernal.

Also born in Ireland, to a comfortably-off farming family, John Desmond Bernal[7] had astonished his Cambridge tutors as an undergraduate by producing unbidden an eighty-page manuscript giving mathematical derivations of the 230 'space groups' of classical crystallography. The diversion of his efforts probably cost him a First, but left them in no doubt of his quick grasp of theoretical concepts, and like Astbury he came to Bragg with their enthusiastic recommendation. Though he never completed a PhD thesis, during his time at the RI he solved the structures of single crystals, notably graphite, designed the X-ray goniometer that all crystallographers used to mount and photograph their crystals for years afterwards, and made further theoretical contributions to the subject. In 1927 he returned to Cambridge as the first Lecturer in Structural Crystallography in the department of mineralogy.

THE SAGE OF SCIENCE

Bernal was a polymath, able to discourse convincingly and at length on any topic from Chinese art to quantum physics. While still an undergraduate he had earned the nickname 'Sage' from his fellow students: the name stuck throughout his life, used by all his friends and colleagues with barely a trace of irony. Nor were his energies confined to intellectual pursuits.

Exchanging devout Roman Catholicism for equally devout Marxism as an undergraduate, he became a leading member of the 'visible college' of scientists and socialists who came to prominence in the 1930s.[8] Always linking thought to action, he was an indefatigable organiser, notably of the Association of Scientific Workers and later its international counterpart, the World Federation of Scientific Workers. His desire for experimentation extended far outside the laboratory: he pursued a private life of unabashed promiscuity, justified to himself and others by his political mission to escape the restrictions of social convention.

Bernal was unusual among scientists in the degree to which he reflected on his experiences and beliefs in both public and private. In his early life he kept diaries charting everything from his scientific and political insights to his sexual conquests, and at the age of only twenty-five began a passionate and idealistic memoir (never published) entitled *Microcosm*. Soon afterwards he produced his first published book, *The World, the Flesh and the Devil: An Enquiry into the Future of the Three Enemies of the Rational Soul* (1929), which accurately predicted a number of scientific developments including the Apollo space programme, and just as inaccurately forecast the triumph of world Communism. A later and much more influential book, *The Social Function of Science* (1939), argued for central planning of science on the Soviet model, with the goal of improving human welfare rather than pursuing knowledge for its own sake.

The Second World War gave Bernal the opportunity to put his own science to the service of society. He was involved in studies of the accuracy of bombing raids and their effects, which influenced both civil defence policy and Bomber Command, and conducted surveys of the Normandy coastline and seabed as part of the preparation for the D-day landings. After the war he aligned himself, like many of his fellow scientists, with opposition to nuclear warfare, coining the phrase 'weapons of mass destruction' at a speech to the British-Soviet Society in London in 1949. His influence might have been greater had it not been for his blindly

7 For more on Bernal, see Andrew Brown, *J.D. Bernal: The Sage of Science* (Oxford, OUP, 2005).
8 See Gary Werskey, *The Visible College: A Collective Biography of British Scientists and Socialists of the 1930s* (London, Allen Lane, 1978).

uncritical support for Soviet Communism, which was unwavering in the face of Stalin's purges, the Lysenko affair and the invasion of Hungary. His accusation that the direction of Western science was dictated by warmongers led to his removal from the Council of the British Association for the Advancement of Science. Despite his valuable service during the war years, he never received any honours in Britain.

The double Nobel Prize-winner Linus Pauling (For.Mem.RS 1949) is one of many who described Bernal as the most brilliant scientist they had ever met. Yet he never personally made the kind of breakthrough that would have set him on the road to Stockholm. With so much to do, and so little time, he rarely pursued a scientific project to its conclusion. Instead, he gathered around him a group of able disciples of both sexes and showered them with ideas. They did not let him down.

PROTEINS AND PRIZES

Dorothy Crowfoot (later Hodgkin[9]) was a slim, soft-spoken, first-class graduate in chemistry from Oxford who came to Bernal's lab in 1932 to begin a PhD. The eldest of four girls, Crowfoot came from a middle-class family who did not see intellectual pursuits as off-limits for women. Her father was a colonial administrator and archaeologist, and her mother, without any formal higher education, became a world expert in ancient textiles. It was she who encouraged Crowfoot's schoolgirl interest in chemistry by giving her W.H. Bragg's collected lectures to read, and his account of crystallography captured her imagination.

Crowfoot excelled in all the practical aspects of crystallography – growing the crystals, mounting them and photographing them – but also had a remarkable ability to visualise the three-dimensional manipulations that the early, trial-and-error stage of the subject demanded. She quickly became Bernal's right hand, conducting preliminary observations on the dozens of crystals that poured into his lab from all over the world. Asked later how she succeeded so early, she modestly replied that there was so much gold lying about, one could not help picking it up.[10]

One day Glenn Millikan, a young scientist and friend of Bernal's, returned to Cambridge from Sweden with a tube of crystals of the digestive enzyme pepsin in his pocket. Like all enzymes pepsin is a protein, one of a class of biological molecules that are the precision tools of the living body. Enzymes are highly specific catalysts that speed the construction and destruction of all the body's constituents; other proteins include keratin and collagen that build strong structures such as hair and skin, antibodies that protect us against disease, and hormones such as insulin. All proteins depend for their function on their molecular structure. With care they can be purified and crystallised just like simple salts (though the crystals tend to be very small). The fact that they crystallise at all implies that their molecules have a regular structure – something that not all chemists believed at the time – and Bernal was convinced that solving these structures would reveal the 'secret of life'.

When he took the pepsin crystals out of the liquid in the tube he found that they quickly lost their crystalline form, so he mounted a crystal with some of the liquid inside a fine glass capillary before putting it in the X-ray beam. He obtained a pattern of spots, the first time anyone had successfully made a single protein crystal diffract. Crowfoot went on to take a further series of photographs of the crystal until they had enough for a letter to *Nature*,[11] describing their preliminary observations. Protein molecules are so large, consisting of thousands of atoms arranged in folded chains, that the relationship between their X-ray reflections and atomic positions is far from straightforward. Trial-and-error methods could not begin to narrow down the range of possible structures that would produce such patterns. Nevertheless, the Bernal and Crowfoot paper heralded the modern era of protein structure analysis.

Already at the forefront of the field at the age of twenty-four, in 1934 Crowfoot returned to Oxford where Somerville College (a women-only college) had given her a fellowship, and embarked on an X-ray study of the

9 For more on Hodgkin, see Georgina Ferry, *Dorothy Hodgkin: A Life* (London, Granta, 1998).
10 Lewis Wolpert and Alison Richards, *A Passion for Science* (Oxford, OUP, 1988).
11 J.D. Bernal and D. Crowfoot, 'X-ray Photographs of Crystalline Pepsin', *Nature*, 133 (1934), 794–5.

protein hormone insulin.[12] She married the historian Thomas Hodgkin, and despite his long absences promoting adult education in the north of England, had given birth to two children by the end of 1941. A supportive college, indulgent in-laws and cheap domestic labour enabled her to keep working with only the briefest of intervals, despite a severe attack of acute rheumatoid arthritis after the birth of her first child. During the Second World War she was recruited to the secret penicillin project, trying to solve the structure of the miraculously effective antibiotic that had been purified from mould by Howard Florey and his colleagues in Oxford's Dunn School of Pathology. Penicillin molecules had only a couple of dozen atoms, but the substance proved difficult to crystallise.

Success followed in 1945 after Kathleen Lonsdale personally brought Hodgkin samples of a more easily crystallisable penicillin derivative from America, where efforts to start industrial production were under way. Hodgkin's structure unequivocally confirmed the presence of a previously unseen ring of atoms in the molecule, known as a beta lactam ring, that was fundamental to the drug's ability to incapacitate bacteria. Although this discovery did not immediately lead to the creation of synthetic antibiotics as the project's industrial partners had hoped, it was one of the first examples of a drug's function being explained in terms of its structure, a principle that underlies all drug discovery programmes today. Lonsdale was delighted, and hoped for the opportunity to exercise her brand-new status as a Royal Society Fellow on Hodgkin's behalf:

I am going to ask a favour; when this work is published, may I communicate it [to the *Proceedings of the Royal Society*]? If … it is possible I think that it would be rather pleasant that a woman Fellow should communicate such a very important paper by another woman, and I would be very proud to do it.[13]

As she had so fervently wished, Hodgkin was herself elected to the Royal Society two years later, aged only thirty-six and by then a mother of three. She went on to solve the structure of the anti-pernicious anaemia factor, Vitamin B12, and in 1960 the Society appointed her its first Wolfson Research Professor. Bernal's prophecy came true when she was awarded the 1964 Nobel Prize for Chemistry, the first (and so far the only) British woman to win a science Nobel. The following year she was appointed to the Order of Merit, the first woman to receive the honour since Florence Nightingale.

While Hodgkin developed Bernal's project in Oxford, another of his students kept it going in Cambridge after Bernal himself had departed for the chair in physics at Birkbeck in 1937. Max Perutz (FRS 1954)[14] came to Cambridge as a wealthy foreign research student, funded by an allowance from his father who ran a textile business in Vienna. He began work on the protein haemoglobin, the pigment in red blood cells that carries oxygen round the body. But with the Anschluss in 1938 his Jewish family lost everything and had to flee for their lives. His parents eventually arrived in Cambridge and became dependent on his support. Fortunately his excellent X-ray photographs of haemoglobin crystals so impressed the new Cavendish Professor of Physics – none other than Bragg junior, soon to be Sir Lawrence to distinguish him from his father – that he found himself taken on in 1939 as Bragg's research assistant with a grant from the Rockefeller Foundation.

As an 'enemy alien', Perutz suffered internment in 1940–41, but on his return was recruited (thanks to Bernal, and to a brief pre-war foray into glaciology) to one of the most audacious scientific projects of the war.

12 Insulin controls levels of glucose in the bloodstream, and has to be given artificially to those with some forms of diabetes.
13 Kathleen Lonsdale to Dorothy, 15 May 1945, Bodleian Library, Hodgkin papers, H.138. In fact the work was published not in the Royal Society journal but as part of a book edited by Hans Clarke and others, *The Chemistry of Penicillin*, published by Princeton University Press in 1949.
14 For more on Perutz, see Ferry, *Max Perutz and the Secret of Life*.

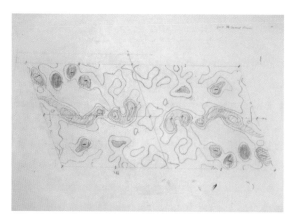

Project Habbakuk,[15] misspelled and misguided, aimed to build a huge fleet of aircraft carriers out of ice to enable planes to refuel as they crossed the Atlantic. Perutz carried out successful experiments on making ice stronger, but the project ran for months before its American partners calculated that construction of the vessels would be hopelessly costly and impractical, and cancelled it. For Perutz, however, its value was incalculable: through it he gained a British passport and the security he had lacked for so long.

More successful wartime scientific projects, such as penicillin, code-breaking and radar, led the government to increase budgets for peacetime research. Perutz's work on haemoglobin, championed by Bragg, seemed sufficiently promising for the Medical Research Council to fund a unit on the Molecular Structure of Biological Systems (later called simply Molecular Biology) in the Cavendish Laboratory, under Perutz's leadership. The crowded but exceptionally well-equipped unit's mix of physics, chemistry, biology and mathematics proved a magnet for curious minds, especially physicists who had become disillusioned with their subject after Hiroshima.

Francis Crick (FRS 1959) was one of these, joining Perutz's unit in 1949 and contributing a new mathematical rigour to his studies of proteins. The restless young American geneticist James Watson (For.Mem.RS 1981) arrived two years later. Informed by fibre diffraction photographs by Maurice Wilkins (FRS 1959) and Rosalind Franklin at King's College London, the two of them discovered the double helix structure of DNA in

1953.[16] The structure was the most important discovery of twentieth-century biology, providing a mechanism that could unite Charles Darwin's theory of evolution and Gregor Mendel's model of heredity. A 'spiral staircase' of two linked chains of complementary pairs of the four nucleotide bases adenine, thymine, guanine and cytosine, it immediately revealed how such a chemically simple molecule could account for life in all its abundant diversity. 'It has not escaped our notice', famously wrote the authors of their classic paper in *Nature*,[17] 'that the specific pairing we have postulated immediately suggests a possible copying mechanism for the genetic material.' Each chain could make a new double helix, enabling cells and organisms to replicate themselves. Crick and Watson also realised that the infinite number of 'words' that could be written in the four-letter alphabet A, T, G and C provided a genetic code to direct the construction of protein chains, though it took the efforts of many scientists until the mid-1960s to crack the code. The discovery ushered in the modern era of biotechnology, in which scientists not only read but edit genetic information to produce animals, plants, medicines or industrial processes tailored to human demands.

Perutz himself and his colleague John Kendrew (FRS 1960) continued with the much more difficult problem of protein structure. In 1953 Perutz discovered that introducing mercury atoms into the haemoglobin molecule could remove ambiguities in the results obtained from such large, irregular molecules. Combining this technique with pioneering computer analysis, Kendrew solved the structure of myoglobin, an oxygen-carrying protein a quarter of the size of haemoglobin, in 1957. Two years later Perutz and his team finally succeeded with haemoglobin. For the first time it was possible to see how the protein chain encoded by a DNA sequence folded itself into a characteristic, compact shape, as specific to its purpose as the nuts, bolts, valves, pistons, sparkplugs and gearwheels of a motor car. Perutz continued to work on haemoglobin for the rest of his life, describing the mechanism by which the 'breathing molecule' seizes and releases oxygen, exploring the evolutionary relationships between haemoglobins of different species, and

---

15 It was conceived by Geoffrey Pyke, a maverick inventor whom Bernal regarded as a genius and persuaded Louis Mountbatten, Chief of Combined Operations, to take seriously.

16 J.D. Watson, *The Double Helix* (London, Weidenfeld & Nicolson, 1968); Robert Olby, *The Path to the Double Helix* (London, Macmillan, 1974).

17 J.D. Watson and F.H.C Crick, 'A structure for deoxyribose nucleic acid', *Nature*, 171 (1953), 737–8.

linking abnormal haemoglobin to disease. Today's structural biologists use essentially the same technique, though with much better X-ray sources and computer analysis, to explore the whole toolbox of molecular machines that make up the living body. These include the enzyme DNA polymerase that builds new DNA chains on the template of a single DNA strand, and the bacterial flagellar motor, a protein complex that rotates the tiny flails that propel bacteria through their fluid worlds.

In 1962 Perutz and Kendrew shared the Nobel Prize for Chemistry, while Crick, Watson and Wilkins received the accolade for Physiology: an extraordinary sweep for one country, let alone a single laboratory. Sir Lawrence Bragg had been instrumental in forwarding all their claims: he heard of the awards while recovering in hospital from an operation for prostate cancer, leading his doctor to tell his wife that he was 'over the worst, but now I think he may die of excitement'.[18] He had left Cambridge in 1953 to take up his father's old job as Director of the Royal Institution. Having failed in his first plan of moving Perutz and Kendrew to the RI with him, Bragg started his own protein structure group there. It included David Phillips (FRS 1967), a young post-doctoral researcher from Cardiff, who led a team that solved the next protein structure, the enzyme lysozyme,[19] in 1965. With his student Louise Johnson (FRS 1990), he was the first to shed light on the molecular interactions that give enzymes their catalytic effect, without which the chemical reactions that power our lives would be impossible.

LEGACIES

Bragg dedicated his last years to restoring the RI, which had gone through a fallow period, to the glory days of Faraday or indeed his own father. Apart from sorting out its finances and establishing a first-class programme of research, he devoted most of his own energies to promoting science literacy. With enormous enjoyment and a knack for the felicitous analogy, he launched a year-round programme of lectures for schools, accompanied by the most spectacular demonstrations his inventive mind could conjure. Not a man for political activism, he took every opportunity through lecturing and broadcasting to present his vision of science as a benign, humanising activity that transcended class, gender and national boundaries. The RI continues this work today.

The triumphant successes of Cambridge molecular biology had been carried out largely in a 'temporary' shed outside the Cavendish Laboratory, known as The Hut. In 1962 they moved to the purpose-built Medical Research Council Laboratory of Molecular Biology, which has continued to expand ever since. Max Perutz chose to be chairman of the lab, not director as was usual in MRC units. He pursued a policy of attracting good people, giving them a share in the resources of the lab, and letting them get on with their research with a minimum of interference while he got on with his. The model also included more or less compulsory tea and coffee breaks in the communal canteen, where even the starriest prima donna would sit down next to the most junior graduate student and discuss science.

It paid off. The tally of Nobel Prize-winners steadily rose, with Fred Sanger (his second), Cesar Milstein, Georges Köhler, Aaron Klug, John Walker, Sydney Brenner, Robert Horvitz, John Sulston and Venkatramen Ramakrishnan joining the list. In 1993 Sulston (FRS 1986) moved to become founding director of the nearby Wellcome Trust Sanger Institute. Its major role in the international Human Genome Project, which published the complete human sequence in 2003, grew directly from Sulston's work at the LMB on sequencing the genome of the nematode worm, work supported by Jim Watson in his role as head of the US Office of Genome Research. Both the LMB and the Sanger Institute continue as international centres of molecular biology, while labs throughout the world are peopled with those who imbibed the LMB philosophy as young researchers. Sulston, supported by the Wellcome Trust, has continued to champion the free availability of biological information and oppose 'land grabs' in the genome for private gain.[20]

Perutz retired as chairman of the LMB in 1979, but never gave up

18 Quoted in Hunter (2004), pp. 232.
19 Found in tears and other secretions, lysozyme provides some protection against bacterial infection.
20 John Sulston and Georgina Ferry, *The Common Thread: A Story of Science, Politics, Ethics and the Human Genome* (London, Bantam, 2002).

research. In his latter years he became a frequent contributor to the *New York Review of Books*, writing witty and lucid essay-reviews on science and scientists. Though he abhorred political extremes of both right and left, he shared Bernal's view of science as a force for good and set out to counter the anti-science movement with his 1989 collection of essays *Is Science Necessary?*[21] His main concern was to promote health and well-being in developing countries, and to that end he advocated birth control, intensive agriculture and nuclear power (later with reservations). Like his more politically motivated colleagues, he argued passionately for an end to nuclear weapons and indeed all forms of warfare:

> A nuclear war would destroy everything that has been built up over centuries without giving us any control over what, if anything, will rise from the ashes. We must work for the application of science to peace and a more just distribution of its benefits to mankind.

As for John Kendrew, after his solution of myoglobin he turned to government advice and scientific organisation. He had a close exposure to nuclear matters during two years' tenure as deputy to the chief scientific adviser of the Ministry of Defence at the time of the Polaris Sales Agreement between Britain and the US. He subsequently became a member of the Council for Scientific Policy, created under the Labour government in 1964, and was knighted. A committed internationalist, he chaired the International Council of Scientific Unions and in 1978 became the founding director of the European Molecular Biology Laboratory, now a flourishing centre for research and training in the subject with a membership of twenty European countries.

On Sir Lawrence Bragg's retirement from the RI, David Phillips moved his group to Oxford. His successors there, Louise Johnson and Dave Stuart (FRS 1996), have in turn headed the life sciences division at the Diamond Light Source, the synchrotron near Didcot in Oxfordshire that since 2007

---

21 Max Perutz, *Is Science Necessary?* (London, Barrie & Jenkins, 1989).

The insulin
molecule from The
Bakerian Lecture,
1972: *Insulin,
Chemistry and
Biochemistry* by
Dorothy Hodgkin.

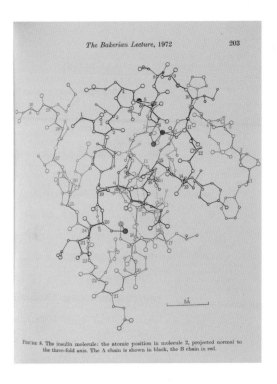

FIGURE 8. The insulin molecule: the atomic position in molecule 2, projected normal to
the three-fold axis. The A chain is shown in black, the B chain in red.

has provided a national source of high-energy X-rays to probe ever more
complex biological molecules and their interactions. Phillips himself spent
his last two decades in scientific administration. As Chairman of the
Advisory Board for the Research Councils from 1983–93, he shouldered
the difficult and thankless task of sharing out an essentially static science
budget among a growing and increasingly high-tech scientific community,
while constantly fighting for better settlements from the government. After
being raised to the peerage, he chaired the House of Lords Select
Committee on Science and Technology.

Dorothy Hodgkin remained a practising scientist well into her eighties,
by which time she was chronically disabled with arthritis. In 1969 her dedicated team of assistants and students finally completed the task she set herself
in 1934, of revealing the structure of insulin. It was a result that depended on
huge advances in technology, including the development of high-speed
computers and innovative ways of programming them. No single individual

could have done all this. Protein crystallography offers a prime example of a style of science that is the antithesis of the 'lone genius' model. It is often said to be a science in which women excel, though most are wary of any suggestion that it is 'women's work'. It is the case, however, that Hodgkin took on many female graduate students who went on to make careers in the field.

Hodgkin was another political idealist and admirer of Communist systems. Like Bernal she had found that her political sympathies made her *persona non grata* in the US during the McCarthy era (and like Bernal she was awarded the Lenin Peace Prize); but unlike him she conducted her politics on a personal level and avoided strident sloganising. She maintained contacts with colleagues in China throughout the Cultural Revolution, and worked indefatigably behind the scenes to bring about their readmission into international scientific organisations. She was a vocal opponent of war and nuclear weapons, a stance that led to her appointment in 1975 as President of the Pugwash Conferences on Science and World Affairs.[22] She was not afraid to use her status as a Nobel Prize-winner in the service of causes she believed in. She personally lobbied education minister Sir Keith Joseph over cuts in the higher education budget, and Prime Minister Margaret Thatcher (who was her former student) on East–West relations; and she insisted on speaking out about the Israeli–Palestinian conflict at a conference of Nobel Prize-winners organised by the French President François Mitterrand in 1988. She is commemorated in the Dorothy Hodgkin Fellowships, launched by the Royal Society to help young scien-

Watercolour of Dorothy Hodgkin by Graham Sutherland, 1980.

tists, especially women, to get on to the academic career ladder.

Bernal never gave up science, taking on the presidency of the International Union of Crystallography in 1963. In the post-war years science policy in both Europe and the US moved a considerable distance in the direction he had mapped out in his 1939 book. The establishment of 'big science' projects such as the Apollo programme, CERN,[23] and the Human Genome Project all required central government planning and support. The Labour Prime Minister Harold Wilson's 'white heat [of the scientific revolution]' speech in 1963, and the UK's science policy White Papers *A Framework for Government Research and Development* (1971), *Realising Our Potential* (1993) and *Excellence and Opportunity* (2000), all stressed wealth creation and the quality of life, though the debate has swung back and forth over whether scientists themselves or their government paymasters should set the agenda. Once again, though, Bernal's take on the interdependence of science and socialism has proved laughably wide of the mark. When he wrote, in 1964, that 'the scientific and computer age is necessarily a socialist one',[24] he could not have envisaged the commercial free-for-all made possible by the World Wide Web.

ENVOI

Like all branches of science, X-ray analysis calls for a combination of imagination and rigorous data collection. Unlike some, it gives hard-earned results that no paradigm shift or new experimental approach can undermine. As Perutz wrote, 'Bragg's structures were not preliminary approximations subject to revision: any student setting out to redetermine the structures of calcite, quartz or beryl will be disappointed.'[25] The knowledge that an exact solution existed gave crystallographers an optimism that kept them going in their darkest hours (insulin took thirty-five years to solve, haemoglobin twenty-two). It is perhaps no coincidence that the scientists in this account were prepared to tackle society's problems in the same hopeful spirit.

22 Pugwash is an international organisation of scientists and others dedicated to research into the dangers of nuclear weapons. It was inspired by the Russell–Einstein Manifesto of 1955.

23 The European Centre for Nuclear Research in Geneva.

24 J.D. Bernal, 'After Twenty-Five Years' in M. Goldsmith and A. McKay (eds), *The Science of Science* (London, Souvenir Press, 1964).

25 Max Perutz, 'How W.L. Bragg Invented X-ray Analysis' in Max Perutz, *I Wish I'd Made You Angry Earlier* (expanded edition) (New York, Cold Spring Harbor Laboratory Press, 2003).

12

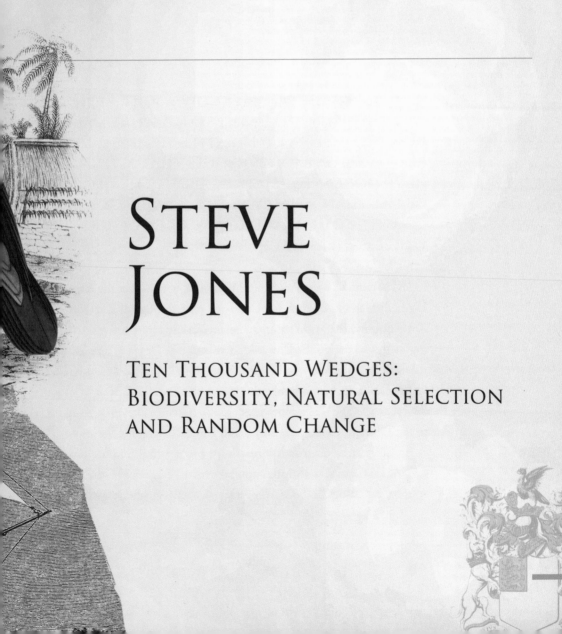

STEVE JONES

Ten Thousand Wedges:
Biodiversity, Natural Selection
and Random Change

Steve Jones is Professor of Genetics at University College London. His popular books include *The Language of the Genes*, *In the Blood* (based on the BBC TV series), *Almost like a Whale: The Origin of Species Updated*, *Y: The Descent of Men*, *Coral* and, most recently, *Darwin's Island*. He writes the 'View from the Lab' column in the *Daily Telegraph*.

H OW MUCH DO WE KNOW ABOUT BIOLOGICAL DIVERSITY? TO UNDERSTAND WHAT MAINTAINS IT MIGHT HELP IN THE BATTLE TO PRESERVE WHAT REMAINS. STEVE JONES ARGUES THAT ALTHOUGH WE HAVE MORE INFORMATION ABOUT THE GEOGRAPHY OF LIFE THAN IN DARWIN'S TIME, WE LACK A THEORY OF WHY SOME PLACES HAVE LOTS OF CREATURES, WHILE OTHERS HAVE FEW.

In 1859, London – with its two million inhabitants – was the largest city on Earth. It was in addition (and in large part through the activities of the Royal Society) the world centre of geological and biological research, its lasting memorial the publication in that year of *The Origin of Species*, the book that gave birth to modern biology. The capital's people were well aware of its fame, and flocked to public displays in the Zoo, the British Museum and Kew Gardens and – as a more select group (Charles Darwin among them) – to the Linnean, Geological, Royal and Royal Geographic Societies.

In 2009 Britain's first city has slipped to a global number seventeen in size, but its status as an international centre of gravity of the intertwined sciences of ecology and evolution has not changed. London still represents, by a considerable margin, the world's largest conglomeration of researchers in this field and remains, as it was in Darwin's time, a global hub for the study of biological diversity. The Natural History Museum (in which the great man's statue, once hidden in the tea room, has been promoted to pride of place) has over twenty million specimens of plant and animal, and Kew plans to store tens of thousands of species of plant as seeds. How many kinds of creature there might be altogether is a matter of guesswork; almost two million have been described, but the total may be – some say – twenty times as great (although that figure depends on just how a 'species' is defined).

Charles Darwin founded the modern sciences of evolution and ecology (although neither word appears in *The Origin*). His book was wrong about plenty of things but impressively right about others. He had an uncanny ability to foresee the difficulties that his new science was likely to face. To him, the nature and origin of species was 'the mystery of mysteries' – as it still is. In a prescient hint of disagreements to come, his writings introduce the tension between the power of directed change (natural selection included) and the importance of accident. That argument pervades the history of evolutionary biology and remains unresolved.

Palaeontology, development, genetics, ecology, demography, species diversity and other parts of evolutionary theory share a history of dissent about the role of chance as opposed to directed forces. Since 1859 there have been many reversals of attitude within each of those fields with – no doubt – more to come. Now, the study of biodiversity is revisiting the controversy, with mixed results.

FROM DELIGHT TO DOUBT

In Darwin's early years, Nature seemed bounteous, complicated, and more or less permanent. For the young naturalist on the *Beagle* the main task was to describe, rather than to explain, the world's variety. His joy in life's abundance is clear: as he wrote on his first steps ashore in South America:

HMS *Beagle* in the Straits of Magellan with Mount Sarmiento in the distance.

The noise from the insects is so loud, that it may be heard even in a vessel anchored several hundred yards from the shore; yet within the recesses of the forest a universal silence appears to reign. To a person fond of natural history, such a day as this, brings with it a deeper pleasure than he ever can hope again to experience … The day has passed delightfully. Delight itself, however, is a weak term to express the feelings of a naturalist who, for the first time, has wandered by himself in a Brazilian forest.

To the delighted Darwin, the tropics – unspoiled by man, filled with sunlight and blessed with sufficient products of the bounteous to allow chains of hungry creatures that prey upon each other – were the centre of the world's diversity. Twenty years later *The Origin*, as it transformed a static view of life into a dynamic one, began to ask why. The book begins with chapters on variation and on the struggle for existence and makes the case that an all-pervasive interaction of the two generates new kinds of creature as the result of an ordered process called natural selection: inherited differences in the chances of reproduction. *The Origin* ends in a hymn to its power with the famous tangled bank: 'clothed with many plants of many kinds, with birds singing on the bushes, with various insects flitting about, and with worms crawling through the damp earth'. It was a vision of what today we refer to as biodiversity.

The original of the paradisiacal bank is only a few hundred yards from Down House, Darwin's home in Kent for the last forty years of his life. To the patriarch of Downe its inhabitants – like those of the Brazilian forest – made up a crowded system of competitors, each squeezed into its own way of life, any vacancy at once filled by a hungry challenger. So finely tuned were their interactions that natural selection was inevitable. Ancestors were replaced by better-adapted descendants and the world was full with no room for passengers. The directive forces of competition, extinction and replacement were essential parts of his theory. 'The face of Nature', he wrote, 'may be compared to a yielding surface, with ten thousand sharp wedges packed close together and driven inwards by incessant blows, sometimes one wedge being struck, and then another with greater force.' Inevitably, the less successful wedges were squeezed out. That vivid image gave rise in time to the well-known 'Red Queen' model of ecology in which competition is the engine of evolutionary change and in which different creatures must run just to stay in the same place.

That view of Nature is still, as in Victorian times, accompanied by the perception that evolution emerges from a series of rules: 'Throw up a handful of feathers, and all must fall to the ground according to definite laws; but how simple is this problem compared to the action and reaction of the innumerable plants and animals which have determined, in the course of centuries, the proportional numbers and kinds ...' Since then, a great variety of laws – definite and less so – has been proposed by ecologists. Many are both linear and prescriptive. Some may have some validity; but most ecologists accept that accident also moulds the diversity of life. As *The Origin* points out, on oceanic islands the number of kinds of inhabitants is scanty and particular groups such as frogs and toads are absent – a result of the hazards of colonisation, denied to creatures that cannot cross the sea and open by chance to only a sample of others. Unpredictable events such as ice ages also helped to shape the distribution of plants and animals. Its author was happy to incorporate such random agents into the evolutionary argument.

The tension between necessity and chance still pervades that science, and its handmaidens, genetics and ecology. Shared disagreements in each of those three fields have appeared and have been (at least temporarily) resolved again and again. The early twentieth century saw a disjunction between genetics and evolution, for it seemed that sudden leaps – the origin of new species – could be explained by the chance appearance of major mutations. Then, population genetics claimed to show that natural selection on variants of minor effect could explain the origin of novel forms of life. We now know, though, that under certain circumstances, large mutations can indeed give rise to new species, as when a change in flower colour leads to a shift in pollinator preference. The perceived importance of selection versus genetic drift – the accidental change of gene frequencies through sampling errors – in maintaining variation has also oscillated. From snail shell patterns to blood groups and to protein variation it was once assumed that most inherited diversity had no influence on fitness; a claim often followed by a belated realisation that in fact the opposite is true. The discovery of extraordinary levels of individual variation in human DNA has caused the pendulum to swing again and most molecular geneticists assume that most such diversity – and perhaps much of the genome – is adaptively irrelevant. Now we know that much of the DNA is transcribed and that changes even in the 'junk' may affect the workings of the creatures that bear it. To balance that, there has been little success in finding the genes involved in important attributes such as human height and weight and the fact that many functional genes have several unrelated phenotypic effects further confuses the search for the action of natural selection. Correlated responses arising from such multiple action or from the involvement of closely linked loci also mean that selection on one trait affects others, as do interactions between apparently unrelated genes. The genome is now seen as a system, filled with non-linear interactions, and speciation as a side-effect of an incompatibility between intricate organisations. Geneticists sometimes need to remind themselves of the stark simplicity of Mendel for reassurance that their subject has any laws at all. Their confusion has a message for those struggling with the perhaps even more complex issue of how species find their place in Nature.

For a brief and golden period at the end of the last millennium, genetics, ecology and evolution seemed to approach a consensus in which the promise of *The Origin* would be fulfilled. Since then, biologists have been forced once again to face the unpalatable truth that life is less simple than seems reasonable to hope. It is increasingly unclear whether patterns of biodiversity and the extent to which the numbers and the relative abundance of species in a community reflect 'definite laws' *à la* Darwin, rather than a chance assemblage of species, of the kind that might be blown on to an island. We do not know why some communities are diverse and some not, some efficient and others less so, some filled with disease and others plagued by predators, and some resilient but others exceeding fragile. Even the consistency that impressed the young naturalist – the vast variety of the unspoiled and generous tropics – appears less impressive than once it did. Ecology, which once saw ordered communities moving through predictable stages to a more or less stable climax, their structure determined by energy flow or predator pressure, now accepts that many may be little more than a random bunch of functionally equivalent creatures and that changes in space or time may often result from accident.

BIODIVERSITY PRESENT AND FUTURE

The term 'biodiversity' was invented in the 1960s and came into widespread use in 1988 as the title for a US National Academy of Sciences forum (Wilson & Peter 1988).* It has attracted plenty of interest for the word is usually accompanied by the qualifier 'threatened'. That statement is familiar or even banal and few doubt that the worst is yet to come. Even so, the grand extinction that marks the new millennium may present an opportunity to understand diversity. Ecology is often derided as a science without a theory but perhaps the upheavals of the past century may reveal more than did the apparently stable patterns of life seen by the early explorers.

Almost nobody denies the crisis that is upon us. Charles Darwin himself, on the last leg of his voyage, had a vision of what the next century would bring. He landed on St Helena in the South Atlantic. It rose 'like a huge black castle from the ocean', with its scenery having 'English, or rather Welsh, character'. The vegetation, too, was decidedly British, with gorse, blackberries, willows and other imports, supplemented by a variety of species from Australia. Many of its inhabitants were invaders. They had driven the natives to extinction. Darwin found the dead shells of nine species of 'land-shells of a very peculiar form' and noted that specimens of one kind 'differ as a marked variety' from others of the same species picked up a few miles away. All apart from one had been replaced by the common English *Helix aspersa*. As he noted, invasion was rife elsewhere, too. European plants were already 'clothing square leagues of surface almost to the exclusion of all other plants' on the La Plata plains of South America, and American natives were spreading through India 'from Cape Comorin to the Himalaya'.

Life in many of the other places visited by HMS *Beagle* is now worse than it was. St Helena had, soon after Darwin's time, forty-nine unique flowering plants and thirteen ferns. Seven have been driven to destruction, two survive only in cultivation and many more are on the edge. The island's giant earwig (the world's largest), its giant ground beetle and the St Helena dragonfly, all common at the time of the ship's visit, have not been seen for years. The St Helena petrel is extinct, and the sole remaining endemic feathered creature, the wire bird, is under threat. Nobody needs to be reminded of the equivalent fate of the Australian fauna, or of the dire state of the Galápagos. The Atlantic forest of Brazil – the site of Darwin's

SPECIMENS COLLECTED BY DARWIN
ON THE VOYAGE OF THE BEAGLE
1831 - 1836

A case displaying various beetle specimens collected by Charles Darwin during the *Beagle* voyage, as well as a map of the ship's route.

apotheosis – retains around twenty thousand kinds of plant, one in twelve of the world's known species, over a thousand vertebrates (including such spectacular creatures as the woolly spider monkey and golden tamarind) and huge numbers of insects, many found nowhere else. The habitat has been reduced to one twentieth of its extent at the time of Columbus. Much of the planet's ecosystem is under equal threat.

In 2002 the World Summit for Sustainable Development set out to 'achieve by 2010 a significant reduction of the current rate of biodiversity loss at the global, regional and national level'. The figures are stark. The International Union for the Conservation of Nature publishes an annual

Red List of species deemed to be in danger. In 2008, 16,928 creatures made it on to that roll of dishonour, six hundred more than in the previous year. The List is biased towards the spectacular and for them the situation is grim. Between 1998 and 2004 (the latest year for which we have figures) the world's birds declined on land, on fresh water and at sea, in the tropics, the temperate zones and the poles. In that brief period, two birds (the Hawaiian crow and Spix's macaw) became extinct. The situation in Europe is also dire (de Heer 2005) with a drop of almost a quarter in the numbers of birds, butterflies and mammals on farmland. One in four of the world's mammals is threatened, with half of the globe's 5,487 species in decline. The numbers of Tasmanian devils have dropped by half over the past decade while whales and dolphins are in almost as bad a shape (Schipper *et al.* 2008). Amphibians are under even greater threat for half of all species face imminent demise. Disease, too, plays a part, for mountain gorillas are under threat from the spread of the virus that causes Ebola fever in humans. The situation is particularly desperate in South East Asia, where almost all species of primate face extinction.

The real danger to diversity in both land and sea is the loss of habitat. The 2005 Millennium Ecosystem Assessment found that almost half of all tropical dry forests and a third of those of the Mediterranean have been replaced by farms and towns, which now cover more than a quarter of the Earth's surface. Asia has lost almost half its forests, even more of its mangroves and its reefs are in a dire state. To balance that, some species are highly invasive. Europe is the source of global threats such as the Austrian pine, the Spanish slug, the German wasp, the Scottish broom and English starling, but its own borders have been crossed by more than ten thousand invaders, among the most troublesome the Canada goose, Argentine ant, Indian strawberry, Chinese mitten crab and New Zealand flatworm.

What drives life's variety and what drives it out? Some extinctions – such as the almost universal loss of island endemics when faced with mainland invaders – do seem to show a certain consistency from place to place.

However, some rules that might seem obvious are not so. Big apes and birds are at more risk than are small ones, but body size has no effect on the fate of carnivores, reptiles, or marine molluscs (Jablonski 2004). And why do some places have many species and others few; and how can some creatures fill the world while others quail before them? Why do some evolve to cope while others give up the ghost? Ecologists have spent years in studying how communities vary in structure and how food, predation, energy flow and sex might change their fate. They have, alas, come to almost no agreement. The problem is that so often revisited by genetics: the difficulty of establishing a reliable scientific framework for an immensely complex system. In ecology, are there any general rules?

THE HIDDEN WORLD OF BIODIVERSITY

The 'species concept' has given rise to some of the most sterile debates in biology: but some clear definition is essential to establishing patterns of biodiversity. Is a bird species – such as one of the two thousand or so distinct kinds of island rail claimed once to have lived on the scattered patches of land across the Pacific – as biologically distinct as the mosquitoes of West Africa once classified under the label of '*Anopheles gambiae*' but now known to encompass several distinct insects? Without an objective statement of what the units are, it is hard to establish real levels of natural richness. Around three hundred thousand plants have been described, and four times that number of animals, but some experts claim that there may be as many as twenty million different kinds among the insects alone, to give weight to the familiar claim that, to a first approximation, all animals are insects. Even among mammals numbers have increased by almost a fifth in the past fifteen years, in part because some taxa have been promoted to species status from lower classificatory levels (Schipper 2008).

New technology also hints that some counts may be far less accurate than they appear. DNA probes make it possible to explore realms of life almost unknown a decade ago. Craig Venter, prominent in the project to map the human genome, has set out to classify the microbes of the sea (Gross 2007). Water from the Atlantic, Pacific, Baltic, Mediterranean and Black Seas was passed through filters to capture organisms of a variety of sizes. Already twenty million new genes and thousands of new protein families, some quite novel, have been found. This may indicate the presence of vast numbers of new species in a habitat which comprises 99 per cent of the whole biosphere. The Sargasso Sea alone has at least 1,800 new varieties of bacteria.

The soil, too, is a hotbed of life. Two shovelfuls of earth taken a metre apart may possess entirely distinct communities. The number of species per gram of soil has been estimated, on the basis of molecular taxonomy, to be between 2,000 and 800,000 depending on what criteria are used (Dance 2008). Two sites in Alaska, one in tundra and one in taiga forest, shared only eighteen kinds of invertebrates (microbes excluded) out of a total of some 1,300. Until our ability to identify the evolutionary units is more dependable the many claims of regular geographic patterns of diversity may be overstated.

BIODIVERSITY AND WHERE TO FIND IT

However it is measured (from species richness, to listings based on information theory, or on weighting the index towards rarer or endemic creatures, or by including data from different ecosystems within a region) there appears at first sight to be a clear tendency for tropical landscapes to be more diverse than those to the north or the south. For terrestrial creatures, part of the global pattern comes from geography: there is relatively more land – and hence more habitat – near the equator than the poles (although the effect does remain when that is corrected for). Sampling effort is also in part to blame: in Darwin's day Britain would have scored top of any biodiversity index – but that was simply because so much was known of the natural history of that undistinguished group of islands. Even on a smaller scale, incomplete sampling confuses real patterns. In mountain ranges, frequently taken as a microcosm of the contrast between the warm tropics and colder poles, species diversity is often claimed to decrease with altitude. However, a survey of more than 400,000 records of 3,000 flower species in the Pyrenees shows that simply by varying the distance between samples almost any pattern of diversity change with height, positive, negative, or hump-shaped, can be generated (Nogues-Bravo 2008).

Recent historical accidents may also have a large influence upon ecological trends. In the Pyrenees altitudinal changes in floral richness are much confused by the fact that farmers have modified lowland habitats more than they have those far above them. One surprise has been to find that what seem to be pristine habitats have long been modified by man: itself a complication when trying to establish natural patterns of variability. Even Darwin's Atlantic forest of Brazil, together with the vast biological storehouse of the Amazon jungle, are partly human constructs, for their structure has been much disturbed by the large indigenous population that lived there in pre-Columbian times and turned parts of it into parkland (Heckenberger *et al.* 2008).

A century and a half of research has improved our knowledge of ecosystems, but many regions of the globe and – more important – many habitats remain relatively unexplored. Sometimes, detailed sampling reveals astonishing patterns of diversity: a single bay on the island of Flores, in the East Indies, has more species of fish than does the entire tropical Atlantic (Briggs 2005). As a result, the geography of diversity has begun to look more complicated than it did. The Conservation International organisation names 34 patches of land as 'hotspots' that contain almost half the world's known plant species and a third of its vertebrates. Together, they represent less than 2 per cent of the terrestrial world. The hottest spots of all are indeed in the tropics – Sundaland, Madagascar, Brazil's Atlantic forest and the Caribbean. Together, in one two-hundredth of the total land surface, they boast a fifth of known plants and a sixth of vertebrates (Sodhi 2008). For mammals, in contrast, the high points of variation include the Andes and the Hengduan Mountains of south-western China (Schipper 2008).

Mediterranean ecosystems such as those of South Africa, of Western Australia, or of the Mediterranean itself, also contain large numbers of creatures although they are well away from the equator. Hotspots are important in conservation, but (Grenyer 2006) there is often little congruence in the distribution of threatened species, particularly when the rarest creatures with the smallest ranges are considered – indeed, if anything they tend to be found in different places. A study of large-scale spatial change in amphibia, birds and mammals across the Western hemisphere suggests that there is some congruence of pattern for birds and amphibians (but much less for all three groups considered together) when areas of high local differentiation are considered. However, the opposite is not true – the three have large regions of relative homogeneity in quite different places (McKnight 2007). In the deep sea, too, there are few consistent associations of biological diversity with depth, latitude, sediment type, or water quality.

WHAT DRIVES BIODIVERSITY?

Many rules of diversity have been proposed. Food, predators, climate, efficiency of energy transfer and complexity of the habitat have all been appealed to as agents underlying community structure. Some cases are convincing, but a closer look reveals a disappointing lack of consistency from one ecosystem to another. Some are under top-down control through the action of predators, while others respond to forces that well upwards from the primary producers and yet more depend on an interaction between the two. It might seem obvious that a complicated place like a rainforest is more productive and more diverse than a peat bog, and a survey of dozens of habitats suggests that the most connected communities may be more efficient. Even that evidence is not always persuasive, for experiments in which particular species have been removed one by one from grassland or pond to see how well the remainder survive give results that are ambiguous at best. A search for order behind local or global patterns of ecological change has not always been a success.

The most productive parts of the world are, it is often said, the most blessed with unique forms of life, perhaps because they have more energy input from sunshine. Whales and dolphins also tend to be most diverse and most abundant in middle latitudes such as the southern Indian Ocean; and although these are not close to the equator they are regions of high productivity (Schipper *et al.* 2008). Metabolic rate may be the main driver, with small or relatively warm-blooded creatures living more speedy lives and generating more species than do their opposites (although the shared geographic patterns of change in warm- and cold-blooded creatures does not fit this notion).

An alternative view emphasises the importance of predators in maintaining community structure and a whole science of food webs attempts to analyse the patterns of eating and being eaten among species in a search for regularity. The re-introduction of wolves to Yellowstone National Park led to more corpses being scattered across the landscape and to increased opportunities for a variety of scavengers, while browsers increase the plant diversity of the pastures upon which they feed. Conservation biologists often believe that large predators help maintain the structure of a community and much effort is devoted to ensuring the survival of such creatures (Sergio *et al.* 2008). Once again a wider look at the dozens of claims made for the importance of predators reveals a depressing lack of consistency. Although the trophic effects of a large predator may be important in some places they do not seem to be important general agents as many creatures seem to live lives rather detached from those of most other species around them. A meta-analysis of twenty food webs (Vermaat *et al.* 2009) hints that they might fall into two classes, highly interconnected or more linear, with fewer links; but the tie between predation, energy flow and community structure is not clear, and may involve further attributes of each species such as how easy they are to eat.

RANDOMNESS AND THE DIVERSITY OF LIFE

In the past few years, a new notion has emerged: that community structure can best be explained with a radical and at first sight absurd assumption that, in effect, all the species involved are equivalent and that their abundance turns on random fluctuations in survival and in reproduction (reviewed in Leigh 2007). This 'neutral model' of ecology has parallels with its equivalent in genetics, in which levels of inherited variation emerge from a balance between random mutation and the accidents of genetic drift. That model has been tested against the real world, and although it sometimes fails, at the level of DNA sequence it retains considerable explanatory power (Clark 2009). In ecology, too, a random model of communities may carry more general conviction than does a series of special cases that explain some patterns in some places but have little predictive power overall.

Darwin accepted random change when he noted that islands contain fewer species than do nearby tracts of mainland and the claim that island life is driven by the accidents of migration and extinction has held up well. The same is true in other populations, on a variety of timescales. Thus, when cataclysms strike, as in the five great extinctions of the past five hundred million years (most associated with comets or great geological upheavals), huge numbers of species of many kinds disappear through mere ill luck, and rules that might help predict their ability to withstand everyday pressures do not much apply (Jablonski 2004). Other geological events quite unrelated to the biological universe such as continental drift also have persistent effects on the diversity of communities. In the same way, the last ice age stripped Northern Europe of most life and the glaciated regions are still depauperate as the result of an ancient historical accident rather than as a response to modern conditions.

The peak of coastal marine species variety is in Indonesia and on the northern coasts of Australia (Renema *et al.* 2008). There, coral reefs flourish. Such places are often appealed to as an epitome of undisturbed and productive nature, in which new kinds of creature can evolve to add to the treasury of life. A closer look at the fossils and the genes shows that in fact the occupants of the reefs have moved across the globe as conditions changed. During the Eocene, marine diversity found its peak in south-west Europe and North Africa, along the Arabian Peninsula and in what is now Pakistan. As these lands were raised from the sea when Arabia crashed into Asia, many of their inhabitants migrated to more congenial places, the present Indo-Australian region included. Most of the animals supposed to have originated there have in fact an ancient and dispersed history. Global disasters of fifty million years ago have done more to shape the geography of today's teeming reefs than have climate, food or sunlight. Evolution, that reminds us, works on a far longer timescale than does ecology.

As in genetics, there are many non-linear interactions in ecology (Andersen *et al.* 2009) and, as in the weather and the stock market, a small disturbance can lead to a sudden and unpredictable change in state. An attempt to shoot foxes to increase the numbers of red grouse prey backfired, for the predators normally caught only the birds most filled with parasites and once they were removed disease spread and killed many more birds than before. In a related case, an attack by one insect herbivore on a leaf often alters its attractiveness to other grazers, while plants that activate a pathway that fights fungal disease may reduce their own ability to combat insect attack with a different biochemical strategy. All these and many more multiple interactions (Strauss & Irwin 2004) emphasise that – as in genetics – many of the connections among species within a community are far from simple.

The importance of randomness first came to attention with the 'paradox of the plankton', the discovery that the apparently homogeneous environment of the sea was host to a vast diversity of drifting creatures all apparently in competition for the same resources, in contradiction to the supposedly fundamental principle of exclusion of species with similar demands (Scheffer *et al.* 2003). The plankton have become even more paradoxical with the discovery of vast numbers of new marine bacteria. The same is true of the world beneath the soil, whose organisms differ wildly from place to place, but generate roughly the same mix of nutrients. Perhaps each of those habitats really is filled with a chance assemblage of ecologically equivalent creatures, each arriving more or less by accident.

That radical notion may have a wider validity, for it seems to apply to some very different terrestrial and freshwater habitats. Fish species diversity across eight hundred tributaries in the entire Missouri–Mississippi river system can be explained by the random loss of species of varying dispersal in a pattern that diffuses from a centre of abundance into streams of smaller and smaller size (Muneepeerakul 2008) with no need for any consideration of the nutrient status of streams, of other species, or of climate. The same is true of patterns of diversity in mature forests.

Temporal shifts, too, hint at an underlying lack of order. In a somewhat heroic experiment (Beninca *et al.* 2008) a series of laboratory containers containing samples of plankton from the Black Sea was cultivated for seven years, in – as far as they could be attained – constant conditions. The abundance of the various species varied dramatically with time, and the relative numbers of each type could not be predicted with any confidence over any period longer than a month (which is, incidentally, the longest period for which the British weather forecast is even slightly dependable). The system was driven by something close to chaos – but, even so, most species persisted at high or low frequency within the containers.

Natural ecosystems can also remain stable until a threshold is reached and then collapse. The effect is familiar to fisheries managers, for a trophic

cascade may be set off by overfishing, with unpredictable results. In the
Black Sea itself, there was, from the 1970s on, a shift from large (and valuable) fish to anchovies that feed on plankton, and then to gelatinous creatures such as jellyfish and ctenophores, which now teem in huge numbers
and have, within a few decades, replaced what seemed a stable ecosystem. A
similar shift in the Pacific from sardines to anchovies and back twice in the
second half of the twentieth century may also have turned on small changes

in climate in a regime poised on the edge of stability that moves unpredictably from one to another. There have been dozens of climate shifts from cool to warm and back again every few hundred or thousand years, in the past hundred thousand years, each of which was no doubt accompanied by sudden upheavals in what might have seemed like stable ecosystems. Even on a much shorter timescale, the numbers of birds and mammals in a particular place when studied for long enough swing wildly for no obvious reason (as in the collapse of the British house-sparrow). Unexpected outbreaks can also destroy whole ecosystems (as in Dutch Elm disease, which appeared from almost nowhere and killed millions of trees). Such fluctuation might maintain a complex community with no external driver, in which case the paradox of the plankton (and, by extension, of land-based ecosystems too) could be explained in terms of random change.

A recent review claims that 'ecological surprises' of this kind have proved to be almost universal (Doak *et al.* 2008). Not only do they reveal our ignorance of the laws behind biodiversity, but they hint that chaos and complexity may be the rule rather than the exception. Darwin himself was well aware of the difficulties of disentangling the patterns of nature. The term 'complexity' appears in *The Origin* almost fifty times, and 'innumerable' and 'endless' almost as often (although 'inextricable web of infinities' makes it just once). The tension between order and disorder remains unresolved and more than a century and a half since that remarkable work we may understand rather less (although we know considerably more) about the patterns of nature than we imagined just a decade ago.

13

PHILIP BALL

MAKING STUFF:
FROM BACON TO BAKELITE

Philip Ball is a science writer. He worked at *Nature* for over 20 years, first as an editor for physical sciences (for which his brief extended from biochemistry to quantum physics and materials science) and then as a consultant editor. He is author of numerous non-fiction works including *Universe of Stone: Chartres Cathedral and the Triumph of the Medieval Mind*, *The Devil's Doctor*, *Elegant Solutions: Ten Beautiful Experiments in Chemistry* and *Critical Mass: How One Thing Leads to Another*. His latest books form a trilogy – *Nature's Patterns: A Tapestry in Three Parts*, individually titled *Shapes*, *Flow* and *Branches*.

FRANCIS BACON'S VISION OF A SCIENCE DRIVEN BY THE URGE FOR 'THE EFFECTING OF ALL THINGS POSSIBLE' HAS BEEN ASTONISHINGLY PRODUCTIVE FOR NEARLY FOUR HUNDRED YEARS. ARE WE GRATEFUL? ASKS PHILIP BALL. ANSWER: YES AND NO. THE REASONS FOR THIS TAKE A BIT OF TEASING OUT.

C.P. Snow's 1959 Rede Lecture is remembered as a critique of the cultural divide then perceived between the scientific and the literary worlds. But there were more than two cultures identified in his discussion. 'I think it is fair to say', he wrote, 'that most pure scientists have themselves been devastatingly ignorant of productive industry, and many still are.'

> It is permissible to lump pure and applied scientists into the same scientific culture, but the gaps are wide. Pure scientists ... wouldn't recognise that many of the problems [of engineering] were as intellectually exacting as pure problems, and that many of the solutions were as satisfying and beautiful. Their instinct ... was to take it for granted that applied science was an occupation for second-rate minds.

Snow wasn't alone in this perception. Writing at much the same time, the English biologist Peter Medawar spoke of Francis Bacon's division of experimental science in the seventeenth century into 'Experiments of Use' and 'Experiments of Light and Discovery'. Bacon's distinction, said Medawar, 'is between research that increases our power over nature and research that increases our understanding of nature, and he is telling us that the power comes from the understanding' – Bacon's famous maxim that 'knowledge is power'. But, Medawar went on:

> Unhappily, Bacon's distinction is not the one we now make when we differentiate between the basic and applied sciences. The notion of *purity* has somehow been superimposed upon it, and in a new usage that connotes a conscious and inexplicably self-righteous disengagement from the pressures of necessity and use. The distinction is not now between the empirically founded sciences and those whose axioms were supposedly known a priori; rather it is between polite and rude learning, between the laudably useless and the vulgarly applied, the free and the intellectually compromised, the poetic and the mundane.

'All this', he added, 'is terribly, terribly English.'

I believe that this situation can't be ignored when looking at the development of the applied sciences over the past several centuries. When several rather austere-sounding books from the post-war years, with titles such as *Metals* [or *Plastics*] *in the Service of Man*, served up to lay audiences a triumphalist celebration of materials technologies, they rather took it for granted that the general public felt indebted to these wondrous advances. But as Snow and Medawar intimated, not even scientists themselves had yet found an accommodation between scientific discovery and its applications. This is scarcely surprising, however, since such ambivalence towards what the Greeks called *techne* – the art of making things – can be discerned throughout history, and pervades not just science and technology but culture in the broad sense.

Many scientists, for instance, will agree with biologist Lewis Wolpert that 'technology is not science'. Science, says Wolpert, 'originated only once in history, in Greece' – although he acknowledges that 'those who equate science with technology would argue differently'. Indeed they do.

The notion that science is distinct from technology would have sat comfortably with the ancient Greek philosophers, most of whom displayed a reluctance to get their hands dirty. Both Plato and Aristotle elected for a top-down approach to understanding the world, launched from the kind of a priori axioms that Medawar mentions. Aristotle, it is true, advocated close observation of nature, and in the Middle Ages Aristotelian natural philosophers such as Roger Bacon and his mentor Robert Grosseteste instigated a methodology in which experiment played a central role. But one must be careful when speaking of 'experimental science' before the Enlightenment, for it often meant demonstrating what one already knew to be the case – and if experiment seemed to contradict axiomatic reason, so much the worse for experiment. In any event, Aristotelianism became rigid dogma in the medieval universities, and Bacon's advocacy of a new, 'experimental philosophy' was a reaction to it: a call for a reformation in how science was conducted.

Meanwhile, what we might now call applied sciences and technologies were commonly conducted by artisans who had no formal university training: metallurgists and alchemists, miners, dye-makers, brewers and bakers, textile makers, barber-surgeons. Their trades were systematically excluded from the academies, where they were often derided as ignorant labourers and recipe-followers (sometimes, it must be said, with good reason).

So it is interesting that, for Wolpert, one of the people confused about the relation between science and technology was Francis Bacon himself. That claim warrants a little examination – for isn't Bacon often credited with the germinal vision of a body of scientific savants like the Royal Society? What was it, exactly, that Bacon would have such an organisation do – science, or something else?

BROTHERHOODS OF SCIENCE

The blueprint for this new philosophy was laid out in Bacon's *Instauratio Magna* (*The Great Instauration*) of 1620. This was a mere fragment, the introductory episode of an unrealised dream to summarise all of human knowledge and to explain how it should be extended and applied. The Latin noun *instauratio* means a renewal or restoration. It has a Biblical connotation, referring to a rebuilding of the House of the Lord like that accomplished in the renovation of Solomon's Temple.

As an addendum to the same volume, Bacon published *Novum Organum* (*The New Organon*), which explains the shortcomings of earlier natural philosophy. Bacon decries both the sterility of academic Aristotelianism, which he compares with spiders weaving tenuous philosophical webs, and the blind fumblings of uninformed practical technologies, which are like the mindless tasks of ants. True scientists, he said, should be like bees, which extract the goodness from nature and use it to make useful things.

Right: Novum Organum frontispiece.

Far right: Francis Bacon. Studio of Paul van Somer.

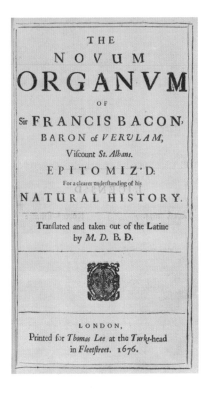

Seven years later, Bacon offered a vision of how this new experimental philosophy might unfold. In *The New Atlantis* he presented a utopian fable in which a group of travellers in the Pacific Ocean encounters a land called Bensalem, run by a sect of scholar-priests in an institution called Salomon's House. Here were Bacon's scientist-bees, engaged in 'the production of wonderful operations'. This is evidently not a scientific body that is content to sit and ponder. It creates marvellous devices and structures: artificial lakes, furnaces, engines, caves where alchemy mimics the natural production of metals. Nature is not merely observed, classified and understood in the manner of some Aristotelian taxonomist – it is dominated, modified, 'improved'. According to the scholars of Salomon's House:

> We make, by art ... trees and flowers to come earlier or later than their seasons; and to come up and bear more speedily than by their natural course they do. We make them also by art greater much than their nature; and their fruit greater and sweeter and of differing taste, smell, colour and figure, from their nature ... We have also parks and enclosures of all sorts of beasts and birds, which we use not only for view or rareness, but likewise for dissections and trials ... We also try all poisons and other medicines upon them ... By art likewise we make them greater or taller than their kind is, and contrariwise dwarf them and stay their growth. We make them more fruitful and bearing than their kind is, and contrariwise barren and not generative. And we also make them differ in colour, shape, activity, many ways. We find means to make commixtures and copulations of different kinds, which have produced many new kinds, and them not barren, as the general opinion is.

Bacon's programme was championed in England during the stormy 1640s by the Prussian exile Samuel Hartlib, one of a clutch of progressive thinkers that included the mathematician William Petty, the chymist Robert Boyle and the Bermudan alchemist George Starkey. During the English Civil War and its aftermath, such ambitions were politically charged: the 'new philosophy' had a distinctly Puritan slant that challenged the traditionalism of the Royalists. But Cromwell's Protectorate was wary of anything that smelled of the utopian, and it was not until the Restoration of Charles II in 1660 that permission was granted for Boyle, Petty and colleagues to found what became, by royal charter two years later, the Royal Society.

Bacon's thinking infused this project. The poet Abraham Cowley, whose pamphlet *The Advancement of Experimental Philosophy* in 1661 was of a distinctly Baconian flavour, wrote an ode to the Royal Society in 1667 in which he hailed Bacon as the liberator who, like Moses, 'led us forth at last' to a 'Promis'd Land'. In fact, in its early days the members of the Royal Society seemed to take so closely to heart Bacon's advocacy of Experiments of Use that its early historian Thomas Sprat complained in 1667 that 'we are not able to inculcate into the minds of many men, the necessity of that distinction of my Lord Bacon's, that there ought to be Experiments of Light, as well as of Fruit'. It was as though they were all intent on creating without delay the technological miracles of a New Atlantis.

PRACTICAL CRAFTS

The aims of the scholars of Salomon's House, Bacon wrote, are 'the knowledge of causes, and secret motions of things; and the enlarging of the bounds of human empire, to the effecting of all things possible'. We are now rather familiar with the former as goals of scientific inquiry. What caused the universe, and what now is causing the 'secret motion' of its accelerating expansion? What are the fundamental forces, and how are they related? How did life begin, and what agencies have governed its trajectory? What are the secret motions of the human mind?

But 'the effecting of all things possible'? You do not have to be one of Snow's anti-scientific snobs to feel a shiver of apprehension at the 'wonders' of Salomon's House, or at the prospect of such subjugation of nature. Today is it painfully evident that we lack much ability to 'control' nature, but possess in abundance a capacity to foul it up. Yes, like Bensalem's scientists we can make 'instruments of war' and 'new mixtures and compositions of gun-powder, wild-fires burning in water, and unquenchable'. We 'make divers imitations of taste, so that they will deceive any man's taste'. We have 'houses of deceits of the senses', 'false apparitions, impostures and illusions'. We do not seem to be any the better off for it.

The debate about where one locates the blame for the excesses and destructiveness of a technological age is an important one, but is certainly not going to be resolved here. I want instead to look at just a few areas of Baconian applied science, to examine where Bacon's vision has in fact taken us and why and how it has acquired the tarnish that Medawar and Snow discerned, whereby engineering becomes simultaneously drab and dangerous to the public view while tolerated by scientists as a somewhat dim and vulgar relation.

MAKING METALS

We have large and deep caves … for the producing also of new artificial metals.

Mining and metallurgy are the first things that the scholars of Salomon's House mention; imagine that! Here is a list, not unlike that attempted in this book, of the great things that science has achieved, and what comes at the top? Cosmology? Genetics? Evolutionary theory? No – metals. That's because Bacon understood the foundations on which his world was built. Political power in the age of the Stuarts depended on metals: on the ability to equip an army and to produce muskets and cannons, and on the control of coinage and bullion. Wealth was measured out in silver, as the Fugger family of Augsburg discovered when it supplied kings and emperors with canny loans from its banking empire in order to gain control of the German silver mines. The foremost technological treatise of the Renaissance was Georgius Agricola's *De Re Metallica* (1556), a summary of mining techniques that remained the standard text for two centuries. It contained woodcuts in which massive machines wrest nature's bounty from the Earth, a truly Baconian picture that foreshadowed the ruthless manufacturing and despoliation of the Industrial Revolution.

But Agricola's book included a staunch defence of mining which reveals a lot about the ambivalent views of his contemporaries. Mining has always been a dirty business – the mines of Rio Tinto in Spain have degraded the environment since the times of Roman occupation. Agricola tells us that people were not blind to this in the late Middle Ages. 'The strongest argument of the detractors', he says, 'is that the fields are devastated by mining operations … Also they argue that the woods and groves are cut down … then are exterminated the beasts and birds … Further, when the ores are washed, the water which has been used poisons the brooks and streams, and either destroys the fish or drives them away.' And he notes that mining is considered to be a profession unsuited to respectable people, a 'degrading and dishonourable' affair once fit only for slaves. In the first century BC the Roman writer Diodorus Siculus wrote that the Egyptian gold mines in the Nubian deserts were manned by 'notorious criminals, captives taken in war, persons against whom the King is incensed', who were worked until 'they drop down dead in the midst of their insufferable labours'. Metals were much prized, but extracting and refining them was a lowly, even despicable task.

That was to change. During the Industrial Revolution, the high price of steel meant that many large engineering projects were carried out that used instead cast iron, which is brittle and prone to failure. The Dee Bridge disaster of 1847 was one such: Robert Stephenson's structure in Chester collapsed as a train passed over it, killing five people. This was why Henry Bessemer's new process for making steel was greeted with jubilation: the details, announced at a meeting of the British Association in 1856, were published in

full in *The Times*. Bessemer himself was lauded not just as an engineer but as a scientist, being elected a Fellow of the Royal Society in 1879.

Bessemer's process controlled the amount of carbon mixed with iron to make steel. That the proportion of carbon governs the hardness was first noted in 1774 by the Swedish metallurgist Torbern Bergmann, who was by any standards a scientist, teaching chemistry, physics and mathematics at Uppsala. Bergmann made an extensive study of the propensity of different chemical elements to combine with one another – a property known as elective affinity, central to the eighteenth-century notion of chemical reactivity. He was a mentor and sponsor of Carl Wilhelm Scheele, the greatest Swedish chemist of the age and a co-discoverer of oxygen.

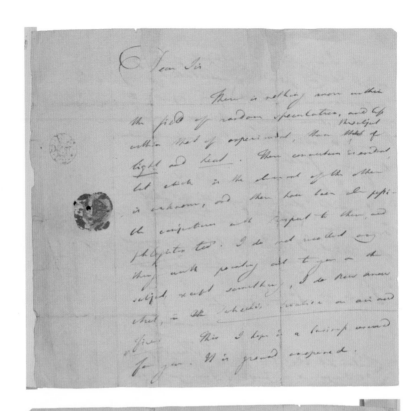

A letter from Joseph Priestley to Thomas Wedgwood on Scheele's discovery of oxygen, 1792.

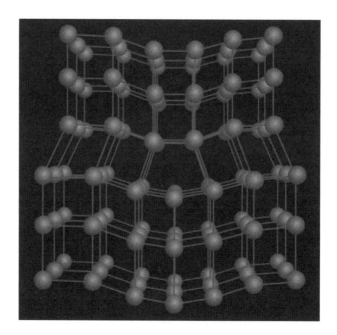

A dislocation in
a crystal lattice.

Oxygen, as a component of air, was the key to the Bessemer process. It offered a way of removing impurities from pig iron and adjusting its carbon content during conversion to steel. A blast of air through the molten metal turned impurities such as silicon into light silica slag, and removed carbon in the form of volatile carbon dioxide. Pig iron contains as much as 4 per cent carbon; steels have only around 0.3–2 per cent. Meanwhile, the heat produced in these reactions with oxygen kept the iron molten without the need for extra fuel (coke was expensive). Basically the same process was invented in Kentucky in the late 1840s by an American inventor, William Kelly, but he had no commercial success with it and went bankrupt in 1857, in the process losing his patent claims to Bessemer.

It was long known that steel can be improved with a spice of other elements. A dash of the metal manganese helps to remove oxygen and sulphur from the iron, and most of the manganese currently produced globally is used for this purpose. Manganese also makes steel stronger, while nickel and chromium improve its hardness. And chromium is the key additive in stainless steel – in a proportion of more than about 11 per cent, it makes the metal rust-resistant. Most modern steels are therefore alloys blended to give the desired properties.

But is this science? Some of the early innovations in steel alloys were chance discoveries, often due to impurities incorporated by accident. In this respect, metallurgy has long retained the air of an artisan craft, akin to the trial-and-error explorations of dyers, glassmakers and potters. But the reason for this empiricism is not that the science of metallurgy is trivial; it is because it is so difficult. According to Rodney Cotterill, a remarkable British physicist whose expertise stretched from the sciences of materials to that of the brain, 'metallurgy is one of our most ancient arts, but is often referred to as one of the youngest sciences'.

One of the principal difficulties in understanding the behaviour of materials such as steel is that this depends on its structure over a wide range of length scales, from the packing of individual atoms to the size and shape of grains micrometres or even millimetres in size. Science has trouble dealing with such a span of scales. One might regard this difficulty as akin to that in the social sciences, where social behaviour is governed by how individuals behave but also how we interact on the scale of families and neighbourhoods, within entire cities, and at a national level. (That's why the social sciences are arguably among the hardest of sciences too.)

The mechanical properties of metals depend on how flaws in the crystal structure, called defects, move and interact. These defects are produced by almost inevitable imperfections in the regular stacking of atoms in the crystalline material. The most common type of stacking fault is called a dislocation. Metals bend, rather than shattering like porcelain, because dislocations can shift around and accommodate the deformation. But if dislocations accumulate and get entangled, restricting their ability to move, the metal becomes brittle. This is what happens after repeated deformation, causing the cracking known as metal fatigue. Dislocations can also get trapped at the boundaries between the fine, microscopic grains that divide a metal into mosaics of crystallites. The arrest of dislocations at grain edges means that metals may be made harder by reducing the size of their grains, a useful trick for modifying their mechanical behaviour.

To understand all of this, one needs a variety of microscopic techniques for investigating metal structure at different levels of magnification. It has also now become possible to simulate the behaviour of vast numbers of atoms on a computer, allowing researchers to relate the properties of dislocations and grains containing thousands or millions of atoms to the packing of constituent particles at the atomic scale.

This sort of insight is making it possible to design metal alloys from the drawing board – figuring out what combinations of elements and arrangements of atoms will supply particular properties, and then attempting to make them. That's true not just for mechanical properties such as strength and hardness but also for electrical and magnetic properties, paving the way for new batteries and super-strong magnets. No one can question that this is hard science, demanding the skills of physics and chemistry as well as the expertise and experience of materials scientists. Among the remarkable metals that have emerged from such research are alloys that can remember shapes, regaining them when bent and then gently warmed; metals that change shape when placed in magnetic fields; metals that don't expand when they get warm (essential for finely engineered devices such as watches); and metals that turn heat into electricity, offering new possibilities in refrigeration. Yet as with so much applied science, the truly 'scientific' aspects of metal engineering tend to be overlooked by the time these substances reach the marketplace: they are just 'stuff', products of a kind of industrial alchemy that passes unquestioned because it is deemed simultaneously prosaic and utterly mysterious.

SYNTHETIC MYTHOLOGY

We have also divers mechanical arts … and stuffs made by them; as papers, linen, silks, tissues …

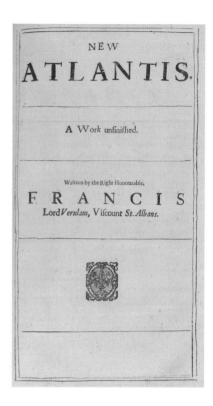

Frontispiece
for Bacon's
New Atlantis.

Bacon's *New Atlantis* is a favourite hunting ground for those who like to find predictions of tomorrow's technologies. With a little imaginative licence, you can find within it intimations of submarines, loudspeakers, even lasers. But even Bacon's fertile mind fails to anticipate that entirely new *classes* of materials might be invented. He does, however, recognise the transformative value of the textile fabrics of everyday life, and it is not hard to imagine him grasping in an instant the idea that approximations to silk might be made from oil, or the genuine article obtained without the aid of spiders and silkworms.

Today, the very notion of 'synthetic' in materials is almost synonymous with plastics: that's to say, with the Protean substances made of long, chainlike polymer molecules with backbones of carbon. Nature's structural fabrics – silk, hair, muscle, horn, wood and so forth – are also essentially carbon-based polymer materials. But whereas they are composed almost

A four-valve TRF (Tuned Radio Frequency) domestic receiver in a circular Bakelite case. It was the first plastic to be used for making radios, and was ideal for the Art Deco-style designs of the 1920s and 1930s.

entirely of just two classes of molecule – proteins and polysaccharides – synthetic plastics have a dazzling diversity of composition.

Plastics open the most revealing window on our relationship with human-made materials and their associated technologies. In many ways, they serve in this regard as a proxy for engineering technologies in general, tracing a complex path between excitement, opportunity, disenchantment, distrust, environmental concerns and even fetishism. Roland Barthes understood this: plastics, he said, are the ultimate representation of tech-nologists' abilities to transform matter: 'the quick-change artistry of plastic is absolute: it can become buckets as well as jewels.' Plastics offer 'the euphoria of a prestigious free-wheeling through Nature' – a poetic description of Bacon's technological utopianism, if ever there was one.

The earliest plastics, invented in the nineteenth century, were semi-natural materials regarded as substitutes for wholly natural ones. Celluloid is made from the cellulose fibres of plants: Christian Schönbein, a Swiss-German chemist who also pioneered the fuel cell and discovered ozone, found in 1832 that cotton fibres could be dissolved in nitric acid to form a glutinous material, cellulose nitrate, that could be moulded and hardened.

John and Isaiah Hyatt, two American brothers, discovered three decades later that castor oil or camphor made this material more malleable and workable, and they marketed it in the 1860s as a kind of imitation ivory, used in billiard balls and false teeth. But it was highly inflammable, even explosive – one form of cellulose nitrate, called gun cotton, was used as an artillery propellant, while celluloid in photographic movie film led to many a reel (and sometimes a cinema) going up in smoke.

A role for polymers as cheap mimics of expensive natural materials was furthered by the serendipitous invention of Bakelite in 1905: this dark resin aped the texture of mahogany. And rayon, another polymer derived from cellulose and marketed from the 1880s, was regarded as a kind of artificial silk – an epithet also attached to nylon, which the American company DuPont sold first for toothbrush bristles and then more lucratively in women's stockings from the late 1930s. Nylon has the better claim: the chemical constitution of its polymer chains is somewhat similar to that of the protein molecules that make up real silk.

So the initial promise of polymers was to provide 'luxury for all': materials resembling those only the wealthy had previously been able to afford. They were egalitarian materials: as Barthes put it, 'they aimed at reproducing cheaply the rarest substances, diamonds, silk, feathers, furs, silver, all the luxurious brilliance of the world'. What's more, the raw ingredients came from cheap oil or, in the case of Bakelite, from a waste product of turning coal into coke. Thus they offered wonders 'for free', and in this sense were a part of the utopian vision that science seemed to promise in the inter-war years. Henry Ford even experimented with an all-plastic car made from extracts of soya beans.

But this vision palled after the Second World War, partly because of shoddy manufacturing. PVC (polyvinylchloride) raincoats had an unpleasant texture and gave off smelly vapours when wet. Polystyrene products were brittle in ways that wood and metal never were. Plastics no longer seemed like cheap luxury, but merely cheap. 'Plastic has climbed down, it is a household material', Barthes announced in the mid-1950s. 'It is the first magical substance which consents to be prosaic.'

And so the plastics industry made that instead its selling point. No longer imitating luxury goods, plastic openly advertised its synthetic nature in garish colours that always looked factory-fresh. These materials were cheap, disposable and convenient: for housewives, much was made of plastics' wipe-clean character, transferable to just about any surface thanks to rolls of adhesive sheeting. The virtue of domestic convenience was exemplified by Teflon, the substance discovered (again serendipitously) at DuPont in the 1930s and later used in 'non-stick' kitchenware.

Historian Jeffrey Meikle of the University of Texas at Austin suggests that plastics thus introduced a 'democratisation of things' in the post-war economic expansion that made a dizzying variety and quantity of goods available to everyone. But this ultimately spawned a backlash against the 'miracle materials', which became emblematic of all that was superficial and wasteful in modern society. And hazardous too: it began with children being suffocated by plastic bags, but during the 1960s and 1970s the dangers started to look far more insidious. The molecular building blocks of PVC were linked to liver cancers among workers in the manufacturing plants, while some of the ingredients used as so-called plasticisers to soften plastics have been implicated as carcinogens and hormone mimics, which disrupt the human endocrine system.

Meanwhile, it became increasingly hard to see a link between these mass products and genuine science. In the World Fairs of the inter-war years, plastics were brought to the public by men in white coats, gazing into test tubes. But could a polymer scientist really belong to the same lofty caste as a geneticist or a particle physicist?

Yet once again, from a scientific and engineering point of view there is an awful lot of complexity to polymer science. These chainlike molecules can get entangled and flow in unusual ways while they are fluid. Engineering specific properties in polymers is a matter of controlling the microstructure, just as it is for metals: modifying the way the chains line up in a more or less orderly manner, say, or controlling their branching. As chemists gradually deduced how to regulate such things, they became capable of synthesising remarkable engineered polymers such as Kevlar, which is strong and tough enough to deflect bullets and tether oil rigs.

In nature this sort of structural tuning is exquisitely managed in protein-based polymers such as silk, a complex collage of tiny crystal-like regions in a disorderly, flexible matrix that creates a material stronger, weight for weight, than steel. Scientists have been attempting to make artificial silk for decades – one of the latest tricks is to produce the silk protein in the milk of genetically engineered goats, a Baconian vision for sure. But a persistent obstacle here is that the superior properties of silk thread arise not just from its chemical composition but from the way the polymer molecules are marshalled, aligned and organised as the threads get spun.

From the utopianism of the 1930s to the bland consumerism of the 1960s and the sleek monochrome minimalism of the 1980s, the mood of the developed world can be gauged from its polymer consumables. Today our bulk plastics are struggling towards a more environmentally friendly image, being biodegradable, made from non-oil-based ingredients, or more easily recycled. Meanwhile, high-tech plastics infiltrate the information technology once monopolised by silicon. Electronic circuits are being written with plastic, manufactured with cheap printing technology instead of demanding expensive high-vacuum conditions. Glowing television screens can be created from all-plastic light-emitting diodes on sheets as thin and flexible as paper.

Even paper itself is being reinvented, partly in plastic, for the information age. It is one of those fabrics that are hard to improve: its cheapness, durability, portability and readability (thanks to the high brightness contrast with ink, whether in bright or dim light) have secured the survival of the book and the newspaper in the digital age. But now the benefits of information technology are being combined with those of paper in a material commonly called e-paper or (to turn the idea on its head) e-ink: a plastic sheet with the lightness and appearance of paper on which the ink can be rearranged electronically. A sheet of the stuff, connected to a microchip loaded with data, is an entire library. These heady possibilities should come with a warning, however, for Bacon was right to say that power stems from knowledge and not mere information.

ENGINEERING LIFE

> We have also means to make divers plants rise by mixtures of earths without seeds; and likewise to make divers new plants, differing from the vulgar; and to make one tree or plant turn into another. We have also parks and enclosures of all sorts of beasts and birds ... By art likewise, we make them greater or taller than their kind is; and contrariwise dwarf them, and stay their growth: we make them more fruitful and bearing than their kind is; and contrariwise barren and not generative. Also we make them differ in colour, shape, activity, many ways ... Neither do we this by chance, but we know beforehand, of what matter and commixture what kind of those creatures will arise.

Of all the 'marvels' in Bensalem, these are surely the most chilling, not least because of the apparent nonchalance with which the priests of Salomon's House tamper with living nature. Here most of all Bacon's treatise takes on a Faustian cast, and it is an easy matter to trace the path from *New Atlantis* to Mary Shelley, whose fantastic fable of life remade tapped into centuries of apprehension about the consequences of scientific hubris.

Of course, we cannot read Bacon's comments now without thinking of biotechnology and genetic engineering, which permit the 'commixture' of creatures: spider genes in goats, plants loaded with the genetic defensive armoury of quite different (even animal) species. We can hollow out animal eggs and load them up with human genomes, and then grow them into embryos. And this is just the beginning. It is probably a matter of a few years before new species are designed on the blackboard and manufactured with genomes synthesised in the laboratory, collections of genes handpicked more fastidiously than anything selective breeding can achieve. These genomes might be transferred into emptied cells, or simply allowed to override the existing genetic instruction manuals of ordinary bacteria. This 'synthetic biology' will represent a new origin of life, after a fashion: the first organisms outside the great chain of being that began almost four billion years ago. And tellingly, such efforts are now framed in terms that relate more to an information age than to the molecular biology of Crick and Watson: organisms, we are told, are being 'reprogrammed' with new 'software', and then 'rebooted' to get them running. The redesign of life 'from scratch' will be accompanied by well-motivated concerns about safety and ethics, but it will also confront us with deeper questions that we have previously preferred to keep at arm's length: What is life? When does it begin? What is 'natural'?

These questions that weigh so heavily for us now might have been regarded as far less burdensome within the uncompromisingly mechanistic worldview of Francis Bacon. Like his contemporaries René Descartes and Thomas Hobbes, he considered all phenomena, whether the workings of the human body or of the stars, to have rational, material causes. Everything was so many atoms, colliding in insensate profusion. Moreover, Bacon's outlook (which accords with that of most scientists and engineers throughout the ages) is essentially optimistic, guided by a belief that the human lot can be improved by technical means. He was eager to free the sciences from religious shackles, to abandon the hierarchy of the Earth and heavens and a reliance on teleological explanations.

By and large, those aspirations still underpin efforts to engineer biology. Some of biotechnology's earliest successes stemmed from an image of living cells, primarily bacteria, as microscopic factories for manufacturing sorely needed drugs. The development in the 1970s of recombinant DNA technology, which enabled genes to be sliced out of the genome of one organism and spliced into that of another, using natural enzymes that conduct such cutting and pasting, enabled human insulin to be derived by fermentation of genetically modified *Escherichia coli* bacteria. Sights are now set not just on pharmaceuticals but on cleaner fuels, greener manufacturing of materials, biological clean-up of environmental contaminants, even 'wet nano-robots' that engage hand-to-hand with disease agents.

There was, as we can see, nothing new in the materialistic conception of life that enabled biotechnologists to view it as amenable to principles of construction and design. And indeed the reception of this 'cut-and-paste' approach to the living world was, all things considered, relatively muted: so long as it declines to re-engineer human beings (and perhaps other higher organisms), biotechnology tends to be seen as just another industrial process, more akin to brewing than to vivisection. While opponents of genetic modification have played on philosophically suspect notions of the 'natural' and 'unnatural', most of the resistance to its introduction has been motivated by concerns about commercial ownership and responsibility, and about public health: issues that might reasonably be raised (and often are) for any new technology. As far as the 'sanctity of life' is concerned, public opinion often shows a solipsistic parochialism. Yet if there is one lesson to be drawn from the controversy in Europe about genetically modified organisms (apart from a reminder of the unwelcome influence of mass media), it is that technologies are less likely to gain easy acceptance until they can demonstrate tangible benefits to potential consumers.

All the same, scientists have been revealingly eager to exploit public sympathy, or at least tolerance, towards 'pure' science in the promotion of biotechnological initiatives with decidedly applied goals. The Human Genome Project was in truth something of a mixture of both, to the extent that the distinction is meaningful at all; but the rhetoric with which the project advertised itself was concerned with uncovering the secrets – in the deeply misleading metaphor, 'reading the book' – of life. The project was entirely dependent on technical advances, and it gave rise to no new theories but rather to an impressive and immensely useful (but only patchily understood) data bank. The frequent comparisons with the Moon landings were more apt than perhaps intended, since both were feats of technical prowess more than they were voyages into the scientific unknown.

Strikingly, then, the extension of engineering ideas to biology has so far been regarded with scarcely more distaste or disdain than is reserved for engineering more generally, and the complaints are often of much the same nature. Few even perceive the philosophical boldness of a word such as 'bioengineering', which is commonly accepted with the indifference one might expect to see accorded to a branch of automotive engineering. Perhaps we are more the heirs to Bacon's vision than we realise. Even concerns about the prospect of the *de novo* creation of life are so far voiced only by rather minor pressure groups, and they too tend to focus on safety issues. Battle lines are only really drawn when biological technologies impinge on human life, as in the cases of stem-cell technology, embryo research and assisted conception. Only here have certain traditional belief systems deemed it necessary to impose assumptions about what life consists in.

Distorted, dogmatic, and dangerous though such assumptions may sometimes be, buried within them are some genuine questions about the ethical responsibilities of the engineer. Opinions may differ on the boundaries of human dignity, but it is surely right that these boundaries feature in any consideration of what we might and might not make. And the desirability of a technological goal is not to be determined simply by a health-and-safety or cost-benefit analysis, but by a careful consideration of the difficult question of whether it seems likely in the long term, on balance, to serve human welfare and well-being. The disturbing aspect of Bacon's utopian scientific writings is often not so much what they consider possible, but how readily he assumes that humankind has the wisdom to handle such power.

Why Engineering Matters

As I write, the Large Hadron Collider, the world's biggest atom-smasher at CERN in Geneva, has switched on with almost unprecedented media jamboree*. Asked about the practical value of it all, Stephen Hawking has said that 'modern society is based on advances in pure science that were not foreseen to lead to practical applications'. It's a common claim, and it subtly reinforces the hierarchy that Medawar identified: technology and engineering are the humble offspring of pure science, the casual cast-offs of a more elevated pursuit.

I don't believe that such pronouncements are intended to denigrate applied science as an intellectual activity; they merely speak into a culture in which that has already happened. Pure science undoubtedly does lead to applied spin-offs, but this is not the norm. Rather, most of our technology has come from explicit and painstaking efforts to develop it. And this is simply a part of the scientific enterprise. A dividing line between pure and applied science makes no sense at all, running as it does in a convoluted path through disciplines, departments, even individual scientific papers and careers. Research aimed at applications fills the pages of the leading journals in physics, chemistry and the life and Earth sciences; curiosity-driven research with no real practical value is abundant in the 'applied' literature of the materials, biotechnological and engineering sciences. The fact that 'pure' and 'applied' science are useful and meaningful terms seduces us sometimes into thinking that they are real, absolute and distinct categories.

This isn't merely a semantic issue. Concerns about a decline in university admissions for science and engineering are more or less universal among the various disciplines, but there is good reason to suspect that the sciences deemed to be more 'pure' retain a greater attraction for the brightest students among those who still gravitate in this direction – even though employment prospects for an engineer are better than for a string theorist (who in recent years has seemed likely to end up on Wall Street). In 1998 the President of the US National Academy of Engineering, William A.

---

* At the time of publication, the hiatus caused by the large Hadron Collider's subsequent malfunction is almost at an end.

Wulf, stated: 'We need to understand why in a society so dependent on technology, a society that benefits so richly from the results of engineering, a society that rewards engineers so well, engineering isn't perceived as a desirable profession.' Yet many of the most pressing global problems – clean energy generation, the management of water resources, securing nuclear non-proliferation, creating less waste and more efficient use of material resources – cry out for technological expertise.

There's no simple formula for the rehabilitation of the engineering, synthetic and technological (in the oldest sense) aspects of science. Celebrating their achievements is all very well, although it remains a conundrum why, for example, the British people seem to hold Isambard Kingdom Brunel in such high esteem without showing much inclination to follow in his footsteps. But no amount of flag-waving can disguise the fact that the practical sciences, the *craft sciences* if you will, have always had and will always have a double-edged nature: along with life-saving drugs, safer transportation, more accessible information and solar power comes pollution, landfills and nuclear weapons. The conventional talk of 'dual-use' technology should rather acknowledge the reality of a thousand uses, guided by as many agendas. As US writer Richard Powers puts it in his 1998 novel *Gain*, an exploration of the social politics of industrial chemistry, 'People want everything. That's their problem.'

Science does itself no favours when it tries to skip away from such complex issues with talk of 'pure knowledge', untainted by the marketplace. That's a privileged position enjoyed by a very few of its practitioners, who even then cannot be sure that their seemingly arcane ideas won't end up guiding the fabrication and operation of some device or other. Science is about making stuff, just as much as it is about understanding stuff. The two go hand in hand, and always have done. Francis Bacon implied as much; but in the twenty-first century, disciplines such as nanotechnology, quantum information technology and synthetic biology are blurring as never before the false distinctions between thinking and doing. So what shall we make tomorrow?

14

Nebula 38

417. Aug.ᵗ 24ᵗʰ 83 38 Nebᵘ. of M. 70 or 80 stars of consider

nitude 7 ft 57. with 298 much mixed

stars invisible to the lower power

619. So. lat. 18ᵗ 86. A very rich cluster of large stars

a size II (M.) Auriga

It is the 38 of the Connois

PAUL
DAVIES

JUST TYPICAL:
OUR CHANGING PLACE
IN THE UNIVERSE

Paul Davies is a British-born theoretical physicist, cosmologist, astrobiologist and best-selling author. He is Director of the Beyond Center for Fundamental Concepts in Science and co-Director of the Cosmology Initiative, both at Arizona State University. He has written 28 books including *The Mind of God*, *About Time*, *How to Build a Time Machine*, *The Fifth Miracle* and *The Goldilocks Enigma*. His latest book, *The Eerie Silence*, is about the search for intelligent life in the universe.

T HE DEVELOPMENT OF COSMOLOGY HAS CONFIRMED OVER AND OVER AGAIN THAT WE DO NOT OCCUPY A CENTRAL POSITION IN THE GREAT SCHEME OF THINGS. BUT AS PAUL DAVIES EXPLAINS, THE STORY OF OUR REALI-SATION THAT WE HOLD NO SPECIAL PLACE IN THE COSMOS COULD YET BE A TALE WITH A TWIST.

When the Royal Society was founded 350 years ago, the Copernican revolution was only a few decades old. Before Copernicus, many people believed the Earth lay at the centre of the universe and mankind was the pinnacle of creation. The discovery that Earth is but one planet among several orbiting the Sun came as a shock and forced human beings to drastically re-evaluate their place in the universe. It is a lesson that has been repeated often in the centuries that followed. The pivotal change that occurred with Copernicus was so far-reaching that scientists refer to 'the Copernican principle' quite generally to mean that our situation in the universe should not be in any way special or privileged. Expressed simply, the Copernican principle asserts that we are typical. Some of the deepest unanswered questions in cosmology and astrobiology in the twenty-first century concern whether and when that principle might break down.

A seventeenth-century representation of the Copernican solar system.

The Copernican principle has been a remarkably reliable guide when applied to astronomy and cosmology, although it got off to a bad start. In the seventeenth century it was widely believed that the other planets and moons in the solar system resembled Earth, even to the extent of being inhabited by plants, animals and sentient beings. Kepler, for example, wrote a treatise about the denizens of Earth's moon. Galileo pioneered the use of the telescope to study the heavens, and it soon became clear that the other planets differ in many respects from Earth; within the solar system, then, Earth turns out to be a very atypical planet. But Galileo also discovered that the Sun is an undistinguished star among a vast number that collectively make up the Milky Way galaxy. Later measurements established

417 Aug.ᵗ 24ᵗʰ 83. 38 Neb. of M. 70 or 80 stars of considerable mag
nitude. 7 ft 57. with 278 much mixed with small
stars invisible to the lower power.

619 Sw. Oct.ʳ 18ᵗʰ 86. A very rich cluster of large stars, all of
a size. 11 (μ) Auriga f — f 2° 39' RA — PD 54° 24'
It is the 38 of the Connoiss. des temps.

693 Sw. Jan 17. 87. A very large cluster of scattered
large stars, extremely rich and beautifull.
21 (σ) Auriga f 4' 10" f 1° 35' RA 5ʰ 14' 20"
PD 54° 24'.

1030 Sw. Feb. 4, 1793. A very large cluster of pretty large stars
very brilliant and very rich. The 38.ᵗʰ of the
Connoiss. des temps. By obs. in 693 Sw. Jan. 17, 1787.
RA 5 14 20 PD 54 24. Cor. since that time by Woll. Cat.
        +24   — 24
Pr. — — 5 14 44   54 23 36

Misc. Oct.ʳ 27. 94. 7 feet Reflector. South of σ Auriga with
120; a pretty rich cluster of small stars.

Rev. Dec.ʳ 30, 99. Nᵒ 38 of the Conoiss. is visible in the finder,
north of 24 Auriga.

Rev. Nov.ʳ 23. 1805. Large 10 feet reflector. A cluster of scattered pretty
large stars of various magnitudes, of an irregular figure;
and pretty rich.
It is in the milky way.

Opposite:
Observations made
by William
Herschel between
1783 and 1805 of
the nebulae and
clusters discovered
by Charles Messier
in 1764.

that the galaxy contains about four hundred billion stars in total, arranged in a disc shape and embellished by spiral arms sprouting from a central spherical bulge. The entire assemblage is about one hundred thousand light years across.

At the turn of the twentieth century, it was widely believed that the Copernican principle might soon fail in two key respects. The first concerned the distribution of stars in the universe. The Dutch astronomer Jacobus Kapteyn made a painstaking analysis and concluded that the Sun lay in a privileged position near the centre of the Milky Way, with the galaxy a sort of 'island universe' surrounded by a seemingly limitless void. But within a decade or two this model was refuted. As far as we can tell, there is after all nothing very special about the location of the solar system. It actually resides in one of the spiral arms about twenty-five thousand light years from the galactic centre – middle suburbia, if you like.

Related to the question of the structure of the galaxy was a controversy concerning the wispy patches of light painstakingly catalogued in the eighteenth century by Frenchman Charles Messier. Some astronomers maintained they were far-flung galaxies in their own right – other 'Milky Ways'. The alternative view was that these nebulae were clouds of glowing gas located within the Milky Way. The dispute was finally settled when telescopes became powerful enough to image individual stars in some of the nebulae, revealing them to be other 'island universes', or galaxies, in their own right, many very similar to the Milky Way. We now know that the Milky Way is in fact a typical galaxy, just as the Sun is a typical star, so the Copernican principle works on an extra-galactic scale too.

At the same time as the true nature of extra-galactic nebulae was being established, similar observations revealed that the other galaxies are in motion with respect to ours and each other, a feature that could readily be deduced from the Doppler shift in the spectral lines of their light. Edwin Hubble in the USA found a systematic pattern to this motion, which can be summarised by saying that the entire universe is expanding: the galaxies are, on average, moving away from each other. Running 'the great cosmic movie' backwards suggests that, some billions of years ago, the matter in the universe was compressed into a small volume of space and was expanding very rapidly, a state of affairs now called the big bang.

With the discovery of other galaxies, the scale of the universe leapt once more. Since the time of Copernicus, the sheer size of the cosmos has dazzled people again and again. The solar system is a few light hours across. The *nearest* large galaxy, Andromeda, is about two million light years away. Hubble observed galaxies ten times further away than this, but saw no end in sight. Hubble's eponymous Space Telescope can now image galaxies more than twelve billion light years away, a volume of space encompassing trillions of galaxies in all. Remarkably, even on the largest scale of size, the Copernican principle again comes through with flying colours. Deep space surveys reveal clusters of galaxies spread with surprising uniformity throughout the universe. It seems we not only live in a typical galaxy, but even our extra-galactic neighbourhood is typical.

The large-scale uniformity of the cosmos is confirmed in another way. The big bang that started off the universe as we know it was intensely hot, and filled space with heat radiation. As the universe expanded so the radiation cooled, but it remains as a fading afterglow of the fiery cosmic birth, detectable today in the form of a background of microwaves coming from all directions of space. The cosmic microwave background radiation has been travelling more or less undisturbed since about 380,000 years after the big bang, which occurred 13.7 billion years ago. It thus carries an imprint of what the universe was like at a very early epoch. Measurements show that, to one part in a hundred thousand, matter and radiation were distributed smoothly throughout space at that time.

The second potential failure of the Copernican principle around 1900 concerned the formation of planets. A popular theory at that time was the so-called encounter hypothesis, according to which the Sun suffered a close approach by another star, which caused blobs of matter to be sucked off and flung into orbit round the Sun. Since such close encounters are highly improbable, the theory predicted that planetary systems will be very rare. In other words, the Sun may be a typical star, but its retinue of planets might be very exceptional.

The problem of the solar system's typicality had to wait far longer for a resolution. It was only in the 1990s that astronomers observed the first extra-solar planets, and with improving techniques the tally has grown to about four hundred. To date, no earthlike planets have shown up, but that is no surprise, because the current instrumentation isn't sensitive enough to detect them. Space-based planet-finding systems should be able to detect other earths, however. There is no good reason why earthlike planets should not exist in abundance throughout our galaxy and others. Although it is not yet

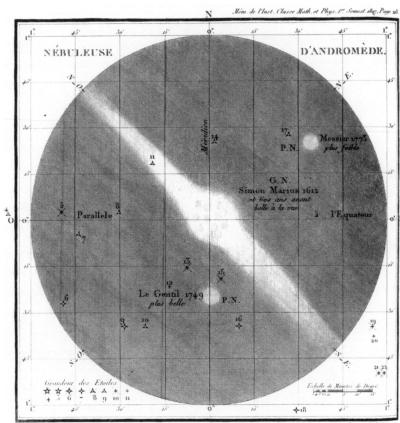

Charles Messier's 1807 sketch of the Andromeda galaxy.

quite certain, it seems therefore that the solar system, and planet Earth, are fairly typical. The Copernican principle may have failed when Earth is compared to our sister planets in the solar system, but within the larger class of all planets, it is probably successful. Of course, success or failure of a typicality hypothesis depends on the level of detail we are interested in. For example, Earth's moon was probably created when a Mars-size body slammed into the proto-Earth shortly after the solar system formed. This cataclysm produced a moon that is unusually large for the size of the planet. It will surely be very rare to find another earthlike planet with a similar-sized moon.

Although the Copernican principle has no basis in physical law – it is more a rule of thumb – it is nevertheless tempting to apply it to other aspects of our circumstances. For example, Earth is host to abundant life. Is that typical of most earthlike planets? Many scientists think so; indeed, the subject of astrobiology is founded on the expectation that life is widespread in the universe. However, there is an obvious complication. We can observe the universe only from a location that supports life, which means we have in a sense selected where we are (or rather, our location has been selected for us). If there was only one planet in the universe with life, we would have to be on it. So we must be cautious in using the typicality argument. In fact, some scientists prefer to invert the reasoning and apply an atypicality, or anti-Copernican, principle.

To illustrate the issues involved, let me discuss not our location in space, but our location in time. In the 1930s, the physicist Paul Dirac and the astronomer Arthur Eddington were struck by a strange relationship in basic physics and cosmology. The hydrogen atom is held together by an electromagnetic force between the proton and electron. There is also a tiny gravitational force of attraction between them. The ratio of these forces is a staggering 10^{40}. How, wondered Dirac and Eddington, did such a large number come out of fundamental physics? (It remains a mystery today.) But the peculiar twist is that the same very large number crops up in a completely different context. The age of the universe – that is, the time since the big bang – is also about 10^{40} when expressed as a ratio using basic atomic units of time. Surely these two very large numbers are not the same by coincidence? Dirac at least thought not. He reasoned that they had to be linked deep down by some law of physics. However, because the age of the universe is not a fixed number – it gets bigger every day! – if there is such a linkage it implies that the ratio of forces must also increase with time, with gravity growing relatively weaker as the universe ages. Dirac developed an elaborate mathematical theory to incorporate this effect, and astronomers set about testing whether the force of gravity is indeed time-dependent.

Dirac's argument, however, contained a hidden Copernican assumption: it supposed that the cosmic epoch at which we find ourselves living isn't special. Therefore an observer seven billion years ago would have found gravity to be twice as strong as it is for us, and an observer fourteen billion years from now would find gravity to be about half as strong as it is today, but in both cases the big number concordance would be the same as it is for us. Clearly the typicality assumption is questionable in this case. In the 1960s, the astrophysicist Robert Dicke pointed out how. The existence of intelligent observers like *Homo sapiens* has two basic prerequisites: suitable chemical elements and a star like the Sun that burns steadily for billions of years while evolution does its stuff. The key element for all earthlife, and probably any form of life, is carbon. Carbon was not coughed out of the big bang; rather, it was made in the cores of massive stars, which then exploded as supernovae and laced the interstellar gases with life-encouraging material. It follows that life would not have been possible until at least one generation of stars had lived and died. On the other hand, after several generations of star burning, the raw material needed for new star formation will dwindle, and stable stars will become a rarity. These considerations therefore bracket the epoch at which life is likely to arise in the universe, to between one and, say, ten stellar lifetimes. Dicke spotted that the lifetime of a star depends on both gravitation and electromagnetism. If by some magic we could make gravity suddenly stronger, the Sun would shrink and get hotter, burn its nuclear fuel faster and die quicker. The strength of the electromagnetic force controls the rate at which heat can diffuse from the energy source (nuclear fusion reactions) in the core of the star, reach the surface, and flow away into space. The balance between these two forces thus turns out to be the dominant factor in determining the star's lifetime. A rough calculation shows that the lifetime of the star, when expressed in atomic units, depends on precisely the ratio of electro-magnetic to gravitational forces flagged by Dirac and Eddington. So the big number 'coincidence' is convincingly explained as a consequence of an observer selection effect. The cosmic epoch at which we are living is indeed typical enough within the range permitted – the solar system is 4.5 billion years old, placing us in the middle range of the 'habitability window' before stars get scarce. However, assuming the universe endures for trillions of years and is not overtaken by a big crunch or similar cosmic catastrophe, the era of 'observership' (at least for observers who evolve naturally) occu-pies an atypical sliver of cosmic history.

How does the Copernican principle play out for the distribution of life across the galaxy and beyond? Until the turn of the twentieth century there was a general belief among scientists that many other life-harbouring worlds existed. Even as late as 1906, the astronomer Percival Lowell was convinced that Mars not only hosted life, but intelligent Martians, who had built a network of canals. During the twentieth century, the mood began to swing against the idea that life is common. Hopes of finding life elsewhere in the solar system began to fade as better telescopes, and then interplanetary space probes, revealed hostile conditions on our sister plan-ets. This mood of scepticism extended to all extraterrestrial life, so that by the 1970s the Nobel Prize-winning biologist Jacques Monod felt able to proclaim in his book *Chance and Necessity*, 'Man at last knows that he is alone in the unfeeling immensity of the universe.' The grounds for this scepticism stemmed from advances in molecular biology, and the growing understanding of life's extraordinary complexity, suggesting to many that its origin must have involved a statistical fluke of stupendous proportions, unlikely to have happened twice. These sentiments were reinforced when, in 1977, two Viking space probes landed on Mars with the express intention of testing for microbes in the soil. Nothing definitive resulted (and certainly no canals were found!). It began to seem as if life on Earth was in fact highly atypical, even unique, in the universe.

Today, the pendulum has swung back again in favour of the idea that life is widespread in the universe. One reason for the renewed optimism is the discovery that terrestrial organisms can flourish under a much wider range of conditions than assumed hitherto. Microbes have been found near deep ocean volcanic vents living at temperatures above 120 °C. Others have been found thriving in acid strong enough to burn human flesh, in the strongly saline waters of the misnamed Dead Sea and in the radioactive waste pools of nuclear reactors. Even the inner core of the Atacama Desert, where the rainfall is essentially zero, supports a low level of bacteria. These discoveries have given hope that microbial life at least might be possible on planets previously thought to be too hostile. In addition, clear evidence for liquid water – thought to be essential for life as we know it – on Mars and Europa (a moon of Jupiter) has rekindled hopes that primitive organisms might yet be found elsewhere in our solar system.

In spite of this new-found optimism, we still lack an accepted theory of life's origin. In 1859, Charles Darwin gave a convincing theory of how life has evolved over billions of years from simple microbes to the richness and diversity of the biosphere we see today, but he pointedly left out of his account how life got started in the first place. 'One might as well speculate about the origin of matter,' he quipped. Nevertheless, he did outline the germ of an idea, by referring to 'a warm little pond' in which all manner of chemicals might accumulate and, driven by the energy of sunlight, would react to form ever more complex molecules. Over an immense period of time sufficient chemical complexity might eventuate that the 'soup' would make the transition from non-living to living (whatever that transition may be – nobody knows).

Darwin's casual suggestion became the 'primordial soup' theory of life's origin, developed by J.B.S. Haldane and Alexander Oparin in the 1920s. The theory was put to an interesting experimental test in 1952, when Stanley Miller, then a student of Harold Urey at the University of Chicago, sought to re-create the conditions on the primeval Earth by putting methane, ammonia, hydrogen and water in a flask and sparking electricity through it for a week. Miller was delighted to discover a red-brown sludge of organic gunk in the flask, from which many amino acids were identified. Amino acids are the building blocks of proteins, and some scientists saw the Miller–Urey experiment as the first step on the road to life down which a simple chemical mixture would be inexorably conveyed by the passage of time. Many pre-biotic soup experiments have since been performed under

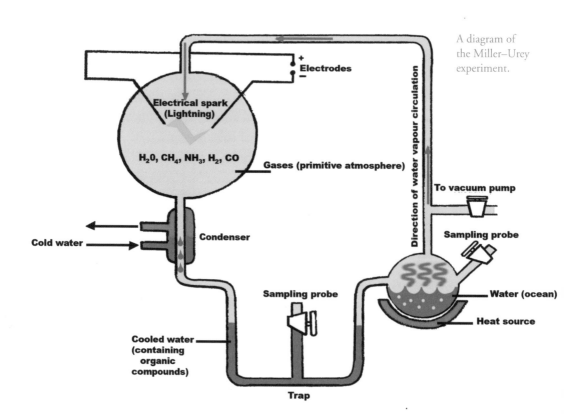

A diagram of the Miller–Urey experiment.

various conditions (we now know that the early Earth did not have an atmosphere quite like that assumed by Miller). It turns out to be easy to make amino acids; in fact, they are even found in meteorites. Much harder, however, is to produce long proteinous chains (peptides), or the building blocks of RNA and DNA. Some scientists are still hopeful that 'more of the same' would create life given enough time, but others are sceptical that simply zapping chemicals willy-nilly with energy will turn a non-living mixture into a living cell. It is often remarked that we may soon be able to make life in the laboratory using existing microbes as a blueprint and reconstructing a new organism piecemeal. (Viruses have already been made that way, but viruses do not satisfy some definitions of life because they lack the ability to reproduce unaided.) While that may be true, and is clearly possible in principle, it would not solve the problem of how Mother Nature performed the trick without fancy equipment, trained biochemists and a clear plan of action.

From the point of view of the Copernican principle, we do not need to know the details of biogenesis, only how probable it is given plausible pre-biotic conditions. Is life on Earth the result of a freak chemical accident, or are there general principles that favour the emergence of organised complexity, and thereby facilitate the formation of life 'against the raw odds' computed from random shuffling of building blocks? Such a 'life principle' (essentially Copernicus' principle for biological systems) is often mooted, but there is no hint of how it might be derived from the known laws of physics and chemistry. Nevertheless, the science of complexity is in its infancy, and it may be that there are general principles of complex organisation that are not yet understood. It is frequently pointed out that the elements needed for life – primarily carbon, but also oxygen, nitrogen, hydrogen, phosphorus and sulphur – are common in the universe, and that even simple organic molecules have been found in interstellar clouds. Sometimes this is used to argue that life must therefore also be common, but that is to confuse a necessary with a sufficient condition. To be sure, these substances are necessary for life, but it may require all sorts of other materials, and special conditions, before the basic building blocks self-assemble into the hugely elaborate structure of a living cell. It's easy to make bricks, but making houses requires far more than throwing a pile of bricks in the air.

If life were discovered on another planet, it would offer support for a life principle. There is, however, a caveat. By common consent, Mars offers the best hope for finding extraterrestrial life in the near future. Unfortunately, it may not settle the matter. Mars and Earth trade rocks blasted off their surfaces by asteroid and comet impacts, and hurled into orbit. A couple of dozen Mars meteorites have been found on Earth so far. During geological history, a prolific traffic of material has taken place between the two planets, mostly Mars to Earth on account of Mars' lower gravity, but some the other way too. It has become clear in recent years that microbes could hitch a ride this way. Cocooned within a rock, a microbe would be shielded from the harsh conditions of interplanetary space, especially the radiation, and could remain viable even after a sojourn of some millions of years orbiting the Sun. It seems inevitable that living terrestrial microbes will have been delivered to Mars this way, especially before 3.5 billion years ago when the bombardment by cosmic debris was far higher than it is today. Conversely, if there was once life on Mars, it will have spread to Earth. The intermingling of the two biospheres complicates the story of life's history though. It may be that life started on Mars and later came to Earth, or vice versa, or that life started from scratch independently on both planets, but became cross-contaminated. Only if there is clear evidence for two independent origins would the discovery lend support to the life/Copernican principle.

While we wait (possibly a very long time) for Mars to be explored for life, and perhaps evidence for a second genesis, there is a way that the life principle can be tested right here on Earth. No planet is more earthlike than Earth itself, so if life does pop up on cue in earthlike conditions, it should have emerged many times over on our home planet. Biologists have long assumed that all life on Earth has descended from a single common origin. Gene sequencing confirms that all known organisms are genetically linked and can be positioned on a universal tree of life. However, the vast majority of species are microbes, and only a tiny fraction of these has even been characterised, let alone sequenced. You can't tell by looking what they are made of. It is entirely possible that some terrestrial microbes are the products of different biogenesis events, in effect 'alien organisms', constituting a type of shadow biosphere. The universal tree of life on Earth might actually be a forest. The identification of a single microbe that is sufficiently alien for us to rule out a common origin with standard life, would have sweeping consequences. It would establish the Copernican principle for biology and point to a universe teeming with life.

And that brings me to the tantalising question of whether we are alone in the universe, as Monod claimed. When it comes to *intelligent* life, the status of the Copernican principle is very uncertain indeed. Even if life has got going on many planets, there is no known law or principle that compels it to evolve intelligence or sentience. The Darwinian mechanism implies that evolution is blind; nature cannot 'look ahead' and strive for the goal of intelligence, or any other trait. So there will be no progressive trend towards sentient beings like ourselves, unless it comes about because natural selection strongly favours certain features and structures, or if there are yet-to-be-discovered principles of organisation at work in nature.

Nevertheless, as always experiment must be the arbiter, and fifty years ago that experiment began with the inception of SETI – the Search for Extraterrestrial Intelligence. A small band of astronomers have been sweeping the skies with radio telescopes in the hope of stumbling across a radio signal from an alien civilisation elsewhere in the galaxy, so far without success. At the time SETI began in 1960, the general feeling was that life, let alone intelligent life, was exceedingly atypical for a planet. The sentiment was summed up by the biologist George Gaylord Simpson in a 1964 article entitled 'On the non-prevalence of humanoids', in which he described SETI as 'a gamble at the most adverse odds with history'. Today, SETI receives far more scientific backing, although the basic facts have changed little since Simpson wrote his article. We still don't know whether the origin of life on Earth was a freak event and whether the evolution of human intelligence was a statistical fluke.

What, then, is our place in the universe as currently understood? As far as we can tell, our planetary system, galaxy and galactic environs are unexceptional out as far as our most powerful instruments can penetrate, over twelve billion light years. But our biological situation remains unresolved. The universe might be teeming with life, or it may turn out that life is very rare – intelligent life more so. It is even conceivable that we are alone in the vastness of space. If so, history will have turned a curious full circle. Before Copernicus, people believed that humans and their planet occupied pole position in the universe. It may yet be that we are privileged after all, in being the only place in the universe with intelligent life.

Is that as far as we can take the Copernican principle, to the edge of the observable universe? As I have commented, each new advance in astronomy has unveiled a universe even larger and more majestic than previously realised, but with instruments like the Hubble Space Telescope we are approaching a fundamental limit due to the finite speed of light. When we see a galaxy 12 billion light years away, we see it as it was 12 billion years ago. Light can have travelled at most 13.7 billion light years since the big bang, so if that explosive event represented the true origin of the universe, then there is an ultimate horizon beyond which we cannot see. That does not mean the universe comes to an end there, any more than a horizon at sea signals the edge of the world. But it does mean we cannot directly observe what lies beyond. An uncritical application of the Copernican principle would suggest that if by some magic we could be transported across the horizon we would find a region of the universe that looked much the same as our region, with

Deployment of the Hubble Space Telescope from the space shuttle *Discovery* on 24 April 1990.

stars, galaxies and galactic clusters uniformly distributed on the largest scale of size. But inevitably this raises the question of how far we can extrapolate. Does this pattern continue to infinity, or is there some variation?

The attempt to construct proper mathematical models of the universe based on the best understanding of gravitation began shortly after Einstein published his general theory of relativity in 1915. For many decades the default assumption was that the Copernican principle applied all the way to infinity (it is called the cosmological principle when applied to gravitational models of the universe). But in the 1970s this conventional wisdom was challenged. The basis for the challenge was the development of a theory of the big bang based on the application of quantum mechanics to the very early stages of the universe. Quantum mechanics is normally reserved for microscopic systems like atoms and molecules, but the theory predicts that, at a sufficiently early time, it would affect the evolution of the universe too. That time is about a hundred trillion-trillion-trillionths of a second after the big bang. According to some variants of the theory, there would not be a single big bang, but a countless number of them scattered randomly throughout space and time. Each quantum event would nucleate a universe with a big bang, 'like bubbles in an uncorked bottle of champagne', to use the words of the physicist Leonard Susskind. The space between the bubbles would expand so rapidly that, even though the bubbles themselves expand, they would rarely intersect. Our own universe would be just one of those bubbles. The entire collection is known as the multiverse. In the most popular multiverse theory, the size of the bubbles is stupendous – about $10^{10,000,000,000}$ km across. So once again, the scale of the universe has leapt dramatically, but by a far larger factor than the jump from pre-Copernican cosmology to the time of Hubble. Now we confront the same Copernican principle on a mega-scale: do we live in a typical bubble? Will the other bubbles be similar to ours?

The evidence from theory suggests no. Physicists are convinced that many features of the laws of physics, such as the masses of subatomic particles, the nature and number of forces, and the density of dark energy (the mysterious stuff that seems to be making the expansion of our universe accelerate) are 'frozen accidents' locked in when the universe cooled from the searing heat of the big bang. If the experiment were done again, so to speak, the masses and forces would come out differently;

there might even be a different number of spatial dimensions. Einstein once famously expressed his distaste for quantum mechanics by declaring that 'God does not play dice with the universe'. In the multiverse theory He plays dice with *universes* (I am tempted to say He plays at randomly blowing bubbles). Taking a God's-eye-view, the multiverse is a patchwork quilt, featuring bubble universes of all hues and textures, distributed across a fantastic range of possibilities. What we had taken to be universal immutable laws of physics turn out to be more like 'local bylaws, valid only in our cosmic patch', to use Martin Rees' evocative description.

A key feature of the multiverse's cosmic smorgasbord is that only a tiny fraction of bubble universes will possess the right laws of physics to permit life and observers to arise. Many prerequisites needed for life, such as abundant carbon, stable stars and a universe neither too hot or chaotic, but cool and inhomogeneous enough to permit galaxies to form, depend very sensitively on the precise values of the parameters that characterise the laws and the initial conditions of the quantum universe-nucleation process. The 'Goldilocks enigma' – why our universe's laws and initial conditions are, amazingly, just right for life – has been a source of puzzlement for a long time. The multiverse theory could explain what otherwise looks suspiciously like a cosmic fix, in terms of an observer selection effect. It is no surprise that we find ourselves living in one of those very rare universes that have bio-friendly laws; we obviously could not inhabit a bio-hostile one.

With the multiverse theory – which, it has to be cautioned, remains extremely speculative and hard to test – the Copernican principle decisively fails. Although we are most likely living in a typical bio-friendly bubble universe, the overall number of life-permitting bubbles is an infinitesimal fraction of the whole multiverse. Earth's address within our bubble universe might well be typical, but our cosmic coordinates in the broader multiverse place us at a very exceptional location indeed.

15

AN INVESTIGATION

OF

THE LAWS OF THOUGHT,

ON WHICH ARE FOUNDED

IAN STEWART

Behind the Scenes:
The Hidden Mathematics
That Rules Our World

Ian Stewart FRS is Professor of Mathematics at the University of Warwick and a leading populariser of mathematics. He has published more than 80 books including *From Here to Infinity*, *Nature's Numbers* and *The Collapse of Chaos*. Recent popular science books include *Why Beauty is Truth*, *Letters to a Young Mathematician*, *The Magical Maze* and the series *The Science of Discworld* (with Terry Pratchett and Jack Cohen). His most recent book is *Professor Stewart's Cabinet of Mathematical Curiosities*. He has also written the science fiction novels *Wheelers* and *Heaven* (with Jack Cohen). Among his many other popular writings, he wrote the monthly 'Mathematical Recreations' column of *Scientific American* for ten years.

T HE STUFF SCIENCE ALLOWS US TO MAKE IS VISIBLE. BUT THE WAYS APPLIED INTELLIGENCE ALLOWS IT TO DO WHAT IT DOES REMAIN HIDDEN FROM MOST OF US – WHEN IT INVOLVES MATHEMATICS. SOMETIMES, AS IAN STEWART REVEALS, IT IS EVEN HIDDEN FROM THE PEOPLE WHO BUILD THE THINGS WHICH EMBODY THE MATHS.

HOW IMPORTANT IS MATHEMATICS IN TODAY'S WORLD?

The role of most sciences is relatively obvious, but mathematics is far less visible than engineering or biology. However, this lack of visibility does not imply that mathematics has no useful applications. On the contrary, mathematics underpins much of today's technology, and is vital to virtually all areas of human activity. To explore how this has happened, and explain why it has gone unnoticed, I'm going to look at two of the great historical figures in British mathematics – both Fellows of the Royal Society – and trace some of the practical consequences of their work. Along the way, we'll see why mathematics is so important, and why hardly anyone outside the subject seems to be aware of that.

My story begins with a strange event, which took place on 4 January 2004, on Mars. A Martian wandering around near Gusev Crater on that particular day would have undergone a life-changing experience. First, a streak of fire high in the sky would have heralded the arrival of an alien artefact, descending rapidly beneath a hemisphere of fabric. Then, as the artefact neared the ground, the fabric would have torn away, allowing it to fall the final hundred metres. And bounce. In fact, it bounced twenty-seven times before finally coming to rest. It would certainly have been a sight to remember.

The bouncy visitor was Mars Exploration Rover A, otherwise known as *Spirit*. After a journey of 487 million kilometres it entered the Martian atmosphere at a speed of 19,000 kilometres per hour. It was still travelling at a healthy 50 kilometres per hour a few seconds before impact when its airbags inflated and it made its touchdown. *Spirit* and its companion *Opportunity* have now spent more than four years exploring the surface of Mars, nearly twenty times as long as originally planned, leading to a wealth of new scientific information about Earth's sister planet. They may not have finished yet.

Much of the credit for this stunning success must go to NASA's engineers and managers, but other disciplines were also essential – among

Artwork of the Mars Exploration Rover *Spirit* landing on and exploring the surface of the planet.

them, mathematics. The spacecraft's trajectories were calculated using Newton's laws of motion and gravity; Einstein's later refinements were not needed. Isaac Newton was elected a Fellow of the Royal Society in 1672, twelve years after the Society was founded. His role in the development of space travel is not hard to identify, even though he died 240 years before the first Moon landing. Less obvious is the influence of a Fellow from the Victorian era, George Boole, whose pioneering ideas in logic and algebra proved fundamental to computer science. His influence can be detected in the error-correcting codes that made it possible for the Rovers (and most other space missions) to send images and scientific data back to Earth. Mathematics, both ancient and modern, is deeply embedded in today's science, and makes vital contributions on a daily basis to many aspects of human society.

The importance of mathematics in the space programme should be evident even to a casual observer. Yet when the Rovers landed, and the American mathematician Philip Davis pointed out that the mission 'would have been impossible without a tremendous underlay of mathematics' – so tremendous, in fact, that 'it would defy the most knowledgeable historian of mathematics to discover and describe all the mathematics that was involved' – he found it necessary to add that 'The public is hardly aware of this.'

This remark was an understatement. In 2007 two Danes with postgraduate mathematics degrees, Uffe Jankvist and Björn Toldbod, decided to uncover the hidden mathematics in the Mars Rover programme. They visited NASA's Jet Propulsion Laboratory at Pasadena, which ran the mission, and discovered that it is not only the general public that lacks awareness of the mathematics used in the Rover mission. Many of the scientists most intimately involved were also unaware of the mathematics being used. Some denied that there was any.

'We don't do any of that,' said one. 'We don't really use any abstract algebra, group theory, and that sort.'

Opposite:
Portrait of
George Boole.

'Except in the channel coding,' one of the Danish mathematicians pointed out.

'They use abstract algebra and group theory in that?'

'The Reed–Solomon codes are based on Galois fields.'

'That's news to me. I didn't know that.'

This story is fairly typical. Few people are aware of the mathematics that makes their world work. Indeed, few are aware that mathematics is involved in their world *at all*. But – as the history of the Royal Society exemplifies – mathematics has long been central to science, and science has long been a major driving force for social change.

What causes this lack of awareness of the importance of mathematics in the modern world? One of the main reasons, as the NASA story shows, is that you don't have to know any mathematics, or even be aware of its existence, to use the technology that it enables. This is entirely sensible – you don't need to understand computer programming to buy CDs over the Internet, and you don't need a degree in engineering to drive a car. However, most computer users are aware that someone had to write the software, and most drivers realise that someone had to design and build the car. With mathematics, it seems to be different.

Why? The story of the Mars Rovers is instructive. JPL scientists did not realise how deeply mathematics was involved in the Rover mission because the mathematical techniques were built into dedicated computer chips and programs. The resulting hardware and software carried out the necessary calculations without human intervention. Moreover, most of the chips and software were designed and manufactured by external subcontractors.

In actual fact, the Rover mission rested on a huge variety of mathematical techniques. These included dynamical systems and numerical analysis to calculate and control the spacecraft's trajectory on its way to Mars, signal processing methods to compress data and eliminate transmission errors caused by electrical interference, even the design and deployment of the

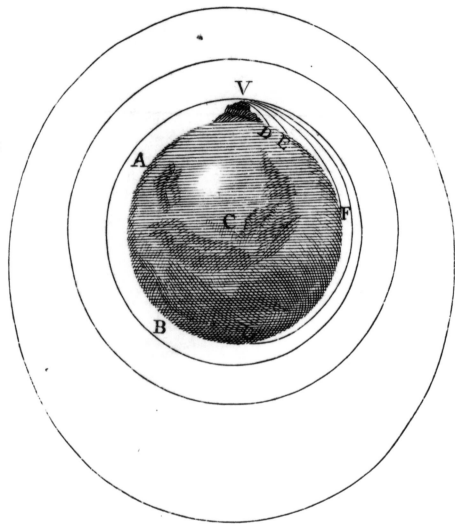

Engraving
published in
Newton's
*A Treatise of
the System of
the World,*
1728, showing
the effects of
gravity.

airbags. These techniques did not come into being overnight, and they were not, initially, developed with the space programme in mind. The work of Newton makes this very clear.

Newton's father was a Lincolnshire farmer, who died three months before his son was born. The boy did not impress some of his schoolteachers, who reported that he was idle and inattentive, but he did impress his head-master, who persuaded Isaac's mother to send him to university. At Cambridge he studied law, but he also read books on physics, philosophy and mathematics. In 1665 the university was closed because of plague, and he returned to Lincolnshire. There, in a few years, he made huge advances in several areas of mathematics and physics, which led to his election as a Fellow of Trinity College.

Newton is famous for many things – his laws of motion, calculus (also discovered by Gottfried Wilhelm Leibniz), the beginnings of numerical analysis. All of this work leaves fingerprints on the Mars Rover mission, but the most significant is the law of gravitation. Every body in the universe, Newton declared, attracts every other body with a force that is proportional to their masses and inversely proportional to the square of the distance between them. When coupled to his laws of motion, the law of gravitation provided accurate descriptions of the motion of the assorted planets and moons of the solar system, and much more. It explained the curious way in which the Moon wobbled on its axis, and the paths of comets. It made the future of the solar system predictable, millions of years ahead.

Newton's motivation was 'natural philosophy', the scientific study of Nature. If he had practical objectives in mind, they were related to things like navigation, and were secondary to understanding what he called 'the system of the world', which was the subtitle to his epic *Principia Mathematica* (*Mathematical Principles of Natural Philosophy*). At that time, the idea that humans might travel to the Moon was considered absurd, when anyone considered it at all. Yet such is the power of mathematics that when spacecraft began to leave the Earth in the 1960s, the tools needed to calculate their orbits and plan their re-entry trajectories through the atmosphere were those developed by Newton and his successors. In particular, since the law of gravitation applies to every particle of matter in the universe, it must apply to spacecraft.

Natural Philosophy has Borne Fruit as Technology

Once pointed out, it's no great surprise that esoteric mathematics can be used in esoteric applications like Martian space probes, even if no one notices … But what does that have to do with the everyday life of the ordinary citizen? Next time you listen to a CD while driving along the motorway in your car, and hit a bump, you may care to ask yourself why the CD player skips tracks only if it's a really *big* bump – big enough to risk damaging your wheel. After all, a CD player is an extremely delicate device, with a tiny laser that hovers a few millionths of a metre away from a plastic disc covered in tiny dots.

The answer goes back to George Boole and the other nineteenth-century mathematicians who founded modern abstract algebra. Boole also hailed from Lincolnshire, being born in Lincoln in 1815; his father was a cobbler who was also interested in making scientific instruments, and his mother was a lady's maid. He did not take a university degree, but his talent for mathematics attracted attention, and in 1849 he became Professor of

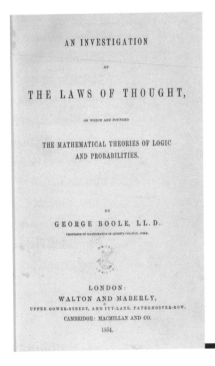

The frontispiece to George Boole's *An Investigation of the Laws of Thought*, 1854.

Mathematics at Queen's College, Cork. His most significant work was his 1854 book *An Investigation of the Laws of Thought*. In it, he reformulated logic in terms of algebra – but a very strange kind of algebra. Most of the familiar algebraic rules, such as $x+y = y+x$, are valid in Boole's logical realm, but there are some surprises, such as $1+1 = 0$. Here 1 means 'true', 0 means 'false', and $x+y$ means what computer scientists now call 'exclusive or': either x is true, or y is true, *but not both*. The first formula says that this statement does not depend on the order in which the two statements x and y are considered. The second says that if x and y are both true, then $x+y$ is false – because the definition of + includes the requirement 'not both'. More elaborate algebraic laws, such as $(x+y)z = xz+yz$, are also true in Boole's system; now the product xy means 'x and y'. So Boole's algebraic rules follow from sensible logical ones.

It is a striking and surprising discovery. Logic, previously thought of as being more basic than mathematics, can actually be *reduced* to mathematics. And the reduction is so natural that the algebra of logic is almost the same as traditional algebra. The new rules do make a difference, but you soon get used to that. Boole knew he was on to something important, but it took a while for most mathematicians to appreciate it. 'Boolean algebra' really took off when digital computers started to appear. Computers are basically logic engines, and Boole is widely recognised as a founder of theoretical computer science.

The link to digital computation is natural, but Boole's influence runs deeper. He was one of the first to realise that algebra need not be about numbers alone: it can be about any mathematical concepts or structures that can be manipulated symbolically according to a fixed system of rules. Boole was one of the earliest thinkers in a long tradition that includes the tragic figure of Évariste Galois, killed in a duel shortly before his twenty-first birthday. Today's abstract algebra, with its key concepts of groups, rings, fields and vector spaces, represents the fruits of their early labours.

These ideas, if I were to explain them in any detail, would seem abstract and impractical – formal games played with symbols, to no clear purpose.

They look like that because they operate on a structural level and focus on deep generalities. But behind the scenes, the abstract algebra that Boole pioneered has taken over most areas of mathematics, because it organises concepts and provokes new ideas. The resulting mathematics can be found, embodied in computer chips, inside most of today's electronic gadgets: CDs, DVDs, digital TVs, mobile phones, iPods, Nintendo Wiis, BlackBerries, SatNav, digital cameras …

Reed–Solomon codes are a typical example. These are the codes that NASA used to detect and correct potential errors in the Rovers' images of the Martian surface as they were beamed across the vastness of the solar system to planet Earth. More familiar devices, such as CD players, also would not work without Reed–Solomon codes. These codes hinge on, and were motivated by, the algebraic legacy of Boole and Galois. They transform the digital data that represents music in a way that makes it easy to spot, and put right, any errors that occur when the CD is being played. Virtually all of today's digital communications are wholly reliant on sophisticated and very modern mathematical coding methods. None of it would work without them. And that turns out to be just the tip of a very large iceberg.

A few weeks ago I looked through a randomly chosen issue of *New Scientist* magazine. Of the fifty or so stories reported, there were a dozen that – to my sensitive eye – involved a significant amount of mathematics. Not one story mentioned this, though a few hinted about 'models' of the process under study. When the contribution of mathematics is hidden that far behind the scenes, it is hardly surprising that the media and the public have little idea of what mathematics is, or what it is good for.

Sometimes mathematics should be kept behind the scenes. When I listen to music in my car, I don't want to have to think about the intricacies of Galois fields. When NASA engineers are firing a space probe's rockets to nudge it into the right entry trajectory to prevent it burning up in the Martian atmosphere, they don't want to be worrying about differential equations. But *someone* has to do the sums, write the program, design the algorithm, invent the concept, or prove the theorem. Someone has to provide the tools for the job and make sure they are reliable. If neither the media, nor the public, nor even practising scientists realise that this hidden mathematics exists, we will stop training mathematicians, and the necessary people will cease to exist too.

To most of us, 'mathematics' is something we did at school, and promptly forgot. Curiously, many of us also think that what we did at school was *the whole of mathematics*: all done and dusted. And pointless, now that we've got computers to do the sums for us. Some of us discover there is more to it than that. Some go on to university, take a science degree, and come to grips with statistics (in biology or medicine), differential equations (physics and engineering), or mathematical logic (computer science). And the mental picture that we get is that there's a certain amount of genuinely *useful* stuff (statistics, differential equations, mathematical logic …) plus a lot of highbrow intellectual fun and games that never has been and never will be useful to anyone living in the 'real world'.

Both of these views of mathematics are caricatures; real mathematics is quite different. Today's mathematics is intimately bound up with two key areas of human knowledge and activity: the natural world, and the society in which we live. Human understanding of our planet, and our universe, rests heavily on the shoulders of mathematics. So does the day-to-day working of our world. Take the hidden mathematics away, and today's world would fall to pieces. That statement applies to a lot of the apparently esoteric parts of the subject, as well as the more obviously applicable ones – partly because mathematics is an interconnected whole, but also because the esoteric concepts are often very general and very powerful. New and unexpected applications are common.

The 'classical' areas of mathematics are mainly those that led up to, or developed from, calculus – continuous mathematics, where everything can be subdivided into pieces that are as tiny as you wish. Most core mathematical physics and classical applied mathematics, such as acoustics or aerodynamics or elasticity theory, are of this kind. An important newcomer is discrete mathematics, which is suited to the digital age. Here the basic ingredients come in indivisible packets; essentially, anything whose natural description uses whole numbers or finite lists of symbols. Straddling both areas is the theory of probability, a mathematical description of uncertainty.

Geometry is also crucial. Despite appearances to the contrary, mathematics is primarily visual, and the formal symbolism tends to be closely related to some kind of mental image. Today's geometric thinking, however, takes a variety of forms, few of them resembling the traditional geometry of Euclid. Modern mathematics rightly places value on generality, when appropriate. That naturally leads to a degree of abstraction, because the focus of attention has to shift from 'what objects are we looking at?' to 'what properties are we assuming?' Logical proof remains central to the enterprise; it's how mathematicians keep themselves and their subject honest. Computers now play an increasingly central role. They seldom solve problems without further thought, but they can create a huge improvement in our understanding when they are used intelligently.

Mathematics, embodied in digital devices, has made technologies possible that seem to verge on magic. In February 2008 my wife and I spent two weeks exploring the private tombs of the Egyptian nobility, from Cairo down to Aswan. We took more than 1,400 photographs with two digital cameras; the whole lot were recorded on three 1-gigabyte memory cards, each the size of a postage stamp. The engineering feats involved are amazing, and they rest on all sorts of advances in materials science, photolithography, even quantum mechanics. Those advances required a lot of mathematics, as it happens, but I want to focus on just one aspect of digital cameras: data compression. The quantity of raw information required to specify 1,400 high-resolution colour pictures is far larger than those three cards can hold. Despite huge advances in miniaturisation, you simply cannot get that amount of data into such a small space.

A Secure Digital (SD) memory card, widely used in digital cameras for storing JPEGs.

Yet the pictures exist. I can print them out, or put them on the computer screen. How do the camera manufacturers cram so much information into so little memory? It may seem like magic, but the magic is mostly invisible mathematics. The clue lies in the names of the image files, which on my camera look something like P1000565.JPG. This tells the computer that the file is formatted using the JPEG standard, issued by the Joint Photographic Experts Group in 1992. This format uses various features of human vision, and typical images, to 'compress' the image data substantially.

In general terms, a computer represents a picture as a list of numbers. The list represents a rectangular array of tiny picture elements, called pixels, and the numbers describe the colour and the brightness of each pixel. If you do the sums, however, you find that there's nowhere near enough space in a memory card to hold all the pictures that undeniably are in there. It's not just like trying to get a quart into a pint pot: more like getting a tanker-load of milk into a pint pot.

This problem is a common one in the digital world, and it is usually tackled by compressing the data – reducing the quantity of information while retaining enough of it to do the job. Just as you can get more luggage into the car if you load it in the right way, so you can get more of the important data into a computer file if you leave out stuff that's not really relevant, or take advantage of certain inbuilt redundancies. For instance, many photographs have a large area of blue sky. Instead of repeating the code for 'blue' thousands of times, once for each pixel, we could tell the computer 'colour everything in this rectangle blue', and specify the rectangle by listing its corners. Suddenly thousands of numbers collapse to a few dozen. That's not how JPEG works, but it shows how redundancies in a list of numbers may make the list compressible. The actual procedure is carefully tailored to what can be done efficiently inside a small camera. The details don't really matter for my main point, but I want you to appreciate that there *are* details, which use several different mathematical ideas. So please indulge me while I tell you just how cunning the process is.

JPEG starts by splitting the data into three separate arrays. One lists how bright each pixel is. The other two take advantage of the fact that the colours perceived by the eye can be specified as points in a plane, the 'colour triangle'. A plane is two-dimensional, so each point can be defined using just two numbers, its horizontal and vertical coordinates. These 'colour components' form the other two lists. The human eye is more sensitive to variations in brightness than in colour, so the two lists of colour components can be shortened – usually they are reduced to one quarter of their original size – by using a coarser list of colours.

The next step uses a trick introduced by the French mathematician Joseph Fourier in 1824 – a year after his election to the Royal Society, as it happens – who at the time was working on the flow of heat. In general terms, Fourier's idea was to represent a pattern of numbers by combining specific patterns with different frequencies – much as the note played by a clarinet is made up from a fundamental 'pure' note and various higher-pitched 'harmonics', all added together in suitable proportions. JPEG uses a similar trick for spatial patterns of numbers, treating each of its three arrays in the same way. First, the array is broken up into 8x8 blocks of pixels. Then each block is transformed into a list of its spatial frequencies in the horizontal and vertical directions. Roughly, this splits the pattern into black-and-white stripes of various thicknesses, and works out how much of each stripe you need to reconstruct the actual image. This step employs a fast Fourier transform, exploiting number-theoretic features of binary numerals to speed up a difficult computation; this is why 8x8 blocks are used, eight being a power of two. The Fourier transform does not compress the data, but rewrites it in a compressible form. The eye is fairly insensitive to high-frequency stripes, so these can be ignored. Medium-frequency stripes can be specified using smaller numbers, which occupy less space on the memory card.

This is not the end: two more tricks are used to squash even more pictures into the same space. If you run through the resulting array of numbers in a zigzag order, from low frequency components to high ones, you typically find runs of repeated numbers, such as 7 7 7 7 7 7 7 7 7. Coding this as '9 consecutive 7's' converts it to 9 7, which is shorter. Finally, another coding method called Huffman coding is used on the resulting file, which compresses it even further.

So JPEG coding is quite complex, with sophisticated mathematical features. You don't need to know how it works to use your digital camera, but without the underlying ideas, that camera could never have been made. Now think of future developments, video cameras, cramming a camera into a mobile phone along with dozens of other applications … We desperately need people who can understand that sort of mathematics.

At any rate, my wife and I were able to take lots of pictures without carrying sacks full of film because a lot of mathematically sophisticated engineers noticed that something that a nineteenth-century Frenchman invented for a completely different reason happened to have an unexpected use. But the hidden mathematics behind our holiday didn't stop there. Without a lot of other mathematics, often with similarly impractical or outmoded origins, we could never have got to Egypt to take the pictures.

Our flight was booked over the Internet and all Internet communications rely on error-correcting codes to ensure that messages are not garbled along the way by electrical interference. Like the codes used by the Mars Rovers, these techniques rely heavily on abstract algebra. The airline's schedules were designed using mathematical methods to improve efficiency – graph theory and linear algebra. Then there was radar, weather-forecasting, even the statistical analysis of different breeds of vegetables that governed the crops from which the airline food was made.

None of this is much use if the aircraft never gets to its intended destination. In the early days of navigation, when the great European explorers were mapping the globe in small wooden sailing ships, navigation was a major consumer of mathematics. Even finding the size and shape of the Earth involved mathematical calculations, as well as experimental observations.

Today we have GPS, the Global Positioning System, which comprises about fifteen satellites orbiting the Earth, sending out signals. A triumph of electronics and engineering, obviously. But mathematics?

Leaving aside the heavy use of mathematics in designing and building launch vehicles and satellites, and in calculating orbital dynamics, let me focus solely on the signalling system that GPS uses. Each satellite transmits a signal, which can be used to work out how far away the satellite is from the GPS receiver (on board the aeroplane, ship, car, yacht, or inside someone's mobile phone). These distances, coupled with knowledge of the positions of the satellites, make it possible to calculate the location of the receiver on the surface of the Earth. That's another highly mathematical step, which I will also ignore.

How do the signals convey distances?

Imagine that the satellite is playing a tune, and that you have access to a second 'copy' of that tune, being sent out from a known source that is in synchrony with the satellite. Because the satellite is further away than the reference source, the signal from the satellite is slightly delayed, by a time equal to the difference in distances divided by the speed of light. The time delay can be measured, very accurately, and the distance is obtained by multiplying that by the speed of light.

Instead of tunes, the signals are sequences of pseudo-random numbers – apparently patternless sequences generated by a fixed mathematical recipe. Both the satellite and the reference source know this recipe, so they can generate and recognise the same signals. So here we find a very practical application of the mathematics of pseudo-random numbers. If you use SatNav in your car, you are a major consumer of the hidden mathematics that runs our world.

Still pursuing the hidden mathematics that made my holiday possible, there is the small matter of designing an aircraft that stays up, one of the heaviest uses of mathematics in the whole enterprise. Nearly all of the analysis of airflow past an aircraft nowadays is done using 'numerical wind-

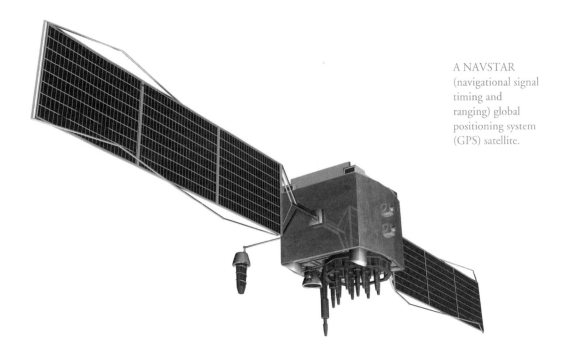

tunnels', which are mathematical simulations. They are much easier to use than physical wind-tunnels, and if anything, more accurate. They have innumerable other applications. They are essential to the design of Formula 1 and NASCAR racing cars, where effective aerodynamics is needed to keep the car on the track and reduce air resistance. If that's not green enough for you, the same techniques improve the fuel efficiency of ordinary road vehicles. Even the dynamics of a football has been analysed mathematically, with useful practical implications about how to make the ball behave unpredictably, which can help it get past the keeper into the goal. Computational Fluid Dynamics also has medical applications to blood flow and heart disease.

This makes the point that mathematics also saves lives. Have you had a medical scan recently? How do you think the scanner works out what's inside you? There's a whole branch of mathematics devoted to such questions. Are you concerned about crime? The FBI uses 'wavelets', a very recent piece of mathematics, to analyse and record fingerprint information to help catch criminals. Other police forces use similar techniques. Do you use oil or natural gas, for heating, cooking, or transport? The oil companies use powerful mathematical techniques to find out what the rocks miles underground look like, based on the echoes from explosions at the surface. Do you use anything with a spring in it – ballpoint pen, video recorder, mattress? The spring-making industry uses mathematics for quality control.

Another huge area that relies on mathematics is science, and science is our most successful method for understanding the natural world. The development of science, and that of mathematics, have gone hand in hand for about five hundred years. Newton invented calculus to understand the movements of the planets. Independently, Gottfried Leibniz developed much the same ideas for purely intellectual reasons. These two sources of mathematical inspiration can be roughly characterised as 'applied' and 'pure' mathematics. The main differences are motivation and attitude, rather than content. The same mathematical concept may appear in the solution of Fermat's last theorem (pure mathematics) or in the construction of a secure code for Internet banking (applied mathematics). Some areas are traditionally considered as being 'pure', others as 'applied', but these are convenient distinctions, not impassable barriers. Today's science is increasingly multi-disciplinary; so is mathematics.

Initially, the main beneficiaries of mathematical techniques were the physical sciences, and these are still the areas in which the use of mathematics is greatest. But the biological and medical sciences are catching up rapidly, and some of the most interesting new problems for research mathematicians are coming out of biology. A century or two from now we will look back at today's Newtons and Booles, and understand how vital their work has been to the development of our society. Provided we do not lose sight of the hidden mathematics that rules our world – because if we do, those advances will never happen.

16

10^{15}

10^{12}

Computational power (bits/second)

10^9

10^6

10^3

Telephone system

Bee

'Smart' missile

Personal computer

Television channel

Radio channel

Bacterialam production

Abacus

Hand calculator

JOHN D. BARROW

SIMPLE REALLY: FROM SIMPLICITY TO COMPLEXITY – AND BACK AGAIN

John D. Barrow FRS is a cosmologist, Professor of Mathematical Sciences, Director of the Millennium Mathematics Project, University of Cambridge, and Gresham Professor of Geometry at Gresham College, London. His many books include *The Anthropic Cosmological Principle, The World Within the World, Pi in the Sky, Theories of Everything, The Origin of the Universe, The Left Hand of Creation, The Artful Universe, Impossibility: The Limits of Science, The Science of Limits, Between Inner Space, Outer Space, The Constants of Nature: From Alpha to Omega* and *Cosmic Imagery: Key Images in the History of Science.* His latest is *100 Essential Things You Didn't Know You Didn't Know.*

MAKING SENSE OF THE WORLD SCIENTIFICALLY HAS OFTEN MEANT SEARCHING FOR SIMPLICITY UNDERLYING THE APPARENTLY COMPLEX. FINE, SAYS JOHN BARROW, EXCEPT WHEN THE COMPLEXITY TURNS OUT TO BE IRREDUCIBLE. OR DOES IT?

Symmetry calms me down, lack of symmetry makes me crazy.

— Yves Saint Laurent

WHAT IS THE WORLD LIKE?

Is the world simple or complicated? As with many things, it depends on who you ask, when you ask, and how seriously they take you. If you should ask a particle physicist you would soon be hearing how wonderfully simple the universe appears to be. But, on returning to contemplate the everyday world, you just know 'it ain't necessarily so': it's far from simple. For the psychologist, the economist, or the botanist, the world is a higgledy-piggledy mess of complex events that just seemed to win out over other alternatives in the long run. It has no mysterious penchant for symmetry or simplicity.

So who is right? Is the world really simple, as the particle physicists claim, or is it as complex as almost everyone else seems to think? Understanding the question, why you got two different answers, and what the difference is telling us about the world, is a key part of the story of science over the past 350 years from the inception of the Royal Society to the present day.

THE QUEST FOR SIMPLICITY

Our belief in the simplicity of Nature springs from the observation that there are regularities which we call 'laws' of Nature. The idea of laws of Nature has a long history rooted in monotheistic religious thinking, and in ancient practices of statute law and social government.[1] The most significant advance in our understanding of their nature and consequences followed Isaac Newton's identification of a law of gravitation in the late seventeenth century, and his creation of a battery of mathematical tools with which to unpick its consequences. Newton made his own tools: with them we have made our tools ever since. His work inspired the early Fellows of the Royal Society, and scientists all over Europe, who followed the advances reported at its meetings and in its published *Transactions* closely during the years of his long Presidency from 1703 to his death in 1727, to bring about a Newtonian revolution in the study of the mathematical description of motion, gravity and light. It gave rise to a style of mathematics applied to science that remains distinctively Newtonian.

Laws reflect the existence of patterns in Nature. We might even define science as the search for those patterns. We observe and document the world in all possible ways; but while this data-gathering is necessary for science, it is not sufficient. We are not content simply to acquire a record of everything that is, or has ever happened, like cosmic stamp collectors. Instead, we look for patterns in the facts, and some of these patterns we have come to call the laws of Nature, while others have achieved only the status of by-laws. Having found, or guessed (for there are no rules at all about how you might find them) possible patterns, we use them to predict what should happen if

1 This civil and theological background can be traced in the study in J.D. Barrow, *The World Within the World* (Oxford, OUP, 1988), of the development of the concept of laws of Nature in ancient societies.

the pattern is also followed at all times and in places where we have yet to look. Then we check if we are right (there are strict rules about how you do this!). In this way, we can update our candidate pattern and improve the likelihood that it explains what we see. Sometimes a likelihood gets so low that we say the proposal is 'falsified', or so high that it is 'confirmed' or 'verified', although strictly speaking this is always provisional, none is ever possible with complete certainty. This is called the 'scientific method'.[2]

For Newton and his contemporaries, the laws of motion were codifications into simple mathematical form of the habits and recurrences of Nature. They were idealistic: 'bodies acted upon by no forces will ...' because there are *no* such bodies. They were laws of cause and effect: they told you what happened if a force was applied. The future is uniquely and completely determined by the present.

Later, these laws of change were found to be equivalent to statements that quantities did not change. The requirement that the laws were the same everywhere in the universe was equivalent to the conservation of momentum; the requirement that they be found to be the same at all times was equivalent to the conservation of energy; and the requirement that they be found the same in every direction in the universe was equivalent to the conservation of angular momentum. This way of looking at the world in terms of conserved quantities, or invariances and unchanging patterns, would prove to be extremely fruitful.

During the twentieth century, physicists became so enamoured of the seamless correspondence between laws dictating changes and invariances preserving abstract patterns when particular forces of Nature acted, that their methodology changed. Instead of identifying habitual patterns of cause and effect, codifying them into mathematical laws, and then showing them to be equivalent to the preservation of a particular symmetry in Nature, physicists did a U-turn. The presence of symmetry became such a

2 In practice, the process of improving central theories of physics usually involves a process of replacing a theory by a deeper and broader version that contains the original as a special, or limiting, case. Thus, Newton's theory of gravity has been superseded by Einstein's theory of general relativity but not replaced by it in some type of scientific 'revolution'. Einstein's theory becomes the same as Newton's when we confine attention to weak gravitational forces and to motions at speeds much less than that of light. Similarly, another limiting process recovers Newtonian mechanics from quantum mechanics. This is why, regardless of the results of our search for the 'ultimate' theory of gravity, structural engineers and sports scientists will still be using Newton's laws in a thousand years' time.

persuasive and powerful facet of laws of physics that physicists began with the mathematical catalogue of possible symmetries. They could pick out symmetries with the right scope to describe the behaviour of a particular force of Nature. Then, having identified the preserved pattern, they could deduce the laws of change that are permitted and test them by experiment.

Since 1973, this focus upon symmetry has taken centre stage in the study of elementary-particle physics and the laws governing the fundamental interactions of Nature. Symmetry is the primary guide into the legislative structure of the elementary-particle world, and its laws are derived from the requirement that particular symmetries, often of a highly abstract character, are preserved when things change. Such theories are called 'gauge theories'. All the currently successful theories of four known forces of Nature – the electromagnetic, weak, strong and gravitational forces – are gauge theories. These theories prescribe as well as describe: preserving the invariances upon which they are based requires the existence of the forces they govern. They are also able to dictate the character of the elementary particles of matter that they govern. In these respects, gauge theories differ from the classical laws of Newton, which, since they governed the motions of all bodies, could say nothing about the properties of those bodies. The reason for this added power of explanation is that the elementary-particle world, in contrast to the macroscopic world, is populated by collections of identical particles ('once you've seen one electron, you've seen 'em all,' as Richard Feynman remarked). Particular gauge theories govern the behaviour of particular subsets of all the elementary particles, according to their shared attributes. Each theory is based upon the preservation of a pattern.

This generation of preserved patterns for each of the separate interactions of Nature has motivated the search for a unification of those theories into more comprehensive editions based upon larger symmetries. Within those larger patterns, smaller patterns respected by the individual forces of Nature might be accommodated, like jigsaw pieces, in an interlocking fashion that places some new constraint upon their allowed forms. So far, this

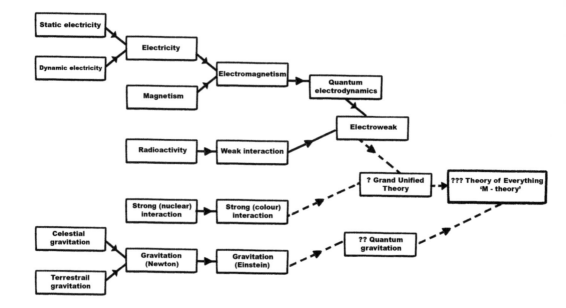

strategy has resulted in a successful, experimentally tested, unification of the electromagnetic and weak interactions, and a number of purely theoretical proposals for a further unification with the strong interaction ('grand unification'), and candidates for a four-fold unification with the gravitational force to produce[3] a so-called 'theory of everything', or 'TOE'. It is this general pattern of explanation by which forces and their underlying patterns are linked and reduced in number by unifications, culminating in a single unified law, that lies at the heart of the physicist's perception of the world as 'simple'. The success of this promising path of progress is the reason that led our hypothetical particle physicist to tell us that the world is simple. The laws of Nature are few in number and getting fewer.

The first candidate for a TOE was a 'superstring' theory, first developed by Michael Green and John Schwarz in 1984. After the initial excitement that followed their proof that string theories are finite and well-defined theories of fundamental physics, hundreds of young mathematicians and physicists flocked to join this research area at the world's leading physics departments. It soon became clear that there were five varieties of string theory available to consider as a TOE: all finite and logically self-consistent, but all different. This was a little disconcerting. You wait nearly a century for a theory of everything then, suddenly, five come along all at once. They had exotic-sounding names that described aspects of the mathematical patterns they contained – type I, type IIA and type IIB superstring theories, SO(32) and E8 heterotic string theories, and eleven-dimensional super-gravity. These theories are all unusual in that they have ten dimensions of space and time, with the exception of the last one, which has eleven. Although it is not demanded for the finiteness of the theory, it is generally assumed that only one of these ten or eleven dimensions is a 'time' and the others are spatial. Of course, we do not live in a nine- or ten-dimensional space so in order to reconcile such a world with what we see it must be assumed that only three of the dimensions of space in these theories became large and the others remain 'trapped' with (so far) unobservably small sizes. It is remarkable that in order to achieve a finite theory we seem to need more dimensions of space than those that we experience. This might be regarded as a prediction of the theory. It is a consequence of the amount of 'room' that is needed to accommodate the patterns governing the four known forces of Nature inside a single bigger pattern without hiving themselves off into sub-patterns that each 'talk' only to themselves rather than to everything else. Nobody knows why three dimensions (rather than one or four or eight, say) became large, or what is the force responsible. Nor do we know if the number of large dimensions is some-thing that arises at random and so could be different – and may be differ-ent – elsewhere in the universe, or is an inevitable consequence of the laws of physics that could not be otherwise without destroying the logical self-consistency of physical reality.

One thing that we do know is that only in spaces with three large dimensions can things bind together to form structures like atoms, mole-cules, planets and stars. No complexity and no life is possible except in spaces with three large dimensions. So, even if the number of large dimensions

---

3 Four fundamental forces are known, of which the weakest is gravitation. There might exist other, far weaker, forces of Nature. Although too weak for us to measure (perhaps ever), their existence may be necessary to fix the logical necessity of that single theory of everything. Without any means to check on their existence, we would always be missing a crucial piece of the cosmic jigsaw puzzle; see J.D. Barrow, *New Theories of Everything: The quest for ultimate explanation* (Oxford, OUP, 2007) and B. Greene, *The Elegant Universe* (London, Jonathan Cape, 1999).

is different in different parts of the universe, or separate universes are possible with different numbers of large dimensions, we would have to find ourselves living where there are *three* large dimensions, no matter how improbable that might be, because life could exist in no other type of space.

At first, it was hoped that one of these theories would turn out to be special and attention would then narrow down to reveal it to be the true theory of everything. Unfortunately, things were not so simple. Progress was slow and unremarkable until Edward Witten, at Princeton, discovered that these different string theories are not really different. They are linked to one another by mathematical transformations that amount to exchanging large distances for small ones, and vice versa in a particular way. Nor were these string theories fundamental. Instead, they were each limiting situations of another deeper, as yet unfound, TOE which lives in eleven dimensions of space and time. That theory became known as 'M-theory', where M has been said to be an abbreviation for Mystery, Matrix, or Millennium, just as you like.[4]

Do these 'extra' dimensions of space really exist? This is a key question for all these new theories of everything. In most versions, the other dimensions are so small (10^{-33} cm) that no direct experiment will ever see them. But, in some variants, they can be much bigger. The interesting feature is that only the force of gravity will 'feel' these extra dimensions and be modified by their presence. In these cases the extra dimensions could be up to one hundredth of a millimetre in extent and they would alter the form of the law of gravity over these and smaller distances. This gives experimental physicists a wonderful challenge: test the law of gravity on submillimetre scales. More sobering still is the fact that all the observed constants of Nature, in our three dimensions, are not truly fundamental, and need not be constant in time or space:[5] they are just shadows of the true constants that live in the full complement of dimensions. Sometimes simplicity can be complex too.

---

4 These mathematical discoveries launched an intensive search for the underlying M theory. But so far it has not been found. Other possibilities have emerged along the way, with the arguments of Lisa Randall and Raman Sundrum that the three-dimensional space that we inhabit may be thought of as the surface of a higher-dimensional space in which the strong, weak, and electromagnetic forces act only in that three-dimensional surface while the force of gravity reaches out into all the other dimensions as well. This is why it is so much weaker than the other three forces of Nature in this picture; see L. Randall, *Warped Passages: Unravelling the Universe's Hidden Dimensions* (London, Penguin, 2006).
5 For a discussion of the status of the constants of Nature and evidence for their possible time variation , see J.D. Barrow, *The Constants of Nature* (London, Cape, 2002).

James Clerk
Maxwell.

ELEMENTARY PARTICLES?

The fact that Nature displays populations of identical elementary particles is its most remarkable property. It is the 'fine tuning' that surpasses all others. In the nineteenth century another of the Royal Society's greatest Fellows, James Clerk Maxwell, first stressed that the physical world was composed of identical atoms which were not subject to evolution. Today, we look for some deeper explanation of the sub-atomic particles of Nature from our TOE. One of the most perplexing discoveries by experimentalists has been that such 'elementary' particles appear to be extremely numerous. They were supposed to be an exclusive club, but they have ended up with an embarrassingly large clientele.

String theories offered another route to solving this problem. Instead of a TOE containing a population of elementary point-like particles, string theories introduce basic entities that are loops (or lines) of energy which have a tension. As the temperature rises the tension falls and the loops vibrate in an increasingly stringy fashion, but as the temperature falls the tension increases and the loops contract to become more and more point-

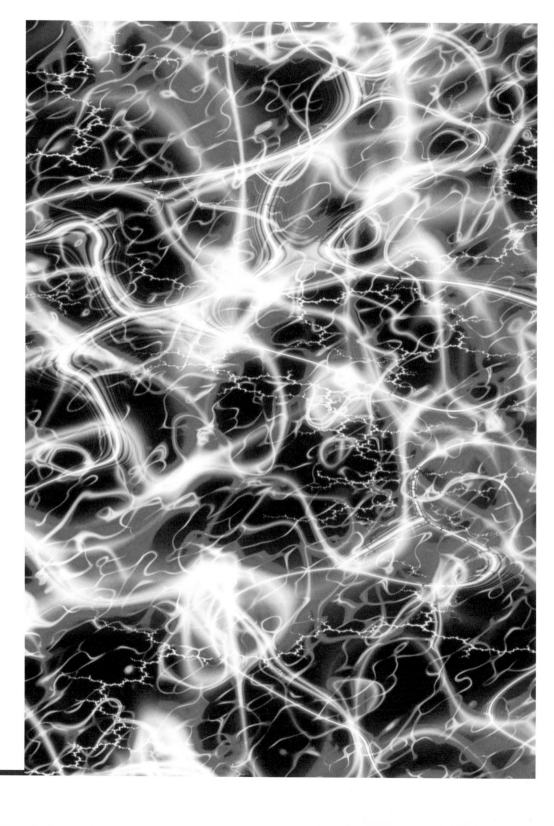

like. So, at low energies the strings behave like points and allow the theory to make the successful predictions about what we should see there as the intrinsically point-like theories do. However, at high energies, things are different. The hope is that it will be possible to determine the principal energies of vibration of the superstrings. All strings, even guitar strings, have a collection of special vibrational energies that they naturally take up when disturbed. If we could calculate these special energies for super-strings, then they would (by virtue of Einstein's famous mass-energy equivalence – $E = mc^2$) correspond to the masses of the 'particles' that we call elementary. So far, these energies have proved too hard to calculate. However, one of them has been found: it corresponds to a particle with zero mass and two units of a quantum attribute called 'spin'. This spin value ensures that it mediates attractions between all masses. It is the particle we call the 'graviton' and it is responsible for mediating the force of gravity. Its appearance shows that string theory necessarily includes gravity and, moreover, its behaviour is described by the equations of general relativity at low energies – a remarkable and compelling feature since earlier candidates for a TOE all failed miserably to include gravity in the unification story at all.

WHY IS THE WORLD MATHEMATICAL?

This reflection on the symmetries behind the laws of Nature also tells us why mathematics is so useful in practice. Mathematics is simply the catalogue of all possible patterns. Some of those patterns are especially attractive and are studied or used for decoration, others are patterns in time or in chains of logic. Some are described solely in abstract terms, while others can be drawn on paper or carved in stone. Viewed in this way, it is inevitable that the world is described by mathematics. We could not exist in a universe in which there was neither pattern nor order. The description of that order, and all the other sorts that we can imagine, is what we call mathematics. Yet, although the fact that mathematics describes the world is not a mystery, the exceptional utility of mathematics is. It could have been that the patterns behind the world were of such complexity that no simple algorithms could approximate them. Such a universe would 'be' mathematical, but we would not find mathematics terribly useful. We could prove 'existence' theorems about what structures exist but we would be unable to predict the future using mathematics in the way that NASA's mission control does.

Seen in this light, we recognise that the great mystery about mathematics and the world is that such *simple* mathematics is so far reaching. Very simple patterns, described by mathematics that is easily within our grasp, allow us to explain and understand a huge part of the universe and the happenings within it.

THE COPERNICAN PRINCIPLE APPLIED TO LAWS

It is often said with hindsight that Nicholas Copernicus taught us not to assume that our position in the universe is special in *every* way. Of course, this does not mean that it cannot be special in *any* way, simply because life is only possible in certain places.[6] Once we start distinguishing between the laws of Nature and their outcomes we should also bring this Copernican view to bear upon the laws of Nature as well as their outcomes.

Universal laws of Nature should be just that – universal – they should not just exist in special forms for some privileged observers at special locations, or who are moving in particular ways, in the universe. Alas, Newton's laws do not have this democratic property. They only have simple forms for privileged observers who are moving in a special way, neither rotating nor accelerating with respect to the distant 'fixed' stars. So there were privileged observers in Newton's universe for whom all the laws of motion look simple.

Newton's first law of motion demands that bodies acted upon by no forces do not accelerate: they remain at rest or move with constant speed. However, this law of motion will only be observed by a special class of observers who are neither accelerating nor rotating relative to the fixed stars. The appearance of these special observers for whom all the laws of motion look simpler violates the Copernican principle.

Imagine that you are located inside a spaceship through whose windows you can see the far distant stars. Put the spaceship in a spin. Through the windows you will see the distant stars accelerating past in the opposite sense to the spin, even though they are not acted upon by any forces. Newton's first law is not true for a spinning observer – a much more complicated law holds. This undemocratic situation signalled that there was something incomplete and unsatisfactory about Newton's formulation of the laws of motion. One of Einstein's great achievements was to create a new theory of gravity in which all observers, no matter how they move, *do* find the laws of gravity and motion to take the same form.[7] By incorporating this principle of 'general covariance', Einstein's theory of general relativity completed the extension of the Copernican principle from outcomes to laws.

OUTCOMES ARE DIFFERENT

The simplicity and economy of the laws and symmetries that govern Nature's fundamental forces are not the end of the story. When we look around us we do not observe the laws of Nature; rather, we see the *outcomes* of those laws. The distinction is crucial. Outcomes are much more complicated than the laws that govern them because they do not have to respect the symmetries displayed by the laws. By this subtle interplay, it is possible to have a world which displays an unlimited number of complicated asymmetrical structures yet is governed by a few, very simple, symmetrical laws. This is one of the secrets of the universe.

Suppose we balance a ball at the apex of a cone. If we were to release the ball, then the law of gravitation will determine its subsequent motion. Gravity has no preference for any particular direction in the universe; it is entirely democratic in that respect. Yet, when we release the ball, it will

6 This is one of the lessons learned from the anthropic principles.
7 Einstein used the elegant fact that tensor equations maintain the same form under any transformation of the coordinates used to express them. This is called the principle of general covariance.

always fall in some particular direction, either because it was given a little push in one direction, or as a result of quantum fluctuations which do not permit an unstable equilibrium state to persist. So here, in the outcome of the falling ball, the directional symmetry of the law of gravity is broken. This teaches us why science is often so difficult. As observers, we see only the broken symmetries manifested as the outcomes of the laws of Nature; from them, we must work backwards to unmask the hidden symmetries behind the appearances.

We can now understand the answers that we obtained from the different scientists we originally polled about the simplicity of the world. The particle physicist works closest to the laws of Nature themselves, and so is especially impressed by their unity, simplicity and symmetry. But the biologist, the economist, or the meteorologist is occupied with the study of the complex outcomes of the laws, rather than with the laws themselves. As a result, it is the complexities of Nature, rather than her laws, that impress them most.

AMBIGUITIES BETWEEN LAWS AND OUTCOMES

One of the most important developments in fundamental physics and cosmology over the past twenty years has been the steady dissolution of the divide between laws and outcomes. When the early quest for a theory of everything began many thought that such a theory would uniquely and completely specify all the constants of physics and the structural features of the universe. There would be no room left for wondering about 'other' universes, or hypothetical changes to the structure of our observed universe. Remarkably, things did not turn out like that. Candidate theories of everything revealed that many of the features of physics and the universe which we had become accustomed to think of as programmed into the universe from the start in some unalterable way, were nothing of the sort. The number of forces of Nature, their laws of interaction, the populations of elementary particles, the values of the so-called constants of Nature, the number of dimensions of space, and even whole universes, can all arise in quasi-random fashion in these theories. They are elaborate outcomes of processes that can have many different physically self-consistent results. There are fewer unalterable laws than we might think.

This means that we have to take seriously the possibility that some features of the universe which we call fundamental may not have explanations in the sense that had always been expected. A good example is the value of the infamous cosmological constant which appears to drive the acceleration of the universe today. Its numerical value is very strange. It cannot so far be explained by known theories of physics. Some physicists hope that there will ultimately be a single theory of everything which will predict the exact numerical value of the cosmological constant that the astronomers need to explain their observations. Others recognise that there may not be any explanation of that sort to be found. If the value of the cosmological constant is a random outcome of some exotic symmetry-breaking process near the beginning of the universe's expansion then all we can say is that it falls within the range of values that permit life to evolve and persist. This is a depressing situation to those who hoped to explain its value. However, it would be a strange (non-Copernican) universe that allowed us to determine everything that we want about it. We may just have to get used to the fact that there are some things we can predict and others that we can only measure. Here is a little piece of science faction to illustrate the point.

Imagine someone in 1600 trying to convince Johannes Kepler that a theory of the solar system won't be able to predict the number of planets in the solar system. Kepler would have had none of it. He would have been outraged. This would have constituted an admission of complete failure. He believed that the beautiful Platonic symmetries of mathematics required the solar system to have a particular number of planets. For Kepler this would have been the key feature of such a theory. He would have rejected the idea that the number of planets had no part to play in the ultimate theory.

Today, no planetary astronomer would expect any theory of the origin

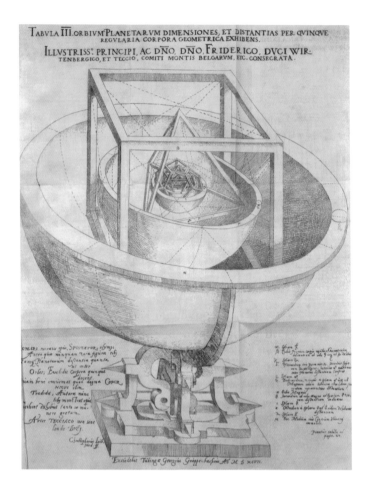

Kepler's spheres. From the *Prodrumus dissertationum cosmo graphicarum* (Prologue to the dissertation on a description of the universe) by Johannes Kepler, 1596.

of the solar system to predict the number of planets. It would make no sense. This number is something that falls out at random as a result of a chaotic sequence of formation events and subsequent mergers between embryonic planetesimals. It is simply not a predictable outcome. We concentrate instead on predicting other features of the solar system so as to test the theory of its origin. Perhaps those who are resolutely opposed to the idea that quantities like the cosmological constant might be randomly determined, and hence unpredictable by the theory of everything, might consider how strange Kepler's views about the importance of the number of planets now seem.

DISORGANISED COMPLEXITIES

Complexity, like crime, comes in organised and disorganised forms. The disorganised form goes by the name of *chaos* and has proven to be ubiquitous in Nature. The standard folklore about chaotic systems is that they are unpredictable. They lead to out-of-control dinosaur parks and frustrated meteorologists. However, it is important to appreciate the nature of chaotic systems more fully than the Hollywood headlines.

Classical (that is, non-quantum mechanical) chaotic systems are not in any sense intrinsically random or unpredictable. They merely possess extreme sensitivity to ignorance. As Maxwell was again the first to recognise in 1873, any initial uncertainty in our knowledge of a chaotic system's state is rapidly amplified in time. This feature might make you think it hopeless even to try to use mathematics to describe a chaotic situation. We are never going to get the mathematical equations for weather prediction 100 per cent correct – there is too much going on – so we will always end up being inaccurate to some extent in our predictions. But although that type of inaccuracy can contribute to unpredictability, it is not in itself a fatal blow to predicting the future adequately. After all, small errors in the weather equations could turn out to have an increasingly insignificant effect on the forecast as time goes on. In practice, it is our inability to determine the weather everywhere at any given time with perfect accuracy that is the major problem. Our inevitable uncertainties about what is going on in between weather stations leaves scope for slightly different interpolations of the temperature and the wind motions in between their locations. Chaos means that those slight differences can produce very different forecasts about tomorrow's weather.

An important feature of chaotic systems is that, although they become unpredictable when you try to determine the future from a particular uncertain starting value, there may be a particular stable statistical spread of outcomes after a long time, regardless of how you started out. The most important thing to appreciate about these stable statistical distributions of events is that they often have very stable and predictable average behaviours.

As a simple example, take a gas of moving molecules (their average speed of motion determines what we call the gas 'temperature') and think of the individual molecules as little balls. The motion of any single molecule is chaotic because each time it bounces off another molecule any uncertainty in its direction is amplified exponentially. This is something you can check for yourself by observing the collisions of marbles or snooker balls. In fact, the amplification in the angle of recoil, θ, in the successive (the $n+1$st and nth) collisions of two identical balls is well described by a rule:

$$\theta_{n+1} = \left(\frac{d}{r}\right) \theta_n$$

where d is the average distance between collisions and r is the radius of the balls. Even the minimal initial uncertainty in θ_0 allowed by Heisenberg's uncertainty principle is increased to exceed $\theta = 360$ degrees after only about 14 collisions. So you can then predict nothing about its trajectory.

The motions of gas molecules behave like a huge number of snooker balls bouncing off each other and the denser walls of their container. One knows from bitter experience that snooker exhibits sensitive dependence on initial conditions: a slight miscue of the cue-ball produces a big miss! Unlike the snooker balls, the molecules won't slow down and stop. Their typical distance between collisions is about 200 times their radius. With this value of d/r the unpredictability grows 200-fold at each close molecular encounter. All the molecular motions are individually chaotic, just like the snooker balls, but we still have simple rules like Boyle's Law governing the pressure P, volume V, and temperature T – the averaged properties[8] – of a confined gas of molecules:

$$\frac{PV}{T} = constant$$

The lesson of this simple example is that chaotic systems can have stable, predictable, long-term, average behaviours. However, it can be difficult to predict when they will. The mathematical conditions that are sufficient to ensure it are often very difficult to prove. You usually just have to explore numerically to discover whether the computation of time averages converges in a nice way or not.[9]

Considerable impetus was imparted to the study and understanding of this type of chaotic unpredictability and its influence on natural phenomena by theoretical biologists like Robert May (later to become the fifty-eighth President of the Royal Society in 2000) and George Oster, together with the mathematician James Yorke. They identified simple features displayed by wide classes of difference equation relating the $(n+1)$st to the nth state of a system as it made the transition from order to chaos.[10]

ORGANISED COMPLEXITIES

Among complex outcomes of the laws of Nature, the most interesting are those that display forms of *organised complexity*. A selection of these are displayed in the diagram on the next page, in terms of their size, gauged by their information storage capacity, which is just how many binary digits are needed to specify them versus their ability to process information, which is simply how quickly they can change one list of numbers into another list.

As we proceed up the diagonal, increasing information storage capability grows hand in hand with the ability to transform that information into new forms. Organised complexity grows. Structures are typified by the presence

---

8 The velocities of the molecules will also tend to attain a particular probability distribution of values, depending only on the temperature, called the Maxwell–Boltzmann distribution after many collisions, regardless of their initial values.

9 This is clearly very important for computing the behaviour of chaotic systems. Many systems possess a *shadowing* property that ensures that computer calculations of long-term averages can be very accurate, even in the presence of rounding errors and other small inaccuracies introduced by the computer's ability to store only a finite number of decimal places. These 'round-off' errors move the solution being calculated on to another nearby solution trajectory. Many chaotic systems have the property that these nearby behaviours end up visiting all the same places as the original solution and it doesn't make any difference in the long-run that you have been shifted from one to the other. For example, when considering molecules moving inside a container, you would set about calculating the pressure exerted on the walls by considering a molecule travelling from one side to the other and rebounding off a wall. In practice, a particular molecule might never make it across the container to hit the wall because it runs into other molecules. However, it gets replaced by another molecule that is behaving in the same way as it would have done had it continued on its way unperturbed.

10 R.M. May, 'Simple Mathematical Models with Very Complicated Dynamics', *Nature*, 261 (1976), 45. Later, this work would be rigorously formalised and generalised by Mitchell Feigenbaum in his classic paper 'The Universal Metric Properties of Nonlinear Transformations', published in *J. Stat. Phys.*, 21 (1979), 669 and then explained in simpler terms for a wider audience in the magazine *Los Alamos Science* 1, 4 (1980).

of feedback, self-organisation and non-equilibrium behaviour. Mathematical scientists in many fields are searching for new types of 'by-law' or 'principle' which govern the existence and evolution of different varieties of complexity. These rules will be quite different from the 'laws' of the particle physicist. They will not be based upon symmetry and invariance, but upon principles of probability and information processing. Perhaps the second law of thermodynamics is as close as we have got to discovering one of this collection of general rules that govern the development of order and disorder.

The defining characteristic of the structures in the diagram below is that they are more than the sum of their parts. They are what they are, they display the behaviour that they do, not because they are made of atoms or molecules (which they all are), but because of the way in which their constituents are organised. It is the circuit diagram of the neutral network that is the root of its complex behaviour. The laws of electromagnetism alone are insufficient to explain the working of a brain. We need to know how it is wired up and its circuits inter-connected. No theory of everything

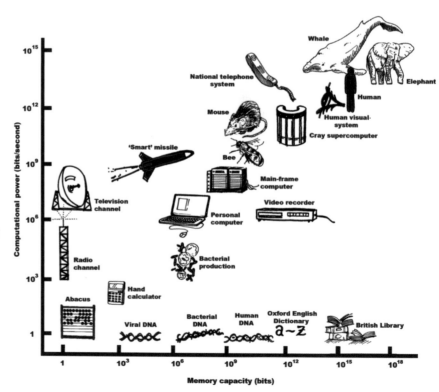

The power and information storage capacity of a variety of complex natural and artificial structures.

that the particle physicists supply us with is likely to shed any light upon the complex workings of the human brain or a turbulent waterfall.

ON THE EDGE OF CHAOS

The advent of small, inexpensive, powerful computers with good interactive graphics has enabled large, complex, and disordered situations to be studied observationally – by looking at a computer monitor. Experimental mathematics is a new tool. A computer can be programmed to simulate the evolution of complicated systems, and their long-term behaviour observed, studied, modified and replayed. By these means, the study of chaos and complexity has become a multidisciplinary subculture within science. The study of the traditional, exactly soluble problems of science has been augmented by a growing appreciation of the vast complexity expected in situations where many competing influences are at work. Prime candidates are provided by systems that evolve in their environment by natural selection, and, in so doing, modify those environments in complicated ways.

As our intuition about the nuances of chaotic behaviour has matured by exposure to natural examples, novelties have emerged that give important hints about how disorder often develops from regularity. Chaos and order have been found to coexist in a curious symbiosis. Imagine a very large egg-timer in which sand is falling, grain by grain, to create a growing sand pile. The pile evolves under the force of gravity in an erratic manner. Sandfalls of all sizes occur, and their effect is to maintain the overall gradient of the sand pile in a temporary equilibrium, always just on the verge of collapse. The pile steadily steepens until it reaches a particular slope and then gets no steeper. This self-sustaining process was dubbed 'self-organising criticality' by its discoverers, Per Bak, Chao Tang and Kurt Wiesenfeld, in 1987. The adjective 'self-organising' captures the way in which the chaotically falling grains seem to arrange themselves into an orderly pile. The title 'criticality' reflects the precarious state of the pile at any time. It is always about to experience an avalanche of some size or another. The sequence of events that maintains its state of large-scale order is a slow local build-up of sand somewhere on the slope, then a sudden avalanche, followed by another slow build-up, a sudden avalanche, and so on. At first, the infalling grains affect a small area of the pile, but gradually their avalanching effects increase to span the dimension of the entire pile, as they must if they are to organise it.

At a microscopic level, the fall of sand is chaotic, yet the result in the presence of a force like gravity is large-scale organisation. If there is nothing peculiar about the sand,[11] that renders avalanches of one size more probable than all others, then the frequency with which avalanches occur is proportional to some mathematical power of their size (the avalanches are said to be 'scale-free' processes). There are many natural systems – like earthquakes – and man-made ones – like some stock market crashes – where a concatenation of local processes combine to maintain a semblance of equilibrium in this way. Order develops on a large scale through the combination of many independent chaotic small-scale events that hover on the brink of instability. Complex adaptive systems thrive in the hinterland between the inflexibilities of determinism and the vagaries of chaos. There, they get the best of both worlds: out of chaos springs a wealth of alternatives for natural selection to sift; while the rudder of determinism sets a clear average course towards islands of stability.

Originally, its discoverers hoped that the way in which the sandpile organised itself might be a paradigm for the development of all types of organised complexity. This was too optimistic. But it does provide clues as to how many types of complexity organise themselves. The avalanches of sand can represent extinctions of species in an ecological balance, traffic flow on a motorway, the bankruptcies of businesses in an economic system, earthquakes or volcanic eruptions in a model of the pressure equilibrium of the Earth's crust, and even the formation of ox-bow lakes by a meandering river. Bends in the river make the flow faster there, which erodes the bank, leading to an ox-bow lake forming. After the lake forms, the river is left a little straighter. This process of gradual build-up of curvature followed by sudden ox-bow formation and straightening is how a river on a flat plain 'organises' its meandering shape.

It seems rather remarkable that all these completely different problems should behave like a tumbling pile of sand. A picture of Richard Solé's, showing a dog being taken for a bumpy walk, reveals the connection.[12] If we have a situation where a force is acting – for the sand pile it is gravity, for the dog it is the elasticity of its leash – and there are many possible equilibrium states (valleys for the dog, stable local hills for the sand), then we can see what happens as the leash is pulled. The dog moves slowly uphill and then is pulled swiftly across the peak to the next valley, begins slowly climbing again, and then jumps across. This staccato movement of slow build-up and sudden jump, time and again, is what characterises the sandpile with its gradual build-up of sand followed by an avalanche. We can see from the picture that it will be the general pattern of behaviour in any system with very simple ingredients.

At first, it was suggested that this route to self-organisation might be followed by all complex self-adaptive systems. That was far too optimistic: it is just one of many types of self-organisation. Yet, the nice feature of these insights is that they show that it is still possible to make important discoveries by observing the everyday things of life and asking the right questions, just like the founding Fellows of the Royal Society 350 years ago. You don't always have to have satellites, accelerators and overwhelming computer power. Sometimes complexity can be simple too.

A realistic system with many possible local equilibrium states and a force which acts to move between them by slow hill-climbing followed by sudden jumps.

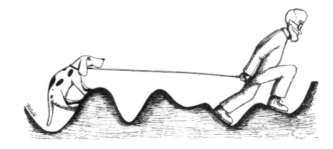

11 Closer examination of the details of the fall of sand has revealed that avalanches of asymmetrically shaped grains, like rice, produce the critical scale-independent behaviour even more accurately because the rice grains always tumble rather than slide.
12 P. Bak, *How Nature Works* (New York, Copernicus, 1996).

17

OLIVER MORTON

GLOBE AND SPHERE, CYCLES AND FLOWS: HOW TO SEE THE WORLD

Oliver Morton is a writer, currently working for *The Economist*. He is the author of *Mapping Mars* and *Eating the Sun*, and currently at work on a book about geo-engineering.

T HE PICTURES OF THE EARTH FROM SPACE BROUGHT HOME BY THE APOLLO ASTRONAUTS TRIGGERED A NEW AWARENESS OF OUR PLANETARY HOME WHICH FED INTO NEW SCIENCE. BUT THE VIEW OF OUR PROBLEMS FROM ASTRONOMICAL DISTANCE IS AN ODD ONE, AS OLIVER MORTON EXPLAINS.

'I know we're not the first to discover this,' Gene Cernan radioed back from about 29,000 kilometres, 'but we'd like to confirm, from the crew of *America*, that the world is round.' Apollo 17 had been thrown up into the night sky over Florida five hours before, but for most of that time the command module *America* and its lunar module *Challenger* had been in low orbit. Only now, having been kicked off to the Moon by the last stage of their Saturn V booster, were the astronauts far enough away to see the planet as a whole. *Challenger* was to land on the edge of the Moon's face as seen from the Earth, rather than near the centre, as previous missions had done, and this meant that Apollo 17 was the first of its kind to head off more or less straight into the Sun, thus allowing Cernan and his crew an unprecedented look back at the shadowless face of the noon-time Earth.

Their photographic record of that view, it is often claimed, is the most

The Earth seen
from Apollo 17,
7 December 1972.

reproduced photographic image in history; given that it is free to use, beautiful and moving, the claim seems not unlikely. Taken from the window of the *America* without the benefit of a viewfinder, the almost-perfectly circular image is dominated by blue oceans and white cloud, an obscuring and captivating pattern which makes the picture clearly and immediately something other than a map. This is a body in space, three-dimensional, a highlight glinting off the ocean, the features at the edge distant and foreshortened. But in this picture, unlike those taken from the Moon itself, there is no doubting what the bewitching body is. The mass of Africa, though centred in a way no traditional map maker would think of, is unmistakable.

When, in late 1946, George Orwell wondered in an essay how he would convince a committed sceptic that the Earth was spherical,[1] he concluded with some reluctance that he would be unable to do better than appeal to the authority of astronomers and to the utility of charts that astronomical observations made possible. Twenty-five years on, the figure of the planet became a matter of direct observation for the select few, and of photographic fact for the rest. There was no longer any need to rely on the

1 George Orwell, 'As I please', *Tribune*, 27 December 1946. In Sonia Orwell and Ian Angus (eds), *Collected Essays, Journalism and Letters of George Orwell, vol. 4: In Front of Your Nose 1945–1950* (London, Secker and Warburg, 1968).

astronomer's authority; by looking at the Apollo pictures one could in effect become an astronomer.

The ability to see the Earth as an astronomer would another planet marked a fundamental shift, the long-term effects of which we still cannot gauge. It has provided valuable new perspectives and treasure troves of data. But no image can reveal everything; and every revelation obscures something. For all that it is an image of the whole, the vision of the Earth from space is necessarily partial. By leaving things out, it makes the Earth too easy to objectify, too easy to hold at a distance, too easy to idealise. It needs to be offset by a deeper sense of the world as it is felt from the inside, and as it extends out of view into past and future. Because of the changes we are putting the planet through, we need as many ways of looking at and thinking about it as we can find. We need ways to see it as a history, a system, and a set of choices, not just a thing of beauty – one which, from our astronomical perspective, we seem already to have left. There are other ways to see the beauty of the world than in the rear-view mirror of progress.

This needs to be stressed in part because the astronomer's gaze is a peculiarly powerful, seductive thing; it is not just thin air that brings dizziness to mountain-top observatories. Its charms are those of photography in general; a form of seeing more removed from direct experience, and frequently from obvious meaning, than any other, its subjects unavailable

The development of the Earth, as represented in Descartes' *Principia Philosophiae*, 1644.

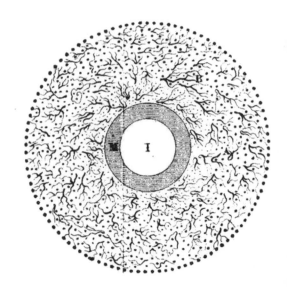

to any cross-examining form of scrutiny. Like photography, astronomy looks, takes joy in looking – but can do no more than keep looking.

In their desire to see and see again, astronomers are particularly well served. Many of the objects of their gaze are eternal and predictable, travelling into our future according to knowable rules. The universe reveals itself in rhythm and return. This is one reason why the visions of astronomy have often stood as an emblem for all the other precise, disinterested but forward-looking observations of science.

Spectacular gains have been made by turning the astronomer's gaze on the Earth. The wetness of clouds, the strength of winds, subtle shifts in the shape of the sea's surface, the thickness of smogs, the colours of the savannah: all are now available on a worldwide scale. Not only can everything be seen: in some of these images, like that icon from Apollo 17, we seem to see everything at once, the Earth entire. It was this completeness which, in the 1970s, gave such images a key role in both the inception and reception of James Lovelock's ideas about Gaia, the self-regulating Earth system – ideas presented, in the subtitle of his first book, as 'A New Look at Life on Earth'.[2]

Such images made clear what Arthur C. Clarke had suggested years before: the archetype for space travel was the *Odyssey*, an adventure completed only in its moment of return. The view that little ship of Apollo brought back gave new reality to the notion, first voiced by Adlai Stevenson in 1965, of 'Spaceship Earth': like the smaller ships, the larger one was a prerequisite for its crew's survival, isolated, fragile. The image of the living Earth as seen from space became a rallying point for environmental activism, an ever-present rebuke to those who would deny the environment's fragility, the finitude of its resources, the limits that it must surely impose on us. It turned the primary concern of the 'space age' from the outward urge of a few to the common heritage of the many.

In 'Globes and Spheres: The Topology of Environmentalism',[3] the anthropologist Tim Ingold voiced an elegant dissent to the way that heritage was represented by those pictures. The global environmental

2 James Lovelock, *Gaia: A New Look at Life on Earth* (Oxford, Oxford University Press, 1979).
3 Tim Ingold, 'Globes and Spheres: The Topology of Environmentalism' in Kay Milton (ed.), *Environmentalism: The View from Anthropology* (London, Routledge, 1993).

movement represented by that objectified, photographed Earth, he argued, was an oxymoron; the environment of a globe is what lies outside it, not what lies within. Thinking about the environment 'from the outside' was a contradictory pursuit that showed a rationalist, map-making mentality taken to its ultimate extreme. To give the planet as a whole precedence over everything it contains, he thought, hid the realities of life as it is lived, and was thus inimical to a deeper-rooted form of environmental awareness. 'The notion of the global environment,' he wrote, 'far from marking humanity's reintegration into the world, signals the culmination of a process of separation.'

The Earth does, as it happens, have an environment in the surrounding sense, a space environment that is both nurturing (a magnetosphere that keeps cosmic rays at bay), a little alarming (near-Earth asteroids, which have in the past caused spectacular calamities and even mass extinctions) and increasingly besmirched (600,000 pieces of space junk and counting). Recognising this cosmic connectedness may in time help to expand our notion of the world we live in, providing new perspectives of its own. But that is not Ingold's point; his point is that to see the earthly environment as something out there and separate is to misunderstand what an environment is.

To agree with Ingold is no to say that everything must be local first and last, nor to deny that there are environmental problems on a planetary scale. It is to say that they are not the planet's problems. They are ours. The drawback of space-age iconography is that it has made the Earth itself the focus of environmental action, the thing at risk, the mother to be celebrated on a consecrated Earth Day. This way of speaking about the planet in peril, of invoking a need to 'save the Earth', suggests either that the needs of people and the needs of the planet are directly opposed – or, at best, that human needs can be reduced to planetary needs.

This line of thinking runs the risk of leading us into futility and sin. Take the futility first. The Earth and its biosphere have, after all, been through far greater, if not faster, fluctuations in temperature than those currently underway. At the hands of those pesky asteroids it has undergone calamities far more sudden. Its seas have frozen; its continents have been licked with flame. Yet even when it has lost species by the bushel, the biosphere has endured, and in the aftermath flourished. Human agriculture, by contrast, is terrifyingly fragile, largely developed during ten millennia of climatic stability, already thin-stretched over too much of the Earth, with ever more people to feed. The late, great comedian George Carlin summed up the true stakes with foul-mouthed pith: 'There is nothing wrong with the planet – the planet is fine. The people are fucked.'

This leads on to the question of morality. To focus on the planet, and not on its people, is wrong; to assume that their interests are identical is to ascribe to the planet attributes it does not possess. It is not an abstracted Earth floating in the velvet vault of space that needs protecting; it's the people inhabiting that world who are at risk of harm, particularly poor people who lack the resources to adapt, to migrate or otherwise to opt out of what is happening to them.

And yet 'planet in peril' rhetoric and attendant catastrophic imagery is everywhere. A quick Google search reveals there to be seven, ten, five, four or eight 'years to save the planet', depending on your headline writer and expert of choice ('Eleven years to save the planet' seems at the moment a rallying cry still up for grabs). It may be that 'planet' here is being used simply to mean the environment on which humanity depends. But this way of talking still acts to raise some abstract notion of the environment above the problems that people actually face, many of which are not environmental. The debates needed to assign priorities to human development, to the reduction of consumerism, to the health of the world's children – important topics all – cannot be reduced to a question about what is good for the planet. Using the planet as a polar bear writ large, a photogenic emblem of the imperilled, obscures more than it illustrates.

At the same time, rather reprehensibly, planet in peril rhetoric trades on a terrible new form of the feeling of the sublime. We are so powerful and

so bad, it says, that we threaten the tough old planet itself; we flatter our human power even while condemning it, seeing ourselves as a problem too big to solve. Thus the old vision of humans as vulnerable to an overpowering nature is reversed. The unstoppable threat is us – and we stand aside, wringing our hands but secretly in awe, as that threat sweeps on.

How better, though, can people see the world than as a fragile blue marble separated from their own experience, cut off from any cosmic continuity by a sharp 360° horizon? And why, given the objective truth of the world as revealed by Apollo, should we even try? To the second question, the answer is that there is more than one way of seeing, just as there is more than one way of speaking. There are times when seeing the Earth as a discrete object, a thing in a picture, is peculiarly helpful; there are times when something else is called for.

Contemporary artists have been confronting this issue for decades. History offers any number of fine traditions of landscape art, in both paint and photography, and invoking a variety of responses in their intended viewers. But more recently something new has arisen: art that seeks not merely to reproduce, or evoke, what it looks like, but to involve it in the artistic process directly, to provide art in which viewers meet the world, rather than just contemplate it – an art that interacts. The British artists Ackroyd and Harvey use the growing of grass as a medium with which to reimagine architecture and photography; Richard Long is fascinated by the traces, material and immaterial, left by walking, and how they can be shared; David Nash grows trees into sculpture while Tim Knowles lets them trace out their own drawings, guided only by the wind; and Andy Goldsworthy imprints and erases ideas on the landscape. As David Nash puts it: 'I think Andy Goldsworthy and I, and Richard Long, and most of the British artists' collectives associated with Land art, would have been landscape painters a hundred years ago. But we don't want to make portraits of the landscape. A landscape picture is a portrait. We don't want that. We want to be in the land.'[4]

---

4 John Grande, 'Real Living Art: A Conversation with David Nash', *Sculpture*, 20:10 (December 2001).

Ingold's response to the inadequacy and contradiction he perceived in the Earth seen as a blue marble was very similar: to look for ways of thinking that put you in the land, rather than just looking at a portrait. He contrasted the outside-in view with cosmologies that look on the world, or cosmos, as a set of spheres experienced from the inside out. Cultures from Ancient Greece to the Inuits have found ways to layer and interpret their worlds in such nests, privileging the local while connecting it to the cosmos. I would not want to suggest that cosmologies should be chosen on aesthetic or practical grounds, assembled piecemeal from those of other cultures, or generated on the basis of what they have to offer. But it seems to me that there is an appealing way of casting the nested-sphere view in the concepts of modern science that does no disservice to that science: transform the spheres into cycles.

Throughout the twentieth century, the Earth sciences have increasingly treated their subject in terms of cycles, whether the oscillations of the atmosphere or the circulation of the core. The past fifty years have seen acceptance of the Milankovitch cycles – subtle variations in the Earth's orbit and attitude – as the causal framework for the ice ages, with ice sheets waxing and waning to their heavenly rhythms. They have seen Earth's magnetic field revealed as a creature that rocks back and forth from North to South, the plaything of dynamic currents circulating in the planet's core. Most fundamental of all, they have seen the uncovering of the great three-dimensional cycles of plate tectonics, in which the slow and mighty over-turning convection of the mantle is coupled to the opening and closing of oceans, the merging and scattering of continents.

In the 1950s Victor Goldschmidt, frequently described as the father of modern geochemistry, put cycles at the centre of that discipline's study of the Earth, defining it in terms of 'the circulation of elements in nature'. Both geochemistry and biogeochemistry remain studies of cycles – in the latter case, quite intimate ones: the carbon dioxide given back to the plants with each animal breath, the nitrogen returned to the world in each drop of urine. The 'Earth Systems Science' that emerged in the 1980s and 1990s, often informed by Lovelock's Gaia, assembled ideas from all these disciplines and subdisciplines into further cycles, cycles made not of matter, but of cause and effect: feedback loops that could stabilise the Earth system or force it into flip-flop oscillations.

Like the components of an astrolabe, the cycles of the Earth system seem to nestle within each other, arranged not by size – they are all, in the end, the size of the planet – but by intimacy and speed, reaching out from the food in our bellies and the wind on our faces to the vastest of vegetable empires and the yet slower, greater mineral realm. Our sweat, once evaporated, spends only days in the sky before falling back as rain. The carbon dioxide we breathe out may be in the air for decades before being eaten up by plants, or take refuge in the oceans for millennia before resurfacing. Other cycles are slower still. While nitrogen compounds can be pumped from sea to sky by microbes, once phosphorus makes its way from soil to the sea it has no easy way back to the atmosphere, and must wait millions of years before, incorporated into sediments, it is lifted up into new mountains to fertilise the soils again. The cycles interpenetrate in such ways all the time, passing through each other in a daunting clockwork of teeth and differentials, their nesting anything but neat, their gearing prone to glitches.

Such a vast machinery seems more daunting to the imagination than a blue marble in space. But while what is circulating, and how it circulates, can be hard and complex questions to fathom, the idea that the world is endlessly recycling itself is an easy perception to train oneself into. The growth of a plant, or the erosion of a gully, are easily seen. And to see a plant grow armed with the knowledge that it does so out of thin air – that is, after all, where the carbon that makes up most of its mass comes from – is to realise that something else must be restoring that nutritive goodness to the atmosphere. To see water cutting into highland rock and washing soil downstream is to realise that, if this is going to go on indefinitely, there must be some way of making new highlands to replace those endlessly whittled away. When Joseph Priestley and James Hutton first had these insights in the eighteenth century they were hard-won breakthroughs. But once known, they become compellingly obvious; it is hard to see how things could be otherwise in a world that endures.

This dynamic image of the Earth is a corollary of one of the most striking aspects of that timeless, static image of the Earth in space: its limits. The Earth is, in material terms, isolated. Very little arrives (those asteroid impacts are few and far between), and only a whisper of gas escapes. Everything else must be endlessly recycled: and so it is. The rain becomes the ocean and the ocean becomes the rain, the mountains are ground down to cover the sea-floors with silt, ancient silts rise up to make new mountains. Nothing stays the same, and yet the system, mostly, persists. Everything is in flux, but nothing is at risk.

And this flux illustrates perhaps the most useful sense of that unhappy phrase, the 'balance of nature'. Nature was not designed to balance, any more than it was designed for anything else. It does not have preferred states with which people meddle at their peril, or that carry some sort of moral weight, or to which it wishes necessarily to be restored. It is precisely to the extent that the Earth is off balance that it works; its rolling cycles are like wheels on slopes. But if there is no static equilibrium, there is balance of another sort – a balance like that of a bank account, its debits and credits constrained always to match over time. For every output there must be an input. Any earthly process not looped back on itself in some way, anything that does not carry the seeds of its own recreation, will either be remarkably slow, or will have run its course long ago, or only just have started. Otherwise it will simply run out of credit.

The existence of the Earth's great recycling can thus be explained by the fact that, in terms of material, it is a closed system. But to explain it this way is immediately to need something more; a source of energy that comes from beyond the system that it powers, and provides the slopes down which the wheels roll. There is work going on in those cycles – pumping, breathing, lifting, grinding – and work can only be done where there are flows of energy. The second law of thermodynamics, the bane of the perpetual-motion-machine designer, means that such flows of energy cannot, themselves, be recycled; the same energy cannot do the same work twice. If work is to be done continuously, fresh energy needs to be provided continuously, and old energy – waste heat – needs to be disposed of. A world closed in one way must be open in another. The Earth depends on there being a beyond.

The Earth's circulating carbon atoms and continents and other constituents depend on three streams of energy from the beyond – and, in the case of the heat of the Earth's interior, from the before as well. Almost all the energy that now comes from within the Earth was put there, in one form or another, at the time of its creation (a tiny amount is now added by the flexing of the planet under the tides of Moon and Sun, but it is the merest smidgen). One stream of energy stems simply from the immense store of heat generated when a planet's worth of gas and dust fell in upon itself, the ingredients smashing into each other in ever larger pieces and at ever greater speeds as the process went on. The Earth thus started off with vast supplies of heat inside it, and a rocky planet, like any other rock, takes a long time to cool down. Stones in a campfire may still be hot the morning after; a stone the size of the Earth can hold heat for billions of years.

Then there is the heat generated since the Earth's creation from energy stored up long before. The chemical elements on Earth that are heavier than helium were created in stars that burned out before the Sun and Earth were born, the vast pressures in their hearts squeezing hydrogen into carbon, silicon, oxygen, nitrogen and iron. When such stellar furnaces explode into supernovae, the energies unleashed become great enough to forge elements even heavier. In the case of elements such as uranium and thorium, those great energies will, in time, leak out. The radioactive elements gathered into the Earth at the time of its creation have steadily meted out the supernova energies stored within them. Thus energy from dying stars helps drive the great internal convection currents which move tectonic plates.

Both these streams of energy, though, are small compared to that which rains down from above. The most easily overlooked and perhaps most fundamental feature of the Apollo 17 picture of the Earth is its brilliant over-the-shoulder illumination. Yes, the Earth floats in pitch-black space – but it floats in sunlight, too. It floats in a torrent of the stuff. The upward flow of ancient heat to the Earth's surface is measured in tens of milliwatts per square metre; the flow from the Sun above is measured in hundreds of watts per square metre. This is the energy that warms the surface and the sky above it, that drives the circulation of atmosphere and ocean. This is the energy of cloud and rain, of sand dune and hurricane. This is the energy which powers the cycles of the biosphere. When plants fix carbon, when bacteria fix nitrogen, when plankton release sulphur from sea-water back into the sky, they do so, directly or indirectly, with solar energy. It is the energy of forest fire and Sunday lunch.

These solar-powered cycles of the biosphere are the ones in which humans are most intimately involved, both as beneficiaries and as rearrangers. Since the development of artificial fertilisers, the nitrogen cycle has come under human control to a remarkable extent, though not in a centralised way. The plough, the field, the roadworks and the building site have increased the rate of erosion far beyond its geological average; the rate at which water flows out of rivers depends on farmers and dam-makers.

And then there is the rate at which ancient sunlight stored in fossil form is used to drive the engines of industry and civilisation. The amount of energy actually liberated in the burning of these fossil fuels is tiny by planetary scales – ten terawatts or so a year, not that much more than the nugatory contribution made by the tides. But the side effects are huge. The carbon dioxide liberated in the burning renders the atmosphere less transparent to the flow of outgoing heat; with the flow thwarted in this way, the temperature at the surface goes up. The resultant warming is, in terms of energy flows, about one hundred times larger than the amount of energy released by the fossil fuels.

Opposite:
An aerial view of the Hoover Dam and Lake Mead in Arizona, June 2009, with the Hoover Dam bypass under construction. The dam was the world's largest concrete structure when completed in 1936.

This great short-circuiting of the geological carbon cycle, though, reveals one of the strengths of seeing the Earth in terms of its turning dynamics. In the purely human realm, cyclic theories of history tend to engender a feeling of hopelessness – the cycle will roll on, regardless. But, perhaps surprisingly, a view of the Earth that focuses on its relentless cycles and the flows of energy that drive them can be empowering. It is a view of the planet in which we are already involved, for good or ill, and to which we can make changes for better or for worse. These are cycles we can use. The Earth seen as a bauble in space is what it is – just a sight, not an experience. The only injunction that is possible faced with that gorgeous globe is 'sustain'. Sustain the gaze; sustain the object. The Earth as an encompassing nest of cycles is a world which we are always already involved with, a Land-art world in which intervention is of the essence. This way of seeing makes things at once more frightening – this is the lived environment of wind in the face and water in the tap at risk, not some idealised representation – and more tractable.

Recognising the openness of the Earth system and the flows of energy that power it offers the clearest way of seeing the solution to the current global environmental crisis. If the manner in which humans currently reap their energy from fossil fuels ties the flow of energy to the material flow of the carbon cycle in a deeply damaging way, we must simply find other flows to tap. Energy is flowing through the winds, in the currents of the oceans, in the rivers, in the growing of the grass. It flows out of the ground and down from the sky. Geothermal plants can speed the flow of heat from the depths; kites in the stratosphere can harvest the endlessly circulating jet streams; mirrors in the deserts can drive turbines with sunshine. There is energy of all sorts flowing through our world; it is not hard to imagine new ways in which that energy can do the work of humanity, new ways to align our needs and the planet's behaviours. And if that capacity for work is harnessed, many other problems can be solved. The carbon cycle could be expanded, the biosphere's capacity for drawing carbon dioxide from the air increased and the greenhouse effect thus diminished. Other waste can be recycled, too, and material resources thus renewed; with a great enough flow of energy from beyond, any closed system can sustain itself with recycling.

The Earth of cycles can hardly be the icon that Apollo's Earth has become; it is more a hum than a sight. But it is a valuable way of thinking of the Earth from inside, of seeing the human and the inhuman as close, interdependent, even indistinguishable. It is an experience that can be taught and shared, and even felt. Stretching from iron core to encompassing cosmos, it has the depth and scale to provide a sublime thrill of its own. True, it offers no gestalt vision to the objective eye. But it can be animated, if you have a mind to. When next you see a picture of the blue marble in space, imagine its clouds coming to life, their whorls beginning to turn like turbine blades. And as you see the scope of the planet's circulation in your mind's eye, let your other mental senses in on the act, too; feel the raw heat of the Sun on the back of your neck as it powers the vision in front of you. Embed the portrait in a vision of process. Turn it into part of something – of a solar system, of an act of the imagination, of a future.

With the right imagination, the world of cycles and the world of the astronomer's gaze can be made to mesh. As mentioned before, the seemingly isolated Earth does in fact have an environment, discovered by astronomy in the abstract, realised as relevant only long after the fact. This environment is the source of the flows of energy that drive the workings of the Earth; it can also be coupled to those workings more directly. The revolutions of the Earth and sky are loosely linked. Orbital cycles carefully calculated by astronomers with no earthly agenda turn out to drive the ice ages. Objects in space affect, and even collide with, the planet from which we watch them. For a long time the possibility of such impacts was deemed of no practical importance, but now it is accepted that they have had great geological significance, and that they merit a certain continued vigilance. As a result of this, 2008 saw the first case of an object on a collision course with the Earth being discovered at an

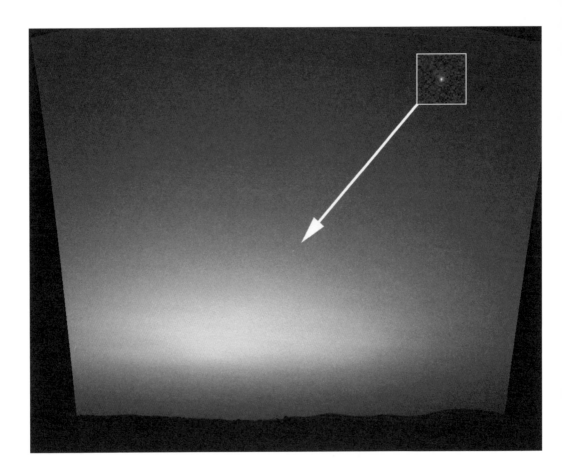

The Earth, photographed from the surface of Mars by the Mars Exploration Rover *Spirit*. The contrast in the image has been increased to make Earth easier to see.

observatory, monitored at the appointed time as a fiery meteor in the sky, and later gathered up in fragments from the ground to which it fell. Asteroid 2008 TC3 was in most ways a small and inconsequential object, but it cut through an important disciplinary distinction. The world of cycles in which we live is not limited to the ball of rock on which we sit; objects elsewhere matter too, and to some extent this must change the way we think about the sky.

And then there is the question of looking further off. Pull back from the Earth just as far as the Moon, and the blue marble loses its features, continents become hard to see, clouds swamping all other detail. From Mars you would need binoculars to even see it had a disc, from Jupiter you would be hard put

.

to make it out with the naked eye. From six billion kilometres away, the greatest range at which the Earth has yet been photographed, *Voyager 1*'s powerful camera saw it as only the palest of blue dots. Yet space scientists now speak of seeing, and learning about, Earth-like planets around other stars.

No telescope currently conceivable could actually produce pictures of such planets as discs in space. But it is possible to look for their cycles, their rhythms. Already such planets are inferred from the way their orbits produce sympathetic wobbles in the movement of their parent's stars, or the regular ways that they pass between those stars and earthly observers. When they become discernible in their own right, less astronomical signs of cycling will be looked for – hints of weather from changes in brightness caused by daily movements of cloud, traces of seasonality as colours shift over the year. Most vital of all, signs of the cycling biosphere will be sought out. As Lovelock pointed out in the late 1960s, biogeochemical cycling has pushed the Earth's atmosphere far from chemical equilibrium. Such disequilibrium may yet, possibly even soon, be seen in the light of planets round other stars. Understanding the Earth's endless recycling sets the stage for measuring the Earthliness of distant specks, and for reading life into a point of light that has no features that could ever be gazed upon.

The Earth is still a beautiful ball floating in space. The Apollo 17 camera did not lie. But by seeming to show everything, that portrait made it too easy to ignore the dynamism its stillness could not show. The Earth is not something put before us, or left behind us. It is around us and within us, turning on itself in every way it can as energy flows through it from the depths of the past and the fires of the Sun. It is not just a spaceship carrying a crew. It is a world, and now aware.

18

THE
END
IS
AT HAND

MAGGIE GEE

BEYOND ENDING: LOOKING INTO THE VOID

Maggie Gee has written eleven novels, a short story collection and a memoir, *My Animal Life*. Her novels include *The Burning Book, Light Years, Where Are the Snows, The Ice People, The White Family, The Flood, My Cleaner* and most recently *My Driver*. She was chair of the Royal Society of Literature from 2004–2008 and is now a vice-president.

S CIENCE REVEALS NEW WORLDS, BUT MAY ALSO BRING NEWS OF THE END OF THE WORLD. THE IDEA HAS A CURIOUS APPEAL, FOR SCIENTISTS AND WRITERS ALIKE, AS MAGGIE GEE EXPLORES.

> Entire nations are uninhabitable. Entire nations have been wiped out. And land cracks and peels in some areas of the globe. In others, deluges of flood water ravage the earth. Welcome to a world six degrees warmer. Welcome to our future.
>
> – From the jacket copy of Mark Lynas' *Six Degrees*, 2007

I

Human beings fear endings, but also crave them. The forbidden thrill of the death-wish stalks many imagined apocalypses, literary, Christian, scientific and filmic; disaster movies do good box office because, in the safety of the present, we can look at the unimaginably terrifying future, and experience the excitement without being annihilated. But our current perils are not just imaginary. Martin Rees' book, *Our Final Century*, suggests that we really are living in dangerous times. In addition to the usual risks to life on Earth, like asteroid impacts, volcanic eruptions and epidemics, twenty-first-century humans have to live with the incidental risks of new technologies – for example, 'bioerror or bioterror', rogue nanoreplicators, mishaps to nuclear power stations – and with the threat of rapidly rising global temperatures due to carbon emissions. So how do twentieth- and twenty-first-century writers and scientists address their sense of an ending?

Of course each generation is assailed by different collective fears and convictions, some validated by events, others not, some with a strong scientific basis, others religious or political. In his two brilliant studies *The Pursuit of the Millennium* and *Cosmos, Chaos and the World to Come*, Norman Cohn traced the recurrence of 'millenarian' cults from the ancient world to the sixteenth century. More recently John Gray in *Black Mass* echoed Cohn's observation that mid-twentieth-century movements like Communism and Nazism also counted on the coming death of the old order. A 1704 letter by Isaac Newton predicting, on skimpy biblical 'evidence', that the world would end in 2060, made news in the twenty-first century partly because it was put on show in the Hebrew University of Jerusalem during the February 2003 run-up to the invasion of Iraq when, as Stephen D. Snoblen has pointed out, apocalyptic fears were already rife. In the nineteenth century, British people's fears of progress often focused on the building of the railways, seen as heralding social revolution, horrifying accidents, 'pollution, destruction, disaster and danger', as Ralph Harrington puts it in his article 'The Neuroses of the Railway'; writers from Elizabeth Gaskell to Charles Dickens were excited by railway terrors; yet in the twenty-first century, we tend to see trains as a low-stress, less

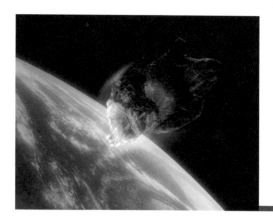

An asteroid heading for Earth. This 'impact event', popular with science fiction writers, is one of the more probable apocalyptic scenarios.

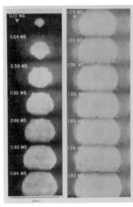

An atomic
explosion from
a paper titled
'The Formation
of a Blast Wave
by a Very Intense
Explosion. II. The
Atomic Explosion
of 1945' by
Geoffrey Taylor,
published in
*Proceedings of the
Royal Society*, 1950.

polluting alternative to planes and cars. Medical doomsday scenarios have proved equally hard to call: AIDS caused, and still causes, a terrible toll of deaths, but has not quite become the all-consuming plague that at one time seemed to threaten us, and nor, so far, have BSE or 'Bird Flu'.

Some collective fears, though, have proved well founded. The 1930s in Europe were marked by fear of totalitarianism; in September 1939, Nazi Germany and Soviet Russia invaded Poland. Other fears continue to stalk us. The cold war and the nuclear arms race brought the shadow of atomic Armageddon. The collapse of the Soviet Union moved it further away, but nuclear proliferation has continued and the famous clock of the Bulletin of Atomic Scientists is still, in 2009, set at five minutes to midnight. In the 1980s and 1990s, laymen got their first glimmer of the global warming that has grown into the pervasive dread of the first decade of the twenty-first century, the thing that most enduringly gives a shape to that vague terror of the end of the world that each generation carries with it from childhood. ('Is it the end of the world?' I asked my mother, aged seven, after the head teacher of our tiny village school told us, during the Suez Crisis, that 'the next few days will decide whether or not the world will go to war'. 'Is it the end of the world?' my teenage daughter asked me, after the Twin Towers fell on 11 September 2001.)

We have just crossed the bar between two millennia and seen a swell of apocalyptic thinking, with the 'Year 2K Bug' cresting the wave. Nick

Davies' book *Flat Earth News* points up the discrepancy between the chaos predicted to follow the digit-change in computers – 'A date with disaster', *Washington Post*, 'The day the world crashes', *Newsweek* – and the actual events: a tide gauge failed in Portsmouth harbour, and a Swansea businessman thought his computer had blown up, only to discover that a mouse had spread droppings on his circuit board. Book titles of the 1990s and early 2000s predicted *The End of Faith*, *The End of Certainty* and *The End of History*, *The End of Food* (two books within two years, in 2006 by F. Pawlick, in 2008 by Paul Roberts), *The End of Oil* and *The End of Fashion*, followed by *The End of the Alphabet*, *The End of America* and *The End of Science*, and finally *The End of Days* and *The End of Time*.

But texts are more complex than titles. Though Paul Roberts' *The End of Food* raises the frightening prospect of 'a perfect storm of food-related calamities' in a globally warmed world, he also suggests practical ways of holding it off: humans might successfully change industrial-scale production, stop demanding ultra-cheap food, use natural fertilisers and practise water conservation. Mark Lynas' book *Six Degrees*, whose colourful blurb prefaces this essay and whose paperback cover shows Big Ben and the Houses of Parliament being neatly toppled by a tidal wave, in fact examines a range of scenarios for global warming, between one degree and six degrees Celsius, and ends with a chapter of suggestions about how his readers can

The paperback
cover of *Six Degrees*
by Mark Lynas.

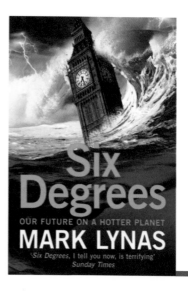

best avoid the worst of them. Fears examined often become less fearful.

Literary writers trying to make sense of our place in history tend to be drawn back constantly to the experience of the present and the physical textures and details we know and love. Stuck in the blank and almost unendurable elevator of time in *Hard-boiled Wonderland and the End of the World*, cult novelist Haruki Murakami metaphorically opens the doors into two alternative universes, both threatened, both situated on the other side of a puzzling schism in history, both marked by a recurrent Arcadian longing for a lost daylight world of physical beauty. More recently acclaimed memoirist Diana Athill published *Towards the End*, her lucid account of why life, however diminished, is still worth living at ninety. Her chosen title both alludes to the end and pushes it away, suspending us in a short but valuable present, the time she has left. She buys and plants a tiny tree-fern even though she knows she will never see it become a tree: the experience of watching it grow is enough.

Regular science fiction and mainstream writers from Mary Shelley onwards have been attracted to the end of the world; it offers drama, heightened emotions and vivid imagery. In the twentieth and twenty-first centuries, Brian Aldiss, Margaret Atwood, J.G. Ballard, Ray Bradbury, Jim Crace, Arthur C. Clarke, Russell Hoban, Anna Kavan, Doris Lessing, Cormac McCarthy, Walter M. Miller, Tim O'Brien, Will Self and Marcel Theroux, among many other novelists, have imagined human life surviving (or sometimes dying) in the grip of great disasters. The final effect of most of these narratives is to make the reader lay down the book with a sense of relief that human civilisation outside its pages still endures. Whether intentionally or not, these books refresh our love of life. But Dr Lee Marsden at the University of East Anglia has drawn my attention to a diametrically opposed trend in the work of Tim Peretti, Tim La Haye, Jerry Jenkins and other figures from the Christian evangelical Right in the USA who are writing an extraordinary sub-genre of apocalyptic fiction that sells in hundreds of thousands and sometimes millions, and is read as literal truth by many of the faithful. These books are based on the premise that we are already living in the 'end times' and can expect 'tribulation, war, famine and pestilence' as a necessary prelude to the ecstasy of the second coming and establishment of Christ's rule on Earth.

Some of my own novels have been described as apocalyptic: my second, *The Burning Book*, written at the apogee of nuclear fears in 1981, ended with nuclear war; two others, *Where Are the Snows* and *The Ice People*, featured runaway climate change; and *The Flood* contains an asteroid strike and a tsunami. But at a conscious level, my strategy is to use the threat of apocalypse to re-focus attention on the short-term miracle of what we have, this relatively peaceful and temperate present where the acts of reading and writing are possible. So in *The Burning Book* and *The Flood* there are, essentially, 'double endings'. I want to offer my readers an active choice. *The Burning Book* ends with all-out nuclear war between America and Russia – but then the narrator steps back, and reminds us that at the time of reading nuclear war is still a fiction. 'Waking again from the book you look out of the window at stillness. The sunlight on the table lying pale and still as peace.' The last section of the book is called 'Against ending', and its final phrases are 'always beginning again, beginning against ending'. *The Flood* uses a similar strategy to *The Burning Book*. The first draft was completed in December 2002, in the long run-up to the 2003 war on Iraq. The people of a city in an imaginary universe are trying to go about their business as usual even though it has been raining for months and the streets are slowly disappearing under the flood waters. President Bare is preoccupied with planning a war against an Islamic country: apocalyptic religion flourishes at home, especially among the poor. The narrative ends with a final tsunami that people have done nothing to prevent. But there's also an epilogue set in the book's first real, named place, Kew Gardens, where the flood has not yet happened. Everyone is there, alive, dancing in their moment, together with the foxes and starlings who are also part of the cast of my dreamed city. *The Flood*'s relationship to subsequent real-life history turned out to be quite unlike *The Burning Book*'s. Britain and America did wage war against an Islamic country, six weeks after I finished the second draft of the novel: a great tsunami did strike Indonesia, Sri Lanka and India the Christmas after *The Flood* was published. Kew Gardens does survive, in real life as it does in metaphor, protecting genetic diversity from all over the world against ending, a vivid botanical carnival of the living moment. I think I wanted to say 'Don't take it all for granted.'

Yet that can't be the whole story. Something in me must be drawn towards disaster. Standing on the cliff edge at Beachy Head in late golden afternoon sunlight, the green of the grass at my feet is glorious, the rocks very far below are white and small as crumbs, the tiny lighthouse is a red-and-white painted toy in front of the sea's crawling glitter, my stomach feels hollow at the brief mile of empty air ahead – yet I like to look, and I am definitely pulled forward towards nothingness before I resolutely pull back and head home across the golden-green slope with its fathers and children flying kites, its jumping dogs, its beautiful restored everydayness.

II

Why do people (or some part of their psyches) long for an ending? Perhaps because continuing down the same path, struggling always to do better, is exhausting and sometimes discouraging, though it is the normal lot of most human beings. Imagining instead a change of state, an abrupt cut-off, offers at least an end to suffering. The great Elizabethan poet Edmund Spenser expressed this longing beautifully in *The Faerie Queen*:

> *Sleep after toil, port after stormy seas,*
> *Ease after war, death after life doth greatly please.*

Perhaps also, in stressful times, people start to crave an ending because its arrival would spare them from the fear of it, and fear, a dynamic emotion alerting us that things are about to get worse, is something human beings

A placard at Speaker's Corner in Hyde Park prophesies the end of the world, 1964.

find peculiarly hard to tolerate. When fear is at its worst, death can start to beckon, slyly whispering that it would be a relief. Readers of thrillers and crime novels, unable to bear the waiting, sometimes skip to the end to know the worst.

I think some individuals, whether artists or scientists or neither, feel the pull of the void more than others. Why should this be? In my own case, I could choose as a defining moment the one in my village school when the head, Mr Norris, perhaps feeling afraid himself or lonely in the midst of us runny-nosed, inarticulate children of all ages from six to eleven, lumped together in one class, said those words about the world going to war that terrified me for months and years afterwards, so that every plane that flew overhead seemed to me the beginning of the end. Yet so far as I know, none of the other children at Watersfield village school became apocalyptic novelists. I would probably have to go back beyond that day to some prior experience of fear to say why I listened to Mr Norris' words with such a painful sense of attunement and recognition: I might also posit some quirk in my own particular neurochemical makeup. Be that as it may, twenty-five years later I did not agree with the two eager psycho-analysts, one of them, as it happened, a dear friend of mine, who turned an agreeable dinner into a battle-ground by trying to convince me that I had written *The Burning Book* as an expression of my infantile desire to destroy the whole world: this helpful interpretation was never going to persuade a card-carrying CND member. But perhaps they were on to something.

I do see some analogy between how I deal with the fear of destruction and how some victims of violence become violent themselves, in order at any rate to play an active rather than a passive role in what is unbearable. It's an attempt to regain a measure of control. When I am writing a story it is I who decide whether the war happens or the tsunami strikes; facing up to these possibilities is arduous and disturbing, yet it is a livelier experience than just waiting in anxiety on the margins of life. Using my role as writer to produce *The Flood* was definitely my way of dealing with my fear and anger about the impending war on Iraq. I wonder if it is the same for scientists working on one facet or other of global warming, or writing about it? Are they putting superficially negative emotions like worry and apprehension to practical use, and so experiencing a kind of victory over circumstance? At the beginning of this chapter I talked about the changing communal fears of human societies and said that only some of them were validated by subsequent events, but I did not add that this is sometimes because fear is a force for good, inspiring effective action. Many computer scientists would argue, contra Paul Davies, that the non-materialising of the Y2K Bug was in fact a validation of the updates they designed. The traits of intellectuals and activists who speculate about disaster – far-sightedess, susceptibility to fear and willingness to tolerate unpalatable facts and the sadness they produce – have not been selected out by evolution, so perhaps they have often enough been thought useful by human beings living in difficult times.

And yet it doesn't make for popularity. I work in these galleys and have dreamed these dreams yet my own heart sinks when I see a title like Lynas'

Six Degrees or even the great James Lovelock's *The Revenge of Gaia,* with the almost comic-book salaciousness of the disaster scene on its front cover. Part of me is repelled by what can seem like gloating. Part of me starts to mutter, 'You don't *know* the living world is going to be wrecked. It isn't yet. You're not *certain.* All this is just extrapolation. Don't wish what we have away.' Despite the ambiguities I have just confessed to, most of me wants very badly not to die just yet, and I am sure the majority of writers and scientists working in this area agree. We too prefer to have fun in the present: we prefer, most of the time, not to think about danger. And yet we cannot for long suppress our half-fearful, half-excited knowledge that we are living at this peculiar and possibly critical point in human history, when, as Martin Rees reminds us, 'within fifty years, little more than one hundredth of a millionth of Earth's age, the amount of carbon dioxide in the atmosphere ... [has begun] to rise anomalously fast', an 'unprecedented spasm ... seemingly occurring with runaway speed'. How are we to live in such anxious times? How to strike a balance between, on the one hand, paying attention to scientific or literary models of possible futures that can draw us ever deeper into possible disaster, and on the other respecting and learning from the quieter practice of, say, a naturalist and ecologist like C.S. Elton, who spent twenty years watching and recording the rhythms and cycles of Wytham Wood? Or a writer like Diana Athill, who, like Elton, focuses her intelligence on what is? Can we learn from Buddhists and lyric poets how to live joyfully in the present at the same time as listening to climate scientists modelling disaster? How often can we afford to stare over the edge of the cliff?

III

Some aspects of ending have a special meaning for the act of writing, perhaps in fiction most of all. A novel is nothing without an ending. In some respects, the end is the most important part. It is vital that the ending should be the right one though; that it satisfies and resolves, and is planned and prepared for. Most ends of human lives, by contrast, are messy, a chapter of missed connections and unwished for accidents, as Julian Barnes' account of his mother's and father's deaths in *Nothing to be Frightened of* elegantly shows. Endings in real life never really end. There are always aftermaths and unintended consequences. But books are places of intended consequences. Fictional ends rest safe in the knowledge that they are final.

Unlike endings in real life, the endings of books can be borne. It is part of the author's job to make the ending bearable for the reader: to help them say goodbye. And in that act, in an ending properly brought off, we help the reader return to life. The end of a good book may make a reader sad, but it is very far from being a death. Whether sad or happy, the ending of

Burning wheatfields, Kansas.

a book should be a complex form of consolation. In *this* world, the invented one, things can end as they were meant to, and in that sense, well. That is one reason why mortal human animals tell stories.

There is another sense of 'end' in the OED on which Paul Muldoon plays in his collected Oxford Lectures on Poetry, *The End of the Poem*: 'the object for which a thing exists; the purpose for which it is designed or instigated'. The narratives in novels do progress inexorably towards the ending. Once arrived there, the reader should be able to look back and see the novel's 'end' in Muldoon's sense: its meaning, or meanings, its purpose. From that viewpoint, everything in the novel should seem both necessary and inevitable. In one sense the ending is also the point of the book.

And this is where life is so very different. Much of Julian Barnes' book about his fear of death centres on his desire not to be caught off guard and outflanked by the unexpected, at the wrong time, with a book unfinished. *Nothing to be Frightened of* reads like his extended attempt in turn to out-think and anticipate death, 'the ruffian on the stair', finally winning at least the aesthetic battle by weaving the unpredictable terror into the smooth texture of his own self-penned story.

There is a ruffianly quality to climate change, too. If and when it comes, it will not be exactly like any of the models. It will catch out governments and individuals. It may be brutal. If we do not do everything we can to lessen its effects, it will cause unprecedented wars and movements of population. We will lose the illusion of control we crave. We may have to give up many of the things we think we cannot do without. We will probably start to value what we have not valued enough only when much of it is already lost. However well prepared we are, we will have to learn very fast and react from day to day. Yet even if governments and electorates are not listening, some scientists are doing their best to inform us. (And so they should: the children of science are technology and industry, whose restless desire to adapt the world to human advantage has helped create this mess in the first place.)

The Royal
Society's 2008
policy report
*Sustainable
biofuels: prospects
and challenges.*

IV

The Royal Society has been active in the climate change debate. As early as
1988, Margaret Thatcher used a Prime Ministerial address to the UK's
national scientific academy to acknowledge the dangers of global warming.
From 1999 to the present, the Royal Society has produced a steady flow of
policy statements, letters to government, workshops, events and guides for
the lay reader on energy policy and global warming. Its policy reports on
environmental issues range from the 1999 *Nuclear Energy: The future
climate*, issued jointly with the Royal Academy of Engineering, to the 2008
Sustainable biofuels. In 2005, 2006, 2007 and 2008 the Royal Society initiated joint statements by the science academies of the G8 +5 countries calling world leaders to urgent action on global warming and saying that 'G8
countries bear a special responsibility for the current high level of energy
consumption and the associated climate change.'

A little further down the Thames, in Somerset House, another Royal
Society, the Royal Society of Literature, holds its meetings and events.
From 2004 to 2008 I was the RSL's Chair of Council, and now I am one

of its Vice-Presidents. What did we, as a body, do to show our concern about climate change during those years?

Er – nothing.

Many things could be said in our defence. True, the RSL is very much smaller, and over 150 years younger, than its scientific sibling the Royal Society. We have little money, few staff, the whole of literature to defend and support, and no expertise in meteorology or energy. Nevertheless, that's not really the point. I think we writers as a group are just like my characters in *The Flood*, still walking through the streets of the drowned city, undeterred by the water rising up to our armpits from trying to get on with our lives as usual. Polls show that a majority of people in the UK believe global warming is a fact, and yet somehow they don't believe it will really affect their lives, and they certainly don't intend to change their own lives radically to help stop it happening. 'Global warming is a problem – but not yet, o Lord, please' is their unconscious prayer. Folk who DO take global warming seriously are thought slightly mad, or over-intense, unlike the sensible majority who just somehow know things will always go on as they do today. 'It's always been like this, so it always will be. Yes, we have had the odd over-hot summer, and springs come earlier, and maybe it seems cloudier and duller than before, but nothing's really going to alter.' And so, out of British politeness, climate change believers keep quiet. It's like religion: don't bring it up. Belief seems like a claim to virtue, a holier-than-thou-ness which will annoy others. Thus some of us, myself included, become cowards, or lazy. Easier to carry on as usual for us too. I am braver in my books, and yet I don't expect to be loved for them. Perhaps climate change scientists have the same problem. None of us want to be bores. None of us want to be laughed at, or groaned over. And writers have a special vanity: we don't want to be thought of as obvious, or preachy, because the subtlety and indirection of contemporary literary language is a cause of pride. In fiction, drama and poetry irony is over-valued, and consequently informativeness, moral depth and emotional truth become qualities not to be assessed, embarrassing to talk about, just as global warming is.

V

I don't blame people for not wanting to peer over the cliff edge. It makes sound sense, in terms of immediate personal survival. Staying happy and optimistic helps people to be healthy. Becoming obsessional about the dangerous future does not help you to navigate the ordinary challenge of each day. Sometimes I myself see young people at global warming conferences almost driven mad by their attempts to live correctly, with a semi-religious belief that they can thus fend off catastrophe, thin and exhausted from taking their bikes on implausible journeys, unable to attend events that are genuinely essential for their work because they refuse to fly, hardly able to eat communally or even shop for food because everything they look at has an environmental cost they feel they cannot pay. I want to stop them and say, 'Be kinder to yourself.' I feel both admiration and pity for their terrible striving. In the end it seems to me we are only animals, and we can only be expected to do our best, not to be angels constantly stretched on the rack. Everyone born surely deserves a little happiness, a little bodily ease and pleasure. The choices always involve benefits and costs, but some of the young are already assuming all the costs and allowing themselves none of the benefits of life on this planet, whereas others, older and much, much richer, have taken all the benefits and paid none of the costs. How are we to strike a balance between self-indulgence and self-flagellation?

In our individual private lives, we will all have to answer that question. If rapid global warming does come, peer pressure will help us make up our minds quite quickly. Already there are the obvious things that everyone can do: walk more, talk about the issue more, drive less, buy less, fly less. In public life, though, scientists and artists play very different roles. To an artist, the scientist's looks harder. Apart from everything else, when scientists make statements to the world, they are vouching, within defined limits, for the truth and solidity of what they say. Artists, on the other hand, are protected by the worn trench-coat of irony. We can place everything we say in distancing quotations; we have a thousand alibis. 'This is fiction: this is a joke: this is a game: this is a confession I am half-ashamed of: this is just personal, take no notice if you don't want to. It's not me, it's just a character. Don't ask me, I'm an entertainer.' It's rather a cushy life we artists have made for ourselves, morally speaking. Scientists have never had the same exemptions.

On the plus side, though, climate scientists at least know clearly what they are doing, and what they can contribute. However many frustrations they have to cope with – financial constraints, deaf or dishonest governments, flawed climate models, inaccurate media reports – they do have a clear part to play. They are useful. Writers very often do not feel useful.

But are we, in fact, useful, and could we be more so? I think the answer is 'Yes' in both cases. Irony, humour and a distancing sense of history can be useful when we apply it critically where it is needed, in this case to the statements of both scientists and politicians. Science means 'knowledge', and it's what writers very often lack. But with great knowledge can come an underestimation of what is still in doubt. Writers are good at casting doubt, and scepticism, in its place, is no bad thing. Great knowledge also brings a degree of power, another thing that writers lack. But again, power can bring with it a blindness to the limits of what it can achieve, a lack of humility. Some of the metaphors used by scientists to express the relationship between human beings and the world they live in are not good metaphors. Some, used so regularly they are barely noticed, in fact embody dangerous untruths. 'Stewardship' is a ubiquitous example. To call human beings 'stewards' of this planet is like accepting that Jack the Ripper is the right man to start a Home for the Care and Protection of Fallen Women. (James Lovelock once said in an interview that it was like putting a goat in charge of a garden.) In 2008, Wallace S. Broecker and that excellent writer Robert Kunzig, author of *Mapping the Deep*, published a survey of climate science called *Fixing Climate*. 'Fix' is a dangerous verb, short, glib and easy. Can human beings really 'fix' the

climate they are currently busy breaking? Do we understand enough even now to do it as easily as Kunzig's peroration – 'the planet is ours to run, and it is up to us to run it wisely' – suggests? These are linguistic quibbles, but perhaps non-scientists can apply their critical intelligence also to the content of some of the remedies suggested by scientists for a globally warmed world. The history of science tells us that once radiation was used as a general tonic, and heroin recommended as a non-addictive alternative to oral morphine. We need a Jonathan Swift to ask sharp questions about the desirability of geothermal engineering along the lines some climate

Opposite:
The Last Man by
John Martin,
1849. Martin
was influenced by
Mary Shelley's
1826 science-
fiction novel of
the same name.

scientists have suggested. Is it really a good idea to seed the ocean with iron to increase the numbers of plankton? Would installing giant mirrors in the sky to reflect sunlight back make sense?

I think writers do have a few special talents we can hope to offer as we look apprehensively into our human future. We can try to defamiliarise the present, make our readers realise afresh how marvellous our living planet is. We can look at scientists' discourse for evidence of solipsism or over-confidence. We suffer from both those traits ourselves, so we should recognise them in others. But in more important ways, both artists and scientists have a similar role to play. Both castes are fortunate to live lives that are not totally taken up with grubbing what we need from the texture of each day as it happens. We are not trying to survive in coal-mines, or struggling to feed livestock. We have the great luxury of being able to look outside this immediate place and time. We can look beyond our own species, too, at the wide web of life which contains us. Unchained from the contingencies of the moment, our imaginations are free to scan the horizon and see the future coming in its many possible forms, and reach out towards it. If we tremble, it is because we are, as Shelley said, 'the antennae of the race'. And if we do not make the attempt, if we sit blindly immured in what we have, we may lose everything. The laboratories and libraries that we need and love to pursue our crafts are some of the first things that would be lost with the collapse of civilisation. Doris Lessing's novel *Mara and Dann*, set tens of thousands of years in the future after a time of great climatic change, imagines the rudimentary survival of only a few broken scraps of writing: a few lines of Shakespeare, though his name is lost. Planes no longer fly; museums have been ransacked and broken long ago. By imagining a darker future, Lessing imparts the golden light of imminent loss to the present. It is just one of the ways in which writers, by daring to look into the void, can help us both to appreciate and evaluate the complex human society that scientists are trying to shore up against ending.

STEPHEN H. SCHNEIDER

CONFIDENCE, CONSENSUS AND THE UNCERTAINTY COPS: TACKLING RISK MANAGEMENT IN CLIMATE CHANGE

Stephen H. Schneider is the Melvin and Joan Lane Professor for Interdisciplinary Environmental Studies and Professor of Biology and Senior Fellow, Woods Institute for the Environment at Stanford University. He has served as an author for the four assessment reports of the Intergovernmental Panel on Climate Change (IPCC), and was a core member for the third and fourth synthesis reports. He shared the 2007 Nobel Peace Prize with other IPCC authors and staff from the previous two decades. His popular books include *The Genesis Strategy: Climate and Global Survival*, *Global Warming: Are We Entering the Greenhouse Century?*, *The Coevolution of Climate and Life* and *Laboratory Earth: The Planetary Gamble We Can't Afford to Lose*. His latest book is *Science as a Contact Sport: Inside the Battle to Save Earth's Climate*.

U NCERTAINTY BEDEVILS COMPONENTS OF THE SCIENCE OF CLIMATE CHANGE. IT WILL NOT BE ELIMINATED FROM MANY ASPECTS ANY TIME SOON, SO THE BEST WAY TO HELP POLICY-MAKERS IS TO TRY AND FORGE A CONSENSUS ABOUT THE DEGREE OF CONFIDENCE THAT CAN BE ASSESSED FOR EACH IMPORTANT CONCLUSION. STEPHEN SCHNEIDER EXPLAINS THE LONG STRUGGLE TO UNDERSTAND HOW TO DO THAT EFFECTIVELY.

Human activities are changing the climate. But how large and how fast will these changes be? What systems will be only partly disturbed and what other systems seriously disrupted? And how can our policy choices reduce the threat they pose to natural and social systems?

The policy problem is hard because the global scale of climate change and its subtly intensifying impacts contrast uneasily with the short-term, local-to-national scales of most management systems. Furthermore, significant uncertainties plague projections of climate change and its consequences.

Such projections stretch the traditional scientific method of directly testing hypotheses because there can be no data for the future before the fact. Any prognostication into that unknown territory is, by definition, a model of the factors that are believed to determine how the future will evolve. But even though we can never fully solve the climate prediction problem, we can go a long way toward bracketing probable outcomes, and even defining possible outliers.

Progress here depends on an international community of scholars, who repeat what others have done with different computer models, make comparisons across models of various designs, compare relevant aspects of simulations to existing observational data to test model performance from 'retrodiction' of past changes, and pioneer new models as data and theory advance. Back in the early 1970s, when a reporter asked how long this model-building and validation process would take to achieve high confidence, I said that our models were 'like dirty crystal balls, but the tough choice is how long we clean the glass before we act on what we can make out inside'. That is still the issue, even as models become more sophisticated and simulate the Earth's conditions increasingly well. What constitutes 'enough' credibility to act is not science per se, but a subjective value judgment on how to gauge risks and weigh costs.

MODELLING FUTURE CLIMATE

How large are the scientific uncertainties, though? People often say that meteorologists' inability to predict weather credibly beyond about ten days bodes ill for climate projection over decades. This misses a key difference between the instantaneous state of the atmosphere – weather – versus its time and space averages – climate. Even though the evolution of atmospheric conditions is inherently chaotic and the slightest perturbation today can make a huge difference in the weather a thousand miles away and weeks hence, large-scale climate shows little tendency to exhibit chaotic behaviour (at least on timescales longer than a decade). Good models can thus make reasonable climate projections decades or even centuries ahead if the processes forcing change are large enough to detect above the background 'noise' of the climate system – the unpredictable part. The Intergovernmental Panel on Climate Change (IPCC)'s laboriously compiled projections combine such modelling with scenarios for greenhouse gas emissions based on different assumptions about economic growth, technological developments, and population increase.[1]

These scenarios, despite major differences in emissions, show paths for global temperature increase that do not diverge dramatically until after the mid twenty-first century. This has led some to declare that there is very little difference in climate change across scenarios, and therefore, emissions reductions can be delayed many decades. That is a big mistake. It takes many decades to replace current polluting energy systems. There is also delay between emissions and temperature change due to the thermal inertia in the climate system caused by the large heat capacity of the oceans. After the mid twenty-first century, there are large differences based on emissions over the next few decades in the projected temperature increases – and the risks of associated dangers – for the late twenty-first century and beyond. Some of these risks imply irreversible changes.

Much of the uncertainty contributing to the ranges of projected future temperature increase derives from the so-called climate sensitivity. How much warming can we expect a given amount of greenhouse gas to cause? It is often estimated as the equilibrium global mean surface temperature increase due to a doubling of atmospheric CO_2 from pre-industrial levels of about 280 parts per million. The IPCC estimates that it is 'likely' (there is a 66–90 per cent chance) that the climate sensitivity is between 2 and 4.5 °C and roughly a 5–17 per cent chance that it is above 4.5 °C (with the remainder being the chance it is less than 2 °C). They also offered a 'best guess' of 3 °C climate sensitivity.

Many studies have produced probability distributions for climate sensitivity with a long right-hand tail, meaning that high climate sensitivity values, while relatively unlikely, still register a probability of a few per cent or more. One example is displayed in figure 1, which shows a very

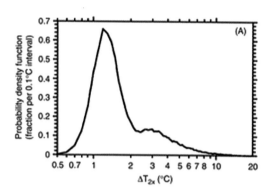
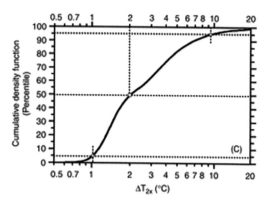

Figure 1: Probability density function (*panel A*) and cumulative density function (*panel C*) for climate sensitivity from Andronova and Schlesinger (2001) generated by scaling observed temperature trends against estimates of radiative forcing over the twentieth century. Owing to uncertainties of the indirect effects of aerosol emissions on the brightness of clouds, the distribution in panel A has a long 'right-hand tail', leaving open some 10 per cent chance of extremely high values for climate sensitivity.[2]

uncomfortable 10 per cent chance that the climate sensitivity is higher than 6.8 °C. The median result – that is, the value that climate sensitivity is as likely to be above as below – is 2.0 °C, while there is a 10 per cent chance the climate sensitivity will be 1.1 °C or less. Like all model dependent studies, the detailed numerical values should not be taken literally, but the overall message must be taken seriously.

Our uncertainty goes beyond scientific understanding of the scale and distribution of climate changes from any single scenario of increasing greenhouse gases to include the trajectory of human development and our adaptive capacity. Moreover, future greenhouse gas emissions are heavily dependent on policy choices worldwide. But we do know that if we wait to act until an increase in undesirable impacts occurs, the inertia in the climate system and in the socioeconomic systems that produce greenhouse gas emissions will have committed us to even more severe impacts stretched out over many decades to centuries.

We cannot eliminate all of the important scientific uncertainties, but we can be more precise about their extent. That, however, is only part of the scientists' job. We also have a responsibility to communicate all of this as well

---

1 IPCC: N. Nakicenovic and R. Swart (eds), *Emissions Scenarios – A Special Report of Working Group III of the Intergovernmental Panel on Climate Change* (Cambridge, Cambridge University Press, 2000).
2 N.G. Andronova and M.E. Schlesinger, 'Objective Estimation of the Probability Density Function for Climate Sensitivity', *Journal of Geophysical Research*, 106 (2001), 22605–12.

as we can. Communicating this complex systems science to policy-makers and the public is difficult. Too often, confusion reigns when an advocate for strong policy cites a well-established severe outcome as the most important consideration, and another advocate from some enterprise institute disliking public control of private decisions cites speculative components of the systems analysis as if that is all there were. Not surprisingly, politicians, media, and just plain folks get frustrated by this 'duelling scientists' mode of presentation, an unfortunate staple of the mainstream media.

Professional training also leads too many scientists to 'bury our leads', as American journalists would put it, rather than finding effective ways to communicate complex ideas. Being straightforward and understandable is a challenge given the strong scientific tradition of full disclosure, which makes us lead with our caveats, not our conclusions. But what I call the 'double ethical bind' – be effective in public communication even if that means there isn't enough space or time to present all of the caveats – is not unbridgeable. It calls for the scientist to develop a hierarchy of products ranging from sound-bites on the evening news to get our findings head-lined on the agenda, to short but meatier articles in semi-popular journals like *Scientific American*, to more in-depth websites, to full-length books in which that smaller fraction of the public or policy worlds that actually want the details about the nature of the processes and how the state of the art has evolved can find them. Yes, it is very time-consuming to produce websites or long books with the details, but it is also necessary for those in complex systems science fields like climate science to simultaneously be effective in public messaging, where all the details are not feasible to communicate, but the longer backup materials can honestly separate the components of the science that are well established from those best characterised as competing explanations and from those which are still speculative.

The Royal Society and my own National Academy of Sciences (if less boldly, I think) have moved into this realm with clear statements of the potential risks of climate change. An evolving series of pronouncements include the joint statement of 2001 of the Royal Society with fifteen other national science academies on the science of climate change.[3] The statement of June 2005 on global response to climate change by the science academies of the G8 nations and of China, India and Brazil stressed that the scientific understanding of climate change is now sufficiently clear to justify prompt action.[4] There followed the May 2007 statement on sustainability, energy efficiency and climate protection of the national science academies of the same countries plus Mexico and South Africa[5] and most recently the June 2009 joint statement calling for the transformation of the G8+5 nations' energy strategies.[6] In addition, I always push at our annual US National Academy membership meetings for us to be more publicly oriented, but it comes slowly. I am glad that our new NAS President, Ralph Cicerone, is committed to communicating quality science in the public interest. It is also encouraging that President Obama's new science adviser, John Holdren, is more in the mould of former UK government adviser and Royal Society President Lord May than some previous science advisers in the US who tended to carry the administration's message to the science community, rather than the other way around, as in the case of May or Holdren.

Along with climate projections, scientists also have to explain how systems science gets done. We cannot usually do traditional 'falsification' controlled experiments. What we can do is assess where the preponderance of evidence lies, and assign confidence levels to various conclusions. Over decades, the community as a whole can 'falsify' earlier collective conclusions – like the sporadic suggestions in the early 1970s that the world would cool. But in systems science it sometimes takes a score of years to even discover that certain data were not collected or analysed correctly, as well as continuing to identify new data, and such discoveries are rarely by individuals but by teams and even assessment groups.

3 'The Science of Climate Change': Joint statement of sixteen national academies of science, 18 May 2001; http://royalsociety.org/displaypagedoc.asp?id=13619

4 'Global Response to Climate Change': Joint science academies' statement, 7 June 2005; http://royalsociety.org/displaypagedoc.asp?id=20742

5 'Sustainability, Energy Efficiency and Climate Protection': Joint science academies' statement, 16 May 2007; http://royalsociety.org/displaypagedoc.asp?id=25576

6 'Climate Change and the Transformation of Energy Technologies for a Low Carbon Future': Joint academies' statement, 11 June 2009; http://royalsociety.org/displaypagedoc.asp?id=34103

THE SCIENCE OF CLIMATE CHANGE

A joint statement issued by the Australian Academy of Sciences, Royal Flemish Academy of Belgium for Sciences and the Arts, Brazilian Academy of Sciences, Royal Society of Canada, Caribbean Academy of Sciences, Chinese Academy of Sciences, French Academy of Sciences, German Academy of Natural Scientists Leopoldina, Indian National Science Academy, Indonesian Academy of Sciences, Royal Irish Academy, Accademia Nazionale dei Lincei (Italy), Academy of Sciences Malaysia, Academy Council of the Royal Society of New Zealand, Royal Swedish Academy of Sciences, and Royal Society (UK).

The work of the Intergovernmental Panel on Climate Change (IPCC) represents the consensus of the international scientific community on climate change science. We recognise IPCC as the world's most reliable source of information on climate change and its causes, and we endorse its method of achieving this consensus. Despite increasing consensus on the science underpinning predictions of global climate change, doubts have been expressed recently about the need to mitigate the risks posed by global climate change. We do not consider such doubts justified.

There will always be some uncertainty surrounding the prediction of changes in such a complex system as the world's climate. Nevertheless, we support the IPCC's conclusion that it is at least 90% certain that temperatures will continue to rise, with average global surface temperature projected to increase by between 1.4 and 5.8°C above 1990 levels by 2100[1]. This increase will be accompanied by rising sea levels, more intense precipitation events in some countries, increased risk of drought in others, and adverse effects on agriculture, health and water resources.

In May 2000, at the InterAcademy Panel (IAP) meeting in Tokyo, 63 academies of science from all parts of the world issued a statement on sustainability in which they noted that "global trends in climate change ... are growing concerns" and pledged themselves to work for sustainability – meeting current human needs while preserving the environment and natural resources needed by future generations[2]. It is now evident that human activities are already contributing adversely to global climate change. Business as usual is no longer a viable option.

We urge everyone - individuals, businesses and governments - to take prompt action to reduce emissions of greenhouse gases. One hundred and eighty-one governments are Parties to the 1992 UN Framework Convention on Climate Change, demonstrating a global commitment to '*stabilising atmospheric concentrations of greenhouse gases at safe levels*'. Eighty-four countries have signed the subsequent 1997 Kyoto Protocol, committing developed countries to reducing their annual aggregate emissions by 5.2% from 1990 levels by 2008-2012.

The ratification of this Protocol represents a small but essential first step towards stabilising atmospheric concentrations of greenhouse gases. It will help create a base on which to build an equitable agreement between all countries in the developed and developing worlds for the more substantial reductions that will be necessary by the middle of the century.

There is much that can be done now to reduce the emissions of greenhouse gases without excessive cost. We believe that there is also a need for a major co-ordinated research effort focusing on the science and technology that underpin mitigation and adaptation strategies related to climate change. This effort should be funded principally by the developed countries and should involve scientists from throughout the world.

The balance of the scientific evidence demands effective steps now to avert damaging changes to the earth's climate.

The Royal Society

Opposite:
The joint
statement of 2001
on the science of
climate change.

BACK TO BAYES

When I first got involved in discussing the range of outcomes in climate
change, I didn't understand Bayesian versus frequentist statistics, but in
fact that was the heart of the matter – how to deal with objectivity and
subjectivity in modelling and in projections.

As Bill Bryson mentions in the Introduction, the English clergyman
and mathematician Thomas Bayes (circa 1702–61) formulated an approach
to probability now called Bayesian inference. His key theorem was
published posthumously in 1764. In essence, it expresses how our knowledge base – and prejudices – establish an *a priori* probability for something
(that is, a prior belief in what will happen based on as much data and theory
as is available). As we further study the system, obtaining more data and
devising better theories, we amend our prior belief and establish a new, *a
posteriori* probability – after the fact. This is called Bayesian updating. Over
time, we keep revising our prior assumptions until eventually the facts
converge on the real probability.

Since we cannot do experiments on the future, prediction is wholly a
Bayesian exercise. This is precisely why the Intergovernmental Panel on
Climate Change produces new assessments every six years or so, since new
data and improved theory allow us to update our prior assumptions and
increase our confidence in the projected conclusions.

That confidence still falls short of certainty for most aspects of the problem. For example, there is only maybe a fifty-fifty chance of sea levels rising
many metres in centuries to come. The conclusion cannot be objective,
since the future is yet to come. However, we can use current measurements
of ice sheet melting. We can compare them with 125,000 years ago, when
the Earth was a degree or two warmer than now and sea levels were four to
six metres (thirteen to twenty feet) higher. Because that ancient natural
warming had a different cause (changed orbital dynamics of Earth around
the Sun) from recent and near future warming caused primarily from
current anthropogenic greenhouse gas increases, we can't say with high

confidence that a few degrees of warming from greenhouse gases will also cause a four-to-six-metre rise in sea levels. But it undoubtedly indicates an uncomfortable Bayesian probability of something similar to that happening in the next few centuries. This indeed was the conclusion of the Synthesis Report of the IPCC's Fourth Assessment in 2007, for exactly those reasons.

Some statisticians and scientists are leery of Bayesian methods. They prefer to stick only with empirical data and well-validated models. But what do you do when you don't have such data? One example is found in clinical trials in cancer treatments, a subject in which I have had a very personal interest. The 'gold standard' is a double-blind trial where half the patients receive a placebo and the other half receive the drug being tested, and neither the patients nor the researchers know who got what. After five or ten years, if there is a statistically significant difference between the recovery rate of drug and placebo, the trial is declared successful. The trial isn't designed to pinpoint individual differences. Even if we knew the odds of recovery for the average person from different treatments, there is a wide spread in individual responses. So medicine should try to tailor treatments to the individual's idiosyncrasies. That makes some doctors – and many insurance companies – nervous. Likewise, some scientists and many policy-makers are nervous about Bayesian inferences based on the best assessment of experts, preferring hard statistics. But as there are no hard statistics on the future, Bayesian methods are all we have. They are certainly better than no assessment at all and hoping that everything will work out fine with no treatment. If we care about the future, we have to learn to engage with subjective analyses and updating – there is no alternative other than to wait for Laboratory Earth to perform the experiment for us, with all living things on the planet along for the ride.

Changing the Culture of Science

While we have refined our models, it has also taken decades to develop the right approach to these scientific realities, and to find the language to convey them properly to policy-makers. In the global climate policy discussion, the most important assessments have been produced by the Intergovernmental Panel on Climate Change, in an extraordinary exercise which involves thousands of scientists reviewing the latest evidence. Ever since the IPCC was founded in 1988, I have pushed hard for a cultural change in the assessments. As I have said, overcoming uncertainties, the traditional approach of what the philosopher Thomas Kuhn[7] called 'normal science', will take an unforeseeably long time. Climate systems science demands a shift to managing uncertainties instead.

That means we scientists, and policy-makers, grappling with climate change impacts are dealing with risk management. As the sea level rise example indicates, outcomes cannot be assessed with high confidence in many important cases, but the probable range can often be estimated.

Risk-management framing is a judgment about acceptable and unacceptable risks. That makes it a value judgment. As with the Bayesian approach to probability, many traditional scientists are uncomfortable with that. I am one of them, but I am more uncomfortable ignoring the problems altogether because they don't fit neatly into our paradigm of 'objective' falsifiable research based on already known empirical data.

Systems science also alerts us to the possibility of 'surprises' in future global climate – perhaps extreme outcomes or tipping points which lead to unusually rapid changes of state. By definition, very little in climate science is more uncertain than the possibility of 'surprises'. But it is nevertheless a real one. Even so, it took several long rounds of assessment just to get IPCC to mention surprises, let alone discuss formal subjective probabilistic treatment of such potentially irreversible, large changes.

John Houghton, former director of the UK Meteorological Office and the IPCC Working Group I leader for the first three assessment reports, was initially very reluctant to get into the surprises tangle. I recall a very clear exchange at a climate meeting in Oxford University in 1993.[8] Houghton thought the public discussion about 'surprises' was too speculative

7 T. Kuhn, *The Structure of Scientific Revolutions* (Chicago, University of Chicago Press, 1962).
8 Oxford Environment Conference: 'Climate Change: Potential for Interactions and Surprise' Oxford University, Oxford, England, 15–16 July 1993.

and would be abused by the media. 'Aren't you just a little bit worried that some will take this surprises/abrupt change issue and take it too far?' he asked. 'I am, John; we have to frame it very carefully,' I replied. 'But I am at least equally worried that if we don't tell the political world the full range of what might happen that could materially affect them, we have not done our jobs fully and are substituting our values on how to take risks for those of society – the right level to decide such questions.'[9]

In the end, despite the worry that discussions of surprises and non-linearities could be taken out of context by extreme elements in the press and NGOs, we were able to include a small section on the need for both more formal and subjective treatments of uncertainties and outright surprises in the IPCC Second Assessment Report (SAR) in 1995.[10] Chapter 11, 'Advancing Our Understanding', was about what to do later, and so was not directly assessed in the more politically sensitive conclusions of the report. Thus, John did not object to the few sentences on those topics in that chapter. As a result, the very last sentence of the IPCC Working Group I 1995 Summary for Policy Makers (SPM)[11] addresses the abrupt non-linearity issue. This made much more in-depth assessment in subsequent IPCC reports possible, simply by noting that 'When rapidly forced, non-linear systems are especially subject to unexpected behaviour.'

A LANGUAGE FOR RISK

Now we had licence to pursue risk assessment of uncertain probability but high consequence possibilities in more depth; but how should we go about it? The basics are that scientists can help policy-makers by laying out the elements of risk, classically defined as *consequence* x *probability*. In other

9 S.H. Schneider, 'The Future of Climate: Potential for Interaction and Surprises' in T.E. Downing (ed.), *Climate Change and World Food Security* (Heidelberg, Springer-Verlag, 1996), NATO ASI Series 137: 77–113.
10 G. McBean, P. Liss and S. Schneider, 'Advancing our understanding' in J.T. Houghton, L.G. Meira Filho, B.A. Callander, N. Harris, A. Kattenberg and K. Maskell (eds), *Climate Change 1995: The Science of Climate Change, Contribution of Working Group I to the Second Assessment Report of the Intergovernmental Panel on Climate Change* (Cambridge, Cambridge University Press, 1996).
11 Houghton et al., 'Summary for Policy Makers' in J.T. Houghton, L.G. Meira Filho, B.A. Callander, N. Harris, A. Kattenberg and K. Maskell (eds), *Climate Change 1995: The Science of Climate Change, Contribution of Working Group I to the Second Assessment Report of the Intergovernmental Panel on Climate Change* (Cambridge, Cambridge University Press 1996).

words, what can happen and what are the odds of it happening?

The plethora of uncertainties inherent in climate change projections clearly makes risk assessment difficult. The inertia in the climate and socio-economic systems and the fact that greenhouse gases emissions will continue to rise, given the absence of strong mitigation policies (or unexpected events like a prolonged recession), indicate that globally most policy-makers have been reluctant to make long-term investments beyond their expected terms in office. But that is changing both in some regions like the EU and even in the US. These kinds of decision-makers are increasingly wary of making what is known as a Type II error – fiddling while the Earth burns. A Type I error is a false positive, which in this case would mean taking action against climate change which subsequently proved relatively needless. Scientists are often leery of making a Type I error when data are scarce for fear of misleading society into unnecessary actions and being blamed for undue alarm. The other kind, a Type II error, is a false negative, and in this case would mean assuming it is preferable to do little or nothing until there is less uncertainty, and subsequently finding that serious climate change ensues unabated with much more damage than if precautionary policies had been undertaken to adapt to and mitigate the effects. So it appears that many scientists are often Type I and our future-oriented decision-makers Type II error avoiders. A less charitable interpretation of those reluctant to invest in precautionary adaptation and mitigation measures is that they know that the really adverse outcomes will likely occur in the future when current decision-makers are not in office and not likely to be held accountable. The short-term incentives are to delay action and pass the risks and the recriminations on to the next generation. None of this is scientific risk assessment, but value judgments on where and how to take risks and make investments in policy hedges – in short, risk management. But risk management is put on a much firmer scientific basis when the managers are schooled in the best risk assessments that state-of-the-art science can produce.

To help decision-makers, the IPCC produced a Guidance Paper on

Uncertainties in 2000[12] which was a foundation for the 2007 Fourth Assessment Report.[13] I prepared the original draft with Richard Moss, now a Senior Scientist, Joint Global Change Research Institute, after convening a meeting in 1996 in which about two dozen IPCC lead authors met with decision analysts to fashion a better way to treat uncertainties in scientific assessments. The final guidance eventually agreed to within the IPCC was a quantitative scale. We would define 'low confidence' as a less than one-in-three chance; 'medium confidence', one-in-three to two-in-three; 'high confidence', above two-thirds; 'very high confidence', above 95 per cent; and 'very low confidence', below 5 per cent.

It took a long time to negotiate those numbers and those words in the Third Assessment Report cycle. There were some people who still felt that they could not apply a quantitative scale to issues that were too speculative or 'too subjective' for real scientists to indulge in 'speculating on probabilities not directly measured'. One critic said, 'Assigning confidence by group discussions, even if informed by the available evidence, was like doing seat-of-the-pants statistics over a good beer.' He never answered my response: 'Would you and your colleagues think you'd do that subjective estimation less credibly than your Minister of the Treasury or the President of the US Chamber of Commerce?'

So we had two things we wanted everyone to use – a set of numbers defining the probability ranges for words such as 'likely', and a set of qualitative phrases for our confidence in the results, going from 'well established' if there were a lot of data and a lot of agreement between theory and data, to 'speculative' without much data and when there wasn't much agreement. We had 'established but incomplete' and 'competing explanations' for the intermediate cases.

And then for the next two years Richard and I became what a journalist later called 'the uncertainty cops'. I read three thousand pages of draft material for the IPCC's Third Assessment Report. People did not always

12 R.H. Moss and S.H. Schneider, 'Uncertainties in the IPCC TAR: Recommendations to Lead Authors for More Consistent Assessment and Reporting' in R. Pachauri, T. Taniguchi and K. Tanaka (eds), *Guidance Papers on the Cross Cutting Issues of the Third Assessment Report of the IPCC* (Geneva, World Meteorological Organization, 2000), pp. 33–51.
13 Intergovernmental Panel on Climate Change (IPCC), *Climate Change 2007: The Fourth Assessment Report of the Intergovernmental Panel on Climate Change* (Cambridge and NY, Cambridge University Press, 2007).

Steam rises from the cooling towers of the Eggborough electricity power station near Selby, England.

use uncertainty terms according to our simple rules. For instance, they would say that because of uncertainties, we can't be 'definitive'. I wrote back, 'What is the probability of a "definitive"?' Early drafts would put the range of outcomes anywhere from a one to five degrees Celsius change in temperature. And then they would say in parentheses 'medium confidence'. That was completely incorrect. It was 'very high confidence', because they were talking about the fact that *between* one and five degrees was a very, very likely place to arrive. But people didn't want to say 'very high confidence' because nobody felt very confident about the state of the science at the level of pinning it down to, say, one degree. So Richard or I would help them to rewrite, and say that we have 'low confidence' in specific forecasts to a precision of a half degree, but we have 'high confidence' that the range is one to five degrees. Simple things like that were needed to achieve consistency of message.

Meanwhile the political chicanery of ideologists and special interests was shamelessly exploiting systems uncertainty by misframing the climate debate as bipolar – 'the end of the world' versus 'it's good for you'. The media compliantly carried it in that frame much of the time, too. But those were and still are, in my view, the two lowest probability outcomes. The confusion that bipolar framing has engendered creates in the public at large a sense that 'if the experts don't know the answers, how can I, a mere lay citizen, fathom this complex situation?' To this, industry-funded pressure groups added the old trick of recruiting non climate scientists who are sceptical of anthropogenic climate change to serve as counterweights to mainstream climate scientists. This spreads doubt and confusion among those who don't look up the credentials of the apparently contending scientists – and that, unfortunately, includes most of the public and too much of the media. The framing of the climate problem as 'unproved', 'lacking a consensus', and 'too uncertain for preventive policy' has been advanced strategically by the defenders of the status quo. This is very similar to the tactics of the Tobacco Institute and its three-decade record of distortion that helped stall policy actions against the tobacco industry, despite the horrendous health consequences and eventually billions of dollars in successful lawsuits against big tobacco.

In the face of such tactics, the IPCC assessment reports are intended to be the best achievable statement of current scientific consensus. But 'consensus' is not necessarily built over conclusions but the *confidence* we have in a host of possible conclusions. With that kind of information policymakers can make risk-management decisions by weighing both the possible outcomes and the assessed levels of confidence – we know it well, sort of know it, or hardly know it at all. Scientists should just say what we do know and don't, and not leave something out because it isn't a well-established consensus yet. It is the job of society, through its officials, to make the risk-management decisions informed by our conclusions and accompanying confidence estimates.

Again, the groups preparing IPCC reports had many hot, contentious discussions on that issue. Working Group I, for example, initially balked at the notion of including subjective estimations, and then embraced it, but then said that they needed to have finer gradations, because they had real data, not just subjective judgments, and they wanted to have a 99 per cent and a 1 per cent. There were also interesting disciplinary differences. Linda Mearns at the National Center for Atmospheric Research, one of the few lead authors in two working groups, helped reconcile the physical scientists in Working Group I who were leery of subjectivity and risk management and the ecologists and social scientists in Working Group II who felt that society, not scientists, should choose how to take risks after *all* the possible conclusions were reported. It took us quite a long time to get both sides to first understand and eventually respect the other point of view. My role was not to endorse one or the other, but rather to be sure all our reporting was explicit about assumptions, so we could have a 'traceable account' of all underlying processes behind important conclusions. That process is building, but is not yet complete across the IPCC or the scientific community in general.[14]

WHERE NEXT?

As I've said, normally science strives to reduce uncertainty through data collection, research, modelling, simulation, and so forth. The objective is to overcome the uncertainty completely – to make known the unknown. Short of that, new information may narrow the range of uncertainty. No doubt further scientific research into the interacting processes that make up the climate system can reduce uncertainty about the response to increasing concentrations of greenhouse gases. This is very unlikely to happen quickly,

14 S.H. Schneider, *Science as a Contact Sport* (Washington DC, National Geographic Press, 2009a), 295 pp.

however, given the complexity of the global climate and the many years of high quality data which will be needed. Meanwhile, even the most optimistic 'business-as-usual' emissions pathway is projected to result in dramatic, dangerous climate impacts. That means making policy decisions before this uncertainty is resolved, rather than using it to justify delaying action.

Risk management also means understanding what is truly uncertain, and what is not. Sometimes critics claim that there should be no strong climate policy until the science is 'settled' and major uncertainties resolved, whereas supporters of strong policies suggest the science is already 'settled enough' and it is time to proceed with action to reduce risks. The science which demonstrates a significant warming trend over the past century is settled; moreover, it is virtually settled that the past several decades of warming have been largely caused by human activity and that much more is being built into the emissions pathways of the twenty-first century. Sounds like the 'settled already' side has won the debate: warming is occurring and human activities are the primary driver of recent changes.

That leaves the uncertainty about how severe warming and its impacts will be in the future, especially when projections for 'likely' warming by 2100 vary by a factor of six. The task then is to manage the uncertainty

> Al Gore may be considered a future-oriented decision-maker. He has done much to suggest to the public that the science of climate change is 'settled enough' and that human activities are the primary driver of recent changes.

ABIGAIL SHERRATT

Blitz spirit? It's needed again to beat global warming, Gore says

Ben Webster Environment Editor
Robin Pagnamenta Energy Editor

Al Gore invoked the spirit of Winston Churchill yesterday when he urged political leaders to follow the example of Britain's wartime leader in the battle against climate change.

The former US Vice-President accused governments around the world of exploiting ignorance about the dangers of global warming to avoid taking difficult decisions.

Speaking in Oxford at the Smith School World Forum on Enterprise and the Environment, sponsored by *The Times*, Mr Gore said: "Winston Churchill aroused this nation in heroic fashion to save civilisation in World War Two. We have everything we need except political will, but political will is a renewable resource."

Mr Gore admitted that it was difficult to persuade the public that the threat from climate change was as urgent as that from Hitler.

"The level of awareness and concern among populations has not crossed the threshold where political leaders feel that they must change," he said. "The only way politicians will act is if awareness raises to a level to make them feel that it's a necessity."

Mr Gore, who brought the issues around climate change to a mass audience with the 2006 documentary *An Inconvenient Truth*, said that the great

Churchill: fought to save civilisation

hope for the future lay in the high level of environmental awareness among young people.

He said sceptics who refused to believe that dramatic cuts in carbon emissions could be delivered should consider the example of the young scientists in the Nasa team who put a man on the Moon on 1969.

"The average age of scientists in the space centre control room was 26, which means they were 18 when they heard President Kennedy say he wanted to put a man on the Moon in ten years. Neil Armstrong did it eight years and two months later." He said

future generations would put one of two questions to today's adults.

"It will either be 'What were you thinking, didn't you see the North Pole melting before your eyes, didn't you hear what the scientists were saying?'. Or they will ask 'How is it you were able to find the moral courage to solve the crisis which so many said couldn't be solved?'."

Sir David King, the Government's former Chief Scientist and now director of the Smith School, also berated politicians. "I do think it's relatively easy for a prime minister to make a speech on climate change, which sounds committed, and very much more difficult for that prime minister to persuade the Treasury to put the finance behind that commitment to make it a reality.

"There is a long distance in government between saying what you think needs to be said and then making budgets available."

Sir David expressed disappointment that no senior British politician had agreed to address a conference attended by top climate scientists, business leaders and the presidents of the Maldives and Rwanda. "I tried to put in a lot of IOUs. But where was Lord Mandelson [the Business Secretary], where was Ed Miliband [the Energy and Climate Change Secretary]? Where was David Cameron? Where was William Hague?"

THE TIMES sseé

WORLD FORUM

ON ENTERPRISE & THE ENVIRONMENT

Al Gore urged political leaders to emulate Britain's wartime Prime Minister

rather than master it, to integrate uncertainty into climate research and policy-making. This kind of risk-management framework is often practised in defence, health, business and environmental decision-making. But the thresholds for action often seem lower. The US has a military arm, of course, and although I may not like everything we do with it, I don't know anybody who says you should get rid of it because a nation has to have security precautions, even against only very low probability – but potentially danger-ous – threats. Well, the climate change threat is not 1 per cent. It's more than 50 per cent for many really significant troubles, and maybe 10 per cent for absolutely catastrophic troubles.

In my personal value frame, it is already a few decades too late for having implemented some policy measures against such risks. Had we begun mitigation and adaptation investments decades ago, when a number of us advocated them,[15] the job of remaining safely below dangerous thresh-olds would be easier and cheaper. Similarly, beyond a few degrees Celsius of warming – at least an even bet if we remain anywhere near our current course – it is likely that many 'dangerous' thresholds will be exceeded. Strong action is long overdue, even if there is a small chance that by luck climate sensitivity will be at the lower end of the uncertainty range and, at the same time, some fortunate, soon-to-be-discovered low-cost, low carbon-emitting energy systems will materialise. For me, that is a high-stakes gamble not remotely worth taking with our planetary life-support system. Despite the large uncertainties in many parts of the climate science and policy assessments to date, uncertainty is no longer a responsible justi-fication for delay.

Adapted in part from S.H. Schneider, *Science as a Contact Sport* (2009a) and S.H. Schneider and M.D. Mastrandrea, 'Managing Climate Change Risk' (2009b).[16]

15 S.H. Schneider, *Science as a Contact Sport* (Washington DC, National Geographic Press, 2009a), 295 pp.
16 S.H. Schneider and M.D. Mastrandrea, 'Managing Climate Change Risk'. Chapter 15 in S.H. Schneider, A. Rosencranz, M.D. Mastrandrea and K. Kuntz-Duriseti (eds), *Climate Change Science and Policy* (Washington DC, Island Press, 2009b).

20

GREGORY BENFORD

Time: The Winged Chariot

Gregory Benford is a Professor of Physics at the University of California, Irvine, and author of over thirty books, mainly science fiction. His novels include *Timescape*, *Cosm*, *Beyond Infinity*, *What Might Have Been*, *The Sunborn* and the six-volume *Galactic Centre* series. His non-fiction includes *Deep Time* and *Beyond Human*.

THE ROYAL SOCIETY IS 350 YEARS OLD, AND STILL GOING. SCIENCE, TOO, WILL GO ON. HOW LONG FOR? WELL, ANSWERING THAT QUESTION NEEDS A PROPER UNDERSTANDING OF TIME – SOMETHING WHICH, AS GREGORY BENFORD EXPLAINS, REMAINS ELUSIVE AFTER ALL THESE YEARS.

But at my back I always hear
Time's winged chariot hurrying near;
And yonder all before us lie
Deserts of vast eternity.

– Andrew Marvell, *To His Coy Mistress*, 1652

Opposite:
Carlo Franzoni's 1810 sculptural figure of Clio, the muse of history, in a winged chariot recording history. Located in National Statutory Hall, the old house chamber of the US Capitol.

When the Royal Society began, time seemed a simple, obvious subject, understood since ancient ages. To Isaac Newton and his colleagues, two long-standing traditions pervaded the idea of time.

The Greeks, like most ancient cultures, saw their world as not completely chaotic, though it could be capricious. Faith in a definite order in nature promised that it could be understood by human reasoning. To them, some physical processes, at least, had a hidden mathematical basis,

and they sought to build a model of reality based on arithmetical and geometrical principles.

Adding to this Western tradition was the Judaic worldview, which had a timeline. God created the universe at some definite moment, arriving fresh and with a fixed set of laws. The Jews thought that the universe unfolds in a sequence running forward, which we now call linear time. Creation enabled evolution, which led forward in linear time to a future we could quite possibly change. This differed greatly from most other ancient cultures, which favoured cosmic cycles, probably by generalising from the march of the year's seasons. In cyclic time everything ends, but eventually returns, so there is eternal recurrence.

These two ideas, time's arrow vs. time's cycle, persist today in physics and also emerge in our art and literature. Physics has constrained time, ordering the music, but the dance between these linear and cyclic views continues.

Four hundred years ago, Europeans assumed a God-created universe that unfolded in orderly ways, in linear time, but that did not mean that the universe always had to be as we see it now. Change was possible, but constrained by physical laws. Einstein once remarked that what most interested him was whether God had any choice in his creation. The Abrahamic religious tradition answered with a resounding *yes*. Further, they insisted on nature's rationality, aided by mathematical principles. These were the only cultures to do so. This driving idea eventually altered the concept of time itself, as the cultural agenda played out in modern science.

The ancient Ouroboros symbol representing cyclicality or eternal recurrence.

Evolving Time

Time has two faces.

First, our sense of it passing seems inevitable, an automatic intuition. Unlike space, in which we can move back and forth, time hammers on relentlessly. This is Intuitive Time.

Second, we frame our position in time, our historical era, by looking at our slowly changing landscapes, and our societies. These alter on the scale that we ourselves see as we age. This is Historical Time.

Both these faces appeal, but they deceive us.

In the 1700s, the philosopher Immanuel Kant saw space and time as elements of a systematic mental framework, structuring our experiences. Spatial measurements tell us how far apart objects are, and temporal measurements show how far apart events occur. This eventually intersected Charles Darwin's idea that many abilities of organisms emerge from evolution by natural selection. Then it follows that time and space are the concepts we and other animals evolved to make the best use of the natural world. In this sense, they emerged from the primordial world where our minds evolved.

But that was not enough. Modern science reveals that time is supple, changeable, and even enigmatic. Further, we stand in a small slice of it, anchored in a moving moment that is an infinitesimal wedge compared with what has gone before, or will come after us. Our telescopes tell us of immensities of space, but other sciences – geology, biology, cosmology – speak of even grander scales of time.

Space and time are so familiar that we forget that they underlie the entire intricate and beautiful structure of scientific theory and philosophy. Perhaps it is not surprising that our first powerful theories built on assumed bedrock, metaphysical intuitions, came to be questioned only later. Clocks in Newton's universe ran everywhere the same. He invoked 'absolute, true and mathematical time' saying that it 'of itself, and from its own nature, flows equably without relation to anything external, and by another name is called duration'. Of the immense expanse of past time Newton had no true idea, for he took as gospel the Genesis story. Space was similarly absolute. Newton avoided the colossal scale of space by supposing that God had fixed up the cosmos so that gravity, the force he was the first to quantify, had not made it collapse – at least, so far.

This view held up well until the nineteenth century. By then even atheist scientists had faith in a lawlike order of nature – not from philosophy, but because it worked. Though this assumption springs from an essentially theological worldview, it gave useful predictions without a god attached. Still, few saw the full implications of regarding time as a subject of study, not belief.

The first collision between religious views and the study of the far past, which we now call Deep Time, came with the newborn science of geology. In 1830, the geologist Charles Lyell proposed that the features of Earth perpetually changed, eroding and re-forming continuously, at a roughly constant rate. This challenged traditional views of a static Earth with rare, intermittent catastrophes. In the eighteenth and nineteenth centuries the vast depth of the eras before humans arose became apparent, through development in geology and evolution's grand perspective. These still had to be licensed by physics, the more secure and quantitative science which sets the stage for the events and processes probed by the other sciences.

When William Smith and Sir Charles Lyell first recognised that rock strata represented successive long eras, they could estimate timescales only very imprecisely, since rates of geologic change varied greatly. Even these early attempts got the sciences into trouble. Creationists, reasoning from the Bible, had been proposing dates of around six or seven thousand years for the age of the Earth based on the Bible. Early geologists suggested millions of years for geologic periods, with some even suggesting a virtually infinite age for the Earth. Geologists and palaeontologists constructed geologic history based on the relative positions of different strata and fossils, estimating the timescales based on studying rates of various kinds of weathering, erosion, sedimentation and lithification. The ages of assorted

Charles Lyell.

rock strata and the age of the Earth were hotly debated. In 1862, the physi-cist William Thomson, whose authority endured – as Lord Kelvin and President of the Royal Society – until the end of the century, set the age of Earth at between 24 million and 400 million years. He assumed that Earth began as a completely molten ball of rock, then calculated how long it took to cool to its present temperature. He did not know of the ongoing heat source from radioactive decay.

Physicists had more prestige, but even then, geologists doubted such a short age for Earth. Biologists could accept that Earth might have a finite age, but even 100 million years seemed much too short for evolution to have yielded such complex plenty. Charles Darwin argued that even 400 million years did not seem long enough.

Until the discovery of radioactivity in 1896, and the development of its geological applications through radiometric dating during the first half of the twentieth century (pioneered by geologists), there were no precise absolute datings of rocks.

Radioactivity introduced another measuring clock. Geologists quickly realised this upset the assumptions used before. They re-examined their estimates. This moved the age into the billions (thousands of millions) of years, sweeping away Archbishop Ussher's biblically inspired dating of Creation to 4004 BC.

Much public ferment paralleled this scientific research and its clash with religion. But by the early twentieth century, opinion settled on an Earth older than a billion years.

Physics, meanwhile, was making hash of the simple view of time that underlay the other sciences. Geology, biology and astronomy would have been happy with Newtonian time, giving them a simple marker of change. The physicists, though, worried about more basic matters.

RELATIVE TIME

In physics, time is, like length, mass and charge, a fundamental quantity – intuitive, given by our basic perceptions. Newton used this view, holding that '*I do not define time, space, place and motion, as being well known to all*' – i.e., obvious. But Einstein showed that it was not.

Nineteenth-century physicists felt that space was the most basic and irreducible of all things. It persisted while time changed, and points made up space – infinitesimal grains close-packed. Einstein's fundamental insight was that space and time, which appear so different to us, are in fact

linked. He argued this using *gedanken* (thought) experiments involving rulers and clocks. These were not just instruments to Einstein; he took them to generate space and time, since they represent it.

He took two basic assumptions. First, the speed of light seems the same to everyone in the universe, whether moving or sunk deep in a gravitational well. This may strike us as odd, but an earlier experiment had found it to be so. Not that Einstein cared; his intuition led him to the conclusion. He proved it valid by using the even deeper second assumption: that the laws of physics had to treat all states of motion on the same footing.

Combining these two assumptions generates the equations of his Special Theory of Relativity. It has astonishing consequences. Moving objects experience a slower passage of time. This is known as time dilation. These transformations are only valid for two frames at *constant* relative velocity. Naïvely applying them to other situations gives rise to such puzzles as the famous twin paradox.

This is a thought experiment in special relativity, in which a twin makes a journey into space in a fast rocket, returning home to find he has aged less than his identical twin, who stayed on Earth. This result appears puzzling because the laws of physics should exhibit symmetry. Since either twin sees the other twin as travelling, each should see the other ageing more slowly. How can an absolute effect (one twin really does age less) come from a relative motion? Hence it is called a paradox.

But there is no paradox, because there is no symmetry. Only one twin accelerates and decelerates, so this differentiates the two cases. Since we each experience minor accelerations, whether on horseback or in a jet plane, we each carry around our own personal scale of time. These are undetectable in ordinary life, but real.

When time stretches, space shrinks. When you rush to catch an aeroplane, the wall clock you see runs a tiny bit slower than your wristwatch. Compensating for the time, the distance to the aeroplane's gate looks closer to you. Time is pricey, though – a second of time difference translates to 300,000 kilometres of space.

The stretching of space and time occurs because they are wired together. More fundamentally, Einstein's work implied that time runs slower the stronger is the gravitational field (and hence the observer's local acceleration). His general relativity theory sees gravity not as a force but as a distortion of space-time.

The rates of clocks on Earth then depend on whether they are on a mountain or in a valley; the valley clock runs slower. This is somewhat like the slowing of clocks as they move past us at high velocity. This gravitational effect is unlike that of the smoothly moving observers on, say, two trains moving by each other, each of whom thinks the other's clock runs slowly. In a gravitational field, the clocks experience different accelerations if they are not at the same altitude. But observers both in the valley and on the mountain agree that the mountain clock runs faster. Experiments checked these results and found complete agreement. Further, particle acceleration experiments and cosmic ray evidence confirmed the predictions of time dilation, where moving particles decay slower than their less energetic counterparts. Gravitational time changes give rise to the phenomenon of gravitational 'redshift', which means that light loses energy as it rises against gravity. There are also well-documented delays in signal travel time near massive objects like the Sun. Today, the Global Positioning System must adjust signals to account for this effect, so the theory has even practical effects.

In empty space, the shortest distance between two points is a straight line. In space-time, this is called a 'world line' that forms the shortest curve between two events. If gravitation curves a space-time, then the straight line becomes a curve, which is the shortest space-time distance between two points. That curvature we see as the curve of a ball when thrown into the distance, a parabola.

This linked with a radical view, pushed by Hermann Minkowski, that neither space nor time is truly fundamental. In relativity, both are mere shadows, and only a union of the two exists in the underlying reality. Minkowski had called Einstein a 'lazy dog' when Einstein was his student. But while reading Einstein's first paper on relativity, he had a brilliant idea, and so laid the foundations for the next great insight. Minkowski's invention was space-time, a joint entity. Einstein later used his intuition to propose that mass curves space-time, and we sense this curvature as gravity.

The fundamental idea of space-time played out in many ways. Time runs faster in space than on a star, because gravity warps space-time. This leads to timewarps that can become severe, when a star implodes and time grinds to a halt. Stars a few times larger than our own can do this, capturing their own light and plunging into an infinitesimal speck we call a black hole. Its gravitation remains with us, though, a timewarp imprinted on empty space. Anyone falling along with the star will see the external world pass through all of eternity, while gravity pulls him into a spaghetti strand. The singularity where all ends up is a 'nowhen' and 'nowhere', signifying that the physical universe as we understand it ceases.

Einstein's singular geometric and kinematic intuition motivated his theory. He assumed that every point in the universe can be treated as a 'centre', whether it is deep in a gravitational well (such as where we live) or in empty space, far from curvatures in space-time induced by gravity. Correspondingly, he reasoned, physics must act the same in all reference frames. This simple and elegant assumption led, after much labour, to a theory showing that time is relative to both where you are and how you are moving. Newton's laws hold well enough in a particular local geometry.

They work in different circumstances, though they must be modified for the environment. Still, this fact can be expressed in the theory itself. This leads to the principle that there is no 'universal clock'. To get things right, we must perform some act of synchronisation between two systems, at the very least.

There is another victim of his intuition. Not only is there not a universal present moment, but also there is no simple division between past, present and future in general – that is, everywhere in the universe. Locally, they do mean something, but not necessarily to those far from us, in a universe that continually expands.

Though you and I on Earth may agree about what 'now' means on the nearest star, Proxima Centauri, an astronaut moving quickly through the solar system who asks this same question when we do will refer to a different moment on Proxima Centauri.

Does this mean that only the present moment 'really' exists? But one person's past can be another's future, so past, present and future must exist in a physical sense, and so be equally real.

Einstein said of the death of an old friend, within months of his own death, 'Now he has departed from this strange world a little ahead of me. That means nothing. People like us, who believe in physics, know that the distinction between past, present and future is only a stubbornly persistent illusion.'

In physics, time is not a sequence of happenings, but a chain that is just there, embedded in space-time. Our lives move along that chain, like a train on a track. Observers differ over whether a given event occurs at a particular time, but there is no universal Now. Instead, an event belongs to a multitude of Nows, depending on others' states of motion or position. Time stretches away into past and future, as we see them, just as space extends away from any place. This is the interwoven thing we call space-time, and it is more fundamental than our particular sense of our local world.

Even more odd possibilities come from these ideas. General relativity allows time travel of a sort, in special circumstances. These may be disallowed by a more fundamental theory, but for now, some puzzling paradoxes emerge from our understanding of time. Presumably events may not happen before their cause, but proving this in general has so far eluded us.

TIME'S MOMENTUM

Time goes, you say? Ah no!
Alas, time stays, we go.
– Henry Austin Dobson, *The Paradox of Time*, 1877

Why do we think that time moves, instead of the fixed, eternal space-time that theory suggests? Because evolution has not selected us to see it that way. Time's flow is a simple way to order the world effectively; that does not mean it is fundamental. Space-time is simple and elegant, but that does not mean it plays well in the rough scramble of life. During a seminar at Princeton University, Einstein remarked that the laws of physics should be simple. Someone asked, 'What if they aren't?' Einstein replied that if so, he was not interested in them.

Yet simplicity may not be the best way to regard time. Time seems to flow because that flow is a holistic concept, not reducible to simple systems like a collision of atoms. In this sense, the paradox of time's flow is an aspect of our minds. We can see time as moving, bringing events to us, or the reverse: we flow through time, sensing a moving moment.

This interlaces with the findings of Sadi Carnot in 1824, when he carefully analysed steam engines with his Carnot cycle, an abstract model of how an engine works. He and Rudolf Clausius noted that disorder, or entropy, steadily increases as machines operate. This means the amount of 'free energy' available continually decreases.

This is the second law of thermodynamics. The continual march of time then defines an arrow of time, defined by the growth of entropy. It is

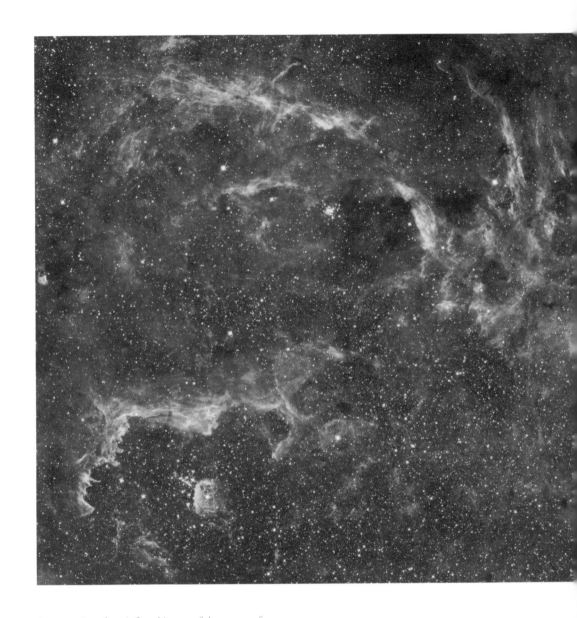

A composite colour infrared image of the centre of our
Milky Way galaxy – the sharpest infrared picture ever
made of the galactic core. At the centre of the image,
ionised gas surrounding the supermassive black hole at the
centre of the galaxy is confined to a bright spiral embedded
within a circumnuclear dusty inner-tube-shaped torus.

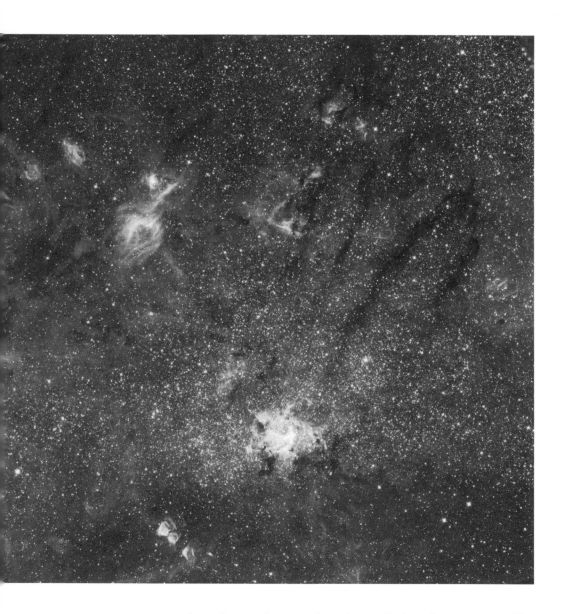

easy enough to observe the arrow by mixing a little milk into your coffee.
Try as you might, you can't reverse it. In the nineteenth century entropy's
increase took its place beside other definitions of time's momentum.
Another definition is the psychological arrow of time, whereby we see an
inexorable flow, dominating our intuitions. The third view, a cosmological

arrow of time, emerged when we discovered the expansion of the universe in the twentieth century.

This dramatic time asymmetry seems to offer a clue to something deeper, hinting at the ultimate workings of space-time. For example, suppose gravity acts on matter – what is the maximum entropy nature can pack into a volume? There is a clean answer: a black hole. In the 1970s Stephen Hawking of Cambridge University, holding the chair Newton had, showed that black holes fit neatly into the second law. Originally the second law described hot objects like steam engines. Applied to black holes, that can also emit radiation and have entropy, the second law shows that a three-million-solar-mass black hole, such as the one at the centre of our galaxy, has a hundred times the entropy of all the ordinary particles in the observable universe. This is astonishing. Collapsed objects are giant repositories of disorder, and thus sinks of the productions of time itself.

These ideas spread throughout science, with varying results. Entropy inevitably increases in thermodynamics, but that seems to fly in the face of our own world, which flourishes with new life forms and increasing order. In contrast to the physical view of time, biologists pointed out that life depends on a 'negative entropy flow' which is local, driven by a larger decrease elsewhere. For us, this 'elsewhere' is the Sun, which supports our entire natural world. The Sun will expand and engulf the Earth in about five billion years. By then we may have a fix for that problem, if we are still around as humans. But then the stars themselves will die out, having burned their core fuels, this will take several tens of billions of years more, and thereafter the universe will indeed cool and entropy will rise throughout.

Increasing entropy implies a 'heat death' as our universe expands. This means the end of time will be cold and dark.

So biological systems do not refute the arrow of time; they define it well during our present, early state of the universe. These realisations ran in parallel through the nineteenth and twentieth centuries, promoting fruitful scientific dialogue.

Deep Time Revisited

The human perception of time has ramified through many sciences. Such fundamental changes in a basic view always echo through culture.

The enormous expansion of our perceptions of time has altered the way we think of ourselves, framed in nature. Palaeontologists track the extinction of whole genera, and in the random progressions of evolution feel the pace of change that looks beyond the level of mere species such as ours. Geologists had told them of vast spans of time, but even that did not seem to be enough to generate the order we see on Earth.

The Darwin–Wallace theory explains our Earthly order as arising from evolution through natural selection. As perhaps the greatest intellectual event of the nineteenth century, it invokes cumulative changes that add up. The fossil record showed that mammals, for example, can take millions of decades to alter significantly. Our own evolution has tuned our sense of probabilities to work within a narrow lifetime, blinding us to the slow sway of long biological time. (And to the fundamental physical space-time, as we discussed.)

This may well be why the theory of evolution came so recently, in an era when our horizons were already quickly expanding; it conjures up spans of time far beyond our intuition. On the creative scale of the great, slow and blunt Darwinnowings, such as we see in the fossil record, no human monument can endure. But our neophyte primate species can now bring extinction to many, and no matter what the clock, extinction is for ever. We live in hurrying times.

Yet we dwell among contrasts between our intuitions and the timescape of the sciences. In their careers, astronomers discern the grand gyre of worlds. But planning, building, flying and analysing a single mission to the outer solar system commands the better part of a professional life. Future technologies beyond the chemical rocket may change this, but there are vaster spaces beckoning beyond which can still consume a career. A mission scientist invests the kernel of his most productive life in a single gesture toward the infinite.

Those who study stars blithely discuss stellar lifetimes encompassing billions of years. In measuring the phases of stellar mortality they employ the many examples, young and old, that hang in the sky. We see suns in snapshot, a tiny sliver of their grand and gravid lives caught in our telescopes. Cosmologists peer at distant galaxies whose light is reddened by the universal expansion, and see them as they were before Earth existed. Observers measure the microwave emission that is relic radiation from the earliest detectable signal of the universe's hot birth. Studying this energetic emergence of all that we can know surely imbues (and perhaps afflicts) astronomers with a perception of how like mayflies we are.

No human enterprise can stand well in the glare of such wild perspectives. Perhaps this is why for some science comes freighted with coldness, a foreboding implication that we are truly tiny and insignificant on the scale of such eternities. Yet as a species we are young, and promise much. We may yet come to be true denizens of Deep Time.

COSMOLOGICAL TIME

Through the twentieth century's developing understanding of stellar evolution, astronomy outpaced even the growing expanses of biological time by dating the age of stars. These lifetimes were several billion years, a fact some found alarming. In the mid-twentieth century some globular clusters of stars even seemed to be older than the universe, a puzzle that better measurement resolved.

However, a still grander canvas awaited. Perhaps the most fundamental aspect of time lies in our description of how it all began, along with the universe itself: cosmology.

There were many 'origin stories' of earlier cultures, but these gave little thought to how the universe came to be, beyond simple stories. Ancient times, until the nineteenth century, preferred eternity to process. As the *Bhagavad Gita* says, 'There never was a time when I was not … there will never be a time when I will cease to be.' Since time and space began together – as both St Augustine and the big bang attest – the *Bhagavad Gita* has a point. The chicken and the egg arrived at the same time.

Yet Newton thought that the universe had to be eternally tuned by God's hand, or else gravitation would cause it to collapse. This view held fairly well until a new theory of gravity and time arrived.

When Einstein developed his theory of general relativity in 1915, physicists believed in a perfectly static universe without beginning or end, like Newton. Though he had a theory of curved space-time, and so could consider all the universe, Einstein inherited this bias. He attempted the first true cosmology – that is, a complete description of the universe's lifetime, from simple assumptions – under the influence of the ancients.

To make his early equations describe a universe unchanging in time, he added a cosmological constant to his theory to enforce a static universe. It had matter in it, which he knew meant that gravitation favoured collapse – but he demanded that it be a time-independent, eternal universe. Analysis soon showed that Einstein's static universe is unstable. A small ripple in space-time or in the mass it contained would make the universe either expand or contract. Einstein had brought his own concepts of time to the issue, and so missed predicting the expanding universe. Soon enough, astronomers' observations showed that our universe is expanding from an earlier, smaller event. After this era, cosmological ideas of time moved beyond him.

Modern cosmology developed along parallel observational and theoretical tracks in the twentieth century. Correct cosmological solutions of general relativity emerged, and astronomers found that distant galaxies were apparently moving away from us. This comes from the expansion of space-time itself, not because we are uniquely abhorrent. Tracking this expansion backward gave a time when space-time approached zero. St Augustine had proposed that God made both space and time, and the big bang told us when that was.

Through the twentieth century, observations of how fast distant galaxies seemed to rush away from us have pushed the age of the universe back to the currently accepted number of 13.7 billion years. By then relativity had altered and even negated our understanding of Intuitive Time, so cosmology's enormous extension of Deep Time only added to the startling changes.

Now astronomers observe that the universal expansion is accelerating, perhaps because of the unknown effects represented by Einstein's added cosmological constant. We now seem to occupy an unusual niche in the long history of this universe, living beyond the early, hot era, yet well before the accelerating expansion will isolate galaxies from each other, then stars, and finally may wrench apart all of matter as space-time stretches ever-faster. Time seems then like a judge, not a mere clock.

***

The essential dilemmas of being human – the contrast between the stellar near-immortalities we see in our night sky, and our own all-too-soon, solitary extinctions – are now even more dramatically the stuff of everyday experience. We now know what a small sliver we inhabit in the long parade of our universe. Who can glimpse these perspectives and not reflect on our mortality? We are mayflies. Yet we now know enough of time and our place in it to reflect upon truly immense issues.

Time is a fundamental, its nature slowly glimpsed. After all this time, we do not fully understand it.

Here, on the level sand
Between the sea and land,
What shall I build or write
Against the fall of night?
– A.E. Housman

Freundschaftlich überreicht
von Ihrem A. Einstein

Über die spezielle und die

allgemeine Relativitätstheorie

(Gemeinverständlich)

Von

A. EINSTEIN

Mit 3 Figuren

Dieses Exemplar ist
das erste, welches die
Druckpresse verlassen hat.
Es wurde mir von
Prof. Einstein zugeschickt unmittelbar nachdem
er es empfangen hatte,
kurz bevor ich nach
Frankreich ins Feld zog.

Hans Museum
Berlin (z.E. Französische
Front. April 1917.

Braunschweig

Druck und Verlag von Friedr. Vieweg & Sohn

MARTIN REES

CONCLUSION:
LOOKING FIFTY YEARS AHEAD

Martin Rees FRS is Professor of Cosmology and Astrophysics and Master of Trinity College at the University of Cambridge. In 2005 he was appointed to the House of Lords and elected President of the Royal Society. He writes and broadcasts regularly about science, and among his books are *Our Final Century: Will the Human Race Survive the Twenty-First Century?* (2003), *Just Six Numbers* (1999) and *Before the Beginning: Our Universe and Others* (1997).

I N 350 YEARS, OUR UNDERSTANDING OF THE UNIVERSE HAS EXPANDED BEYOND THE DREAMS OF THE FOUNDERS OF THE ROYAL SOCIETY. BUT SCIENTISTS NEVER REACH FINALITY, WRITES MARTIN REES. NEW KNOWLEDGE AND NEW APPLICATIONS WILL MAKE A VITAL CONTRIBUTION TO HUMANITY IN THE COMING DECADES.

The Royal Society's founders were inspired by the English philosopher and statesman Francis Bacon. For Bacon, science was driven by two imperatives: the search for enlightenment, and 'the relief of man's estate'. Christopher Wren, Robert Hooke, Robert Boyle and the other 'ingenious and curious gentlemen' who regularly convened in Gresham College were enthusiasts for what we would now call 'curiosity-driven' research. But they engaged also with the practical life of the nation. Indeed, in 1664 John Evelyn reported on the optimum management of forests to ensure a steady supply of good oak for the navy's ships. And the first issue of *Philosophical Transactions* – the world's oldest surviving scientific periodical – contained a paper by Christiaan Huygens on improvements to the pendulum clock and how to get it patented.

Bacon's dichotomy is still germane today: a former President of the

Royal Society, George Porter, encapsulated it by the maxim 'there are two kinds of science, applied and not yet applied'. There can be no better aim, for the next fifty years, than to sustain the curiosity and enthusiasm of our founders, while also achieving the same broad engagement with society and public affairs as they did.

The Society aims, above all, to support and recognise the creative individuals on whom scientific advance depends. What issues will engage such people in 2060, when the Society celebrates its 400th anniversary? Will we continue to push forward the frontiers, enlarging the range of our consensual understanding?

What Will We Understand in 2060?

It is sometimes claimed that the big ideas have been discovered already, and that it only remains to fill in the details and apply what is already known. But nothing could be more wrong. Science is an unending quest: as its frontiers advance, new mysteries come into focus just beyond those frontiers. Most of the questions now being addressed simply couldn't have been posed fifty years ago (or even twenty); we can't conceive what problems will engage our successors.

A prime aim is to understand our world – and, in my own field of astronomy, to probe what lies beyond it. Just as geophysicists have come to understand the processes that made the oceans and sculpted the continents, so astrophysicists can understand our Sun and its planets – and even the other planets that may orbit distant stars. Astronomy is the grandest environmental science. And our exploration is just beginning. There are still domains where, in the fashion of ancient cartographers, we must inscribe 'here be dragons'.

Armchair theory alone cannot achieve much. We are no wiser than Aristotle was. It is technical advances that have enabled astronomers to probe immense distances, and to trace the evolutionary story back before our solar system formed, back to an epoch long before there were any stars, when everything was initiated by an intensely hot 'genesis event', the so-called big bang. The first microsecond is shrouded in mystery, but everything that happened since then – the emergence of our complex cosmos from amorphous beginnings – is the outcome of processes that we are starting to grasp in outline. And our cosmic horizons are still expanding. What we've traditionally called our universe could be just one island – just one patch of space and time – in an infinitely larger cosmic archipelago.

Could there be, far beyond our Earth, other forms of life – perhaps even more complex and advanced than humans? Here again we're flummoxed. Until we find out how life began on Earth we can't understand how likely it is that life may have started elsewhere – nor where to focus our search. However, as Paul Davies describes, there is now some progress: exciting new ideas, and new ways to seek signs of life beyond our home planet. Perhaps we'll one day 'plug in' to a galactic community. On the other hand, searches for extraterrestrial intelligence may fail. Earth's intricate biosphere may be unique. Either way, the search for alien life – exobiology – will surely be one of the most exciting scientific frontiers in the next fifty years.

An undoubted intellectual peak of twentieth-century science was the quantum theory, which describes how atoms behave, and how they combine with each other to make the complex chemistry of the everyday world. The second 'peak' was Einstein's general relativity. More than two hundred years earlier, Isaac Newton had achieved the first major 'unification' by showing that the force that makes apples fall is the same as the gravity that holds planets in their orbits. Newton's mathematics is good enough to fly rockets into space and steer probes around planets. But Einstein transcended Newton: his general theory of relativity could cope with very high speeds, and strong gravity, and offered deeper insight into gravity's nature.

A synthesis of these two great theories – an overarching theory that links the cosmos and the microworld, and applies the quantum principle to space, time and gravity – is unfinished business for the twenty-first century.

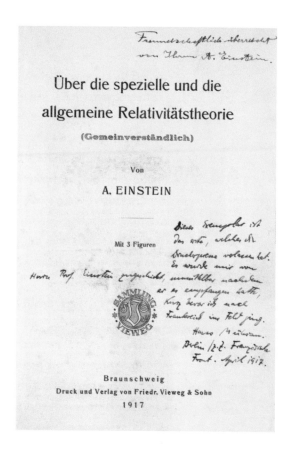

Success will require new insights into what might seem the simplest entity of all: 'mere' empty space. Space itself may have a rich structure – on scales a trillion trillion times smaller than an atom, and also on scales far larger than the entire universe we know.

Einstein was not a first-rate mathematician, despite his deep physical insights. He was lucky that the geometrical concepts he needed had already been developed by the German mathematician Georg Riemann a century earlier. The cohort of young quantum theorists led by Erwin Schrödinger, Werner Heisenberg and Paul Dirac were similarly fortunate in being able to apply ready-made mathematics.

But the twenty-first-century counterparts of these great physicists – those seeking to mesh general relativity and quantum mechanics in a unified theory – are not so lucky. The most favoured theory posits that all subatomic particles are made up of tiny loops, or strings that vibrate in a space with ten or eleven dimensions. String theory involves intensely complex mathematics that certainly can't be found on the shelf and offers a creative stimulus to 'real' mathematicians.

Einstein himself worked on an abortive unified theory till his dying day. In retrospect it is clear that his efforts were premature – too little was then known about the forces and particles that govern the subatomic world. Cynics have said that he might as well have gone fishing from 1920 onwards. But there's something rather noble about the way he persevered and 'raised his game' – reaching beyond his grasp. (Likewise, Francis Crick, the driving intellect behind molecular biology, shifted, when he reached sixty, to the 'Everest' problems of consciousness and the brain even though he knew he'd never get near the summit.)

Einstein averred: 'The most incomprehensible thing about the universe is that it is comprehensible.' It is remarkable that atoms on Earth are the same as in distant stars. And that our minds, which evolved – along with our intuitions – to cope with life on the African savannah, can grasp the highly counterintuitive laws governing the quantum world and the cosmos.

Scientists can never reach finality. Let me recall something that puzzled Isaac Newton three hundred years ago. He could explain why the planets traced out ellipses around the Sun, but the initial 'set-up' of the solar system was a mystery to him. Why were the orbits of the planets all close to a single plane, the ecliptic, whereas the comets plunged in from random directions? In his book *Opticks* he writes: 'blind fate could never make all the planets move one and the same way in orbits concentrick'. 'Such a wonderful uniformity' must, he claimed, be the result of providence. This coplanarity of the orbits, however, is now understood: it's a natural outcome of the solar system's origin as a spinning protostellar disc. Indeed, we can trace things back far further still – to the initial instants of the big bang.

But this 'flashback' to Newton reminds us that, in conceptual terms, things are not qualitatively different from his time. However much the causal chain may have been lengthened – however much further back we can trace our origins than he could – we still at some stage have to say 'things are as they are because they were as they were'.

The phrase 'theory of everything', often used in popular books to denote a unification of the fundamental forces, has connotations that are not only hubristic but very misleading. Such a theory would actually offer absolutely zero help to 99 per cent of scientists. There is another open frontier: the study of things that are very complicated. This is the frontier on which most scientists work. They aren't impeded at all by ignorance of subnuclear physics or the big bang. They are challenged and perplexed by complexity – by the way atoms combine to make all the intricate structures in our environment, especially those that are alive.

There are nonetheless reasons to hope that simple underlying rules might govern some seemingly complex phenomena. John Conway is one of the most charismatic figures in mathematics. His research deals with a branch of maths known as group theory. But he reached a wider audience with his 'game of life'. In 1970 Conway (then based in Cambridge) wanted to devise a game that would start with a simple pattern and use basic rules to evolve it again and again. He began experimenting with the black and white tiles on a Go board and discovered that by adjusting the simple rules and the starting patterns, some arrangements produced incredibly complex results seemingly from nowhere. The simple rules merely specify when a white square turns into a black square and vice versa. But when applied over and over again, they create a fascinating variety of complicated patterns. Objects emerged that seemingly had a life of their own as they moved around the board. Some of them can reproduce themselves. The real world is like that – simple rules allow complex consequences.

The sciences are sometimes likened to different levels of a tall building: logic in the basement, mathematics on the ground floor, then particle physics, then the rest of physics and chemistry, and so forth, all the way up to psychology, sociology – and the economists in the penthouse. But the analogy is poor. The superstructures, the 'higher level' sciences dealing with complex systems, aren't imperilled by an insecure base, as a building is. There are laws of nature in the macroscopic domain that are just as much of a challenge as anything in the micro world, and are conceptually autonomous: for instance, those that describe the transition between regular and chaotic behaviour, and which apply to phenomena as disparate as dripping water pipes and animal populations.

Problems in chemistry, biology, the environment and human sciences remain unsolved because scientists haven't elucidated the patterns, structures and interconnections – not because we don't understand subatomic physics well enough. In trying to understand how water waves break, and how insects behave, analysis at the atomic level doesn't help. An albatross may return predictably to its nest after wandering thousands of miles in the Southern ocean. But its behaviour couldn't be predicted, even in principle, by regarding it as an assemblage of atoms and solving Schrödinger's equation. Finding the sequencing of the human genome – discovering the string of molecules that encode our genetic inheritance – is one of the greatest achievements of the last decade. But it is just the prelude to the far greater challenge of post-genomic science: understanding how the genetic code triggers the assembly of proteins, and expresses itself in a developing embryo.

It may seem topsy-turvy that cosmologists can speak confidently about galaxies billions of light years away, whereas theories of diet and child rearing – issues that everyone cares about – are still tentative and controversial. But astronomy is, quite genuinely, far simpler than the human sciences. Stars are simple: they're so big and hot that their content is broken down into simple atoms – none match the intricate structure of even an insect, let alone a human. Our everyday world presents twenty-first-century Einsteins with intellectual challenges just as daunting as those of the cosmos and the quantum.

The 'Relief of Man's Estate'

The Royal Society's founders, though fascinated by weird animals, air-pumps and telescopes, were also engaged with the practical issues of their time – the rebuilding of London, navigation and the exploration of the New World. Our horizons have expanded. But the same engagement is imperative in the twenty-first century: there are more people than ever on our planet, all empowered by ever more powerful technology.

Technology advances in symbiosis with science. Computers, for instance, owe their burgeoning power to progress in materials science (and in mathematics too, as Ian Stewart's chapter reminds us). The silicon chip was perhaps the most transformative single invention of the last century. It has allowed miniaturisation, spawning mobile phones and an Internet with global reach – promoting economic growth, while being sparing of energy and resources.

It was physicists who developed the World Wide Web, and the international scientific community has benefited immensely. Astronomers or geneticists can quickly download any body of data and analyse it. And the Internet has hugely benefited our colleagues in the developing world who formerly depended on slow and inefficient postal services.

A visual representation of the Internet.

A few years ago, three young Indian mathematicians invented a faster scheme for factoring large numbers – something that would be crucial for code-breaking. They posted their results on the web. Such was the interest that within just a day, twenty thousand people had downloaded the work, which was the topic of hastily convened discussions in many centres of mathematical research around the world.

There is a stark contrast here with the struggles of an earlier Indian mathematician to achieve recognition. In 1913 Srinivasa Ramanujan, a clerk in Mumbai, mailed long screeds of mathematical formulae to G.H. Hardy in Cambridge. Hardy had the percipience to recognise that Ramanujan was not the typical green-ink scribbler who finds numerical patterns in the bible or the pyramids. He arranged for Ramanujan to come to Cambridge, and did all he could to foster his genius. (Ramanujan became an FRS. But culture shock and poor health led him to an early death.)

Advances in information technology amaze us by their rapidity – iPhones would have seemed magic thirty years ago. Each mobile phone today – indeed, each washing machine – has more computing power than NASA could deploy on the Apollo programme. We can't of course guess what twenty-first-century inventions will seem 'magic' to us today. Scientists have a poor record as forecasters. Ernest Rutherford averred that nuclear energy was moonshine; Ken Olson, founder of Digital Equipment Corporation (DEC), said, 'There is no reason anyone would want a computer in their home'; and an earlier Astronomer Royal said space travel was utter bilge. I have no crystal ball and won't add to this inglorious roll call.

Francis Bacon pointed out that the most transformative advances are the least predictable. He cited gunpowder, silk and the mariner's compass, and contrasted them with (for instance) the techniques for printing, which progressed incrementally.

Incremental steps from today's technology will, perhaps within a decade, offer each of us (at least in the developed world) high-bandwidth communication with everyone else, and instant access to all recorded

Illustrating how far the silicon chip has transformed our world, mobile phones continue to become slimmer with increasing computing power.

knowledge, all music and all visual art. As the genome is better understood, the read out of our genetic code may tell us how (and perhaps when) we are most likely to die. Computer networks will continue to become ever more powerful and pervasive.

Computers may, within less than fifty years, achieve a wide range of human capabilities. Of course, in some respects this has happened already. The most basic pocket calculators can hugely surpass us at arithmetic. IBM's 'Deep Blue' beat Garry Kasparov, the world chess champion. But not even the most advanced robot can recognise and handle the pieces on a real chessboard as adeptly as a five-year-old child. There's a long way to go before interactive human-level 'robotic intelligence' is achieved.

An arena where advanced robots will surely have clear advantages over humans is outer space. By mid-century, the entire solar system will have been explored by flotillas of tiny robotic craft. And, even if people haven't followed them, 'fabricators' may perform large construction projects, using raw materials that need not come from Earth.

Future robots may relate to their surroundings (and to people) as adeptly as we do, through our eyes and other sense organs. Indeed, their far faster 'thoughts' and reactions could give them an advantage over us. Everyone's lifestyle and work patterns will then surely be transformed. Robots will be perceived as intelligent beings, to which (or to whom) we can relate, at least in some respects, as we would to our fellow-humans. Moral issues then arise. We generally accept an obligation to ensure that other human beings (and at least some animal species) can fulfil their 'natural' potential. Will we have the same duty to sophisticated robots, our own creations? Should we feel guilty about exploiting them? Should we fret if they are underemployed, frustrated, or bored?

'Deep Blue' didn't work out its strategy like a human player: it exploited its computational speed to explore millions of alternative series of moves and responses before deciding an optimum move. Likewise, machines may make scientific discoveries that have eluded unaided human brains – but by testing out millions of possibilities rather than via a theory or strategy. However, the programmer will get the acclaim – just as, in Olympic equestrian events, the medal goes to the rider, not the horse.

Some kind of mental prosthetics may become essential if theorists are to make headway in the most difficult fields. Meteorology and astronomy have been hugely boosted by the ability to simulate a 'virtual universe'. A unified theory of the physical forces, or a theory of consciousness, might be beyond the powers of unaided human brains, just as surely as quantum mechanics would flummox a chimpanzee.

Another speculation – and a 'wild card' in population projections – is that the human lifespan could be substantially extended. Some Americans, worried that they'll die before this nirvana is reached, bequeath their bodies to be 'frozen' on their death, hoping that future generations will resurrect them or download their brains into a computer. For my part, I'd rather end my days in an English churchyard than a Californian refrigerator.

But flaky futurologists aren't always wrong. Students can derive more stimulus from first-rate science fiction than from second-rate science. We should keep our minds open, or at least ajar, to wacky-seeming concepts – just as the Royal Society's first Fellows did 350 years ago.

A Hazardous World

One thing we can be sure of, however: there will be an ever-widening gulf between what science allows us to do and what it is prudent or ethical to do – more doors that science could open but which are best kept closed. In respect of (for instance) human reproductive cloning, genetically modified organisms, nanotechnology and robotics, regulation will be called for, on ethical as well as prudential grounds.

But the social and geopolitical context in which these issues will be debated fifty years hence is even harder to forecast than the science itself. The upheavals of the present century will surely be as turbulent as those in the last.

An overwhelming challenge for governments will be to ensure food, energy and resources for a rising and increasingly empowered population, and to avoid catastrophic environmental change or societal disruption. By 2060 there will, barring a global catastrophe, be far more people than today. Fifty years ago the world population was below 3 billion. It has more than doubled since then, to 6.8 billion today. And it's projected to reach around 9 billion by mid-century. By then, it will be in Asia – not Europe nor the US – that the world's physical and intellectual capital will be concentrated.

More than half of the world's people live in countries where fertility has now fallen below replacement level. If this trend quickly extended world-wide, then the global population could gradually decline after mid-century – a development that would surely be benign.

Another firm prediction is that, half a century from now, the world will be warmer than today – though by how much is uncertain, as Stephen Schneider's chapter explains. Shifts in weather patterns (especially in rainfall) impact most grievously on those least able to adapt, and on countries that have themselves contributed minimally to global CO_2 emissions. The prospects seem especially gloomy in Africa, where there will be a billion more people by mid-century than there are today and the birth rate remains high. Climate change aggravates the challenge of feeding this growing population. What should make us more anxious is the significant

Many Asian cities, such as Shanghai, have grown rapidly since the 1990s in terms of both population and economic might.

probability of triggering a grave and irreversible global trend: rising sea levels due to the melting of Greenland's icecap; runaway release of methane in the tundra, and so forth.

Collective human actions are ravaging the biosphere and threatening biodiversity. There have been five great extinctions in the geological past. Humans are now causing a sixth. The extinction rate is a thousand times higher than normal and is increasing. In the words of Robert May, my immediate predecessor as Royal Society President, 'we are destroying the book of life before we have read it'. Our Earth harbours millions of species that have not yet even been identified – mainly insects and bacteria.

Biodiversity is often proclaimed as a crucial component of human well-being and economic growth. It manifestly is: we're clearly harmed if fish stocks dwindle to extinction; there are plants in the rainforest whose gene pool might be useful to us. But for many of us, these 'instrumental' and anthropocentric arguments aren't the only compelling ones. Preserving the richness of our biosphere has value in its own right, over and above what it means to us humans.

Overall, our lives are getting safer and healthier. But in our ever more interconnected world, there are new threats whose consequences could be so widespread that even a tiny probability is disquieting. Infectious diseases are a resurgent hazard. In the coming decades there could be an 'arms race' between ever-improving preventative measures, and the growing virulence of the pathogens that could plague us – the latter augmented by risks of 'bioerror' or 'bioterror'. The spread of epidemics is aggravated by rapid air travel, plus the huge concentrations of people in megacities with fragile infrastructures.

We're all precariously dependent on elaborate networks – electricity grids, air-traffic control, the Internet, just-in-time delivery and so forth. It's crucial to optimise the resilience of such systems against accidental malfunction – or against wilful disruption by individuals or small groups empowered by technology. The global village will have its village idiots.

Scientific and technical effort has never been applied optimally to human welfare. Some subjects have had the 'inside track' and gained disproportionate resources. Huge funds are still devoted to new weaponry. On the other hand, environmental protection, renewable energy, and so forth deserve more effort. Indeed, US President Barack Obama has urged that the development of clean carbon-free energy should have the priority accorded to the Apollo programme in the 1960s.

THE ROLE OF ACADEMIES AND 'CITIZEN SCIENTISTS'

In confronting global societal challenge in the twenty-first century – these 'threats without enemies' – we can derive inspiration from some of the scientists who worked on the Manhattan Project to create the first atomic bomb. Among them were some of the great intellects from the 'heroic age' of nuclear science – Hans Bethe and Rudolf Peierls, for instance. These individuals set us a fine example. Fate had assigned them a pivotal role in history. When war ended, they returned with relief to peacetime academic pursuits. But they didn't say that they were 'just scientists' and that the use made of their work was up to politicians. They continued as engaged citizens – promoting efforts to control the power they had helped unleash. They maintained an informed commitment for the rest of their lives – none more than Joseph Rotblat, the founder of the Pugwash Conferences.

In his valedictory address as Royal Society President in 1995, Michael Atiyah reminded us that 'the ivory tower is no longer a sanctuary' and that scientists have a special responsibility. We feel there is

US President Barack Obama touring Nellis Solar Power Plant in Nevada, May 2009.

'They maintained an informed commitment for the rest of their lives'. (*Left to right*) Hans Bethe, Rudolf Peierls and Joseph Rotblat.

something lacking in parents who don't care what happens to their children in adulthood, even though this is largely beyond their control. Likewise, scientists shouldn't be indifferent to the fruits of their ideas – their intellectual creations. They should try to foster benign spin-offs – and of course help to bring their work to market when appropriate. But they should campaign to resist, so far as they can, ethically dubious or threatening applications. And they should, as 'citizen scientists', be prepared to engage in public debate and discussion. The challenges of the twenty-first century are more complex and intractable than those of the nuclear age.

In the UK, an ongoing dialogue with parliamentarians on embryos and stem cells has led to a generally admired legal framework. On the other hand, the GM crops debate went wrong because scientists came in too late, when opinion was already polarised between eco-campaigners on the one side and commercial interests on the other. We have recently done better on nanotechnology, by raising the key concerns 'upstream' of any legislation or commercial developments. The Society can draw on collective expertise to clarify key issues – and perhaps identify them before others can.

Epilogue

It's sometimes wrongly imagined that cosmologists and evolutionists must be serenely unconcerned about next year, next week and tomorrow. I conclude with a 'cosmic perspective' which actually strengthens my own concerns about the here and now.

The stupendous timespans of the evolutionary past are, through the work of Darwin and the geologists, now part of common culture. But most people still regard humans as necessarily the culmination of the evolutionary tree. That hardly seems credible to an astronomer. Our Sun formed 4.5 billion years ago, but it's got 6 billion more before the fuel runs out. It will then flare up, engulfing the inner planets and vaporising whatever remains on Earth. And the expanding universe will continue – perhaps for ever – destined to become ever colder, ever emptier. As Woody Allen said, 'eternity is very long, especially towards the end'.

Any creatures witnessing the Sun's demise 6 billion years hence, here on Earth or far beyond, won't be human – they'll be as different from us as we are from bacteria. As Charles Darwin himself recognised, 'not one living species will transmit its unaltered likeness to a distant futurity'. Post-human evolution – here on Earth and far beyond – could be as prolonged as the Darwinian evolution that's led to us – and even more wonderful. Life from this planet could spread through the entire Galaxy, evolving into a teeming complexity beyond what we can conceive.

However, even in this 'concertinaed' timeline – extending billions of years into the future, as well as into the past – the present century may be a defining moment. It's the first in our planet's history where one species – ours – has Earth's future in its hands, and could not only jeopardise itself but foreclose life's immense potential.

Suppose some aliens had been watching our planet for its entire history, what would they have seen? Over nearly all that immense time, 4.5 billion years, Earth's appearance would have altered very gradually. The continents drifted; the ice cover waxed and waned; successive species emerged, evolved and became extinct.

But in just a tiny sliver of the Earth's history – the last one millionth part, a few thousand years – the patterns of vegetation altered much faster than before. This signalled the start of agriculture. The pace of change accelerated as human populations rose.

Then there were other changes, even more abrupt. Within fifty years, little more than one hundredth of a millionth of the Earth's age, the carbon dioxide in the atmosphere began to rise anomalously fast. The planet became an intense emitter of radio waves (the total output from all TV, cell-phone and radar transmissions). And something else unprecedented happened: small projectiles launched from the planet's surface and escaped the biosphere completely. Some were propelled into orbits around the Earth; some journeyed to the Moon and planets.

If they understood astrophysics, the aliens could confidently predict that the biosphere would face doom in a few billion years when the Sun flares up and dies. But could they have predicted this unprecedented 'fever' less than halfway through the Earth's life?

If they continued to keep watch, what might these hypothetical aliens witness in the next hundred years? Will a runaway spasm be followed by silence? Or will the planet itself stabilise? And will some of the objects launched from the Earth spawn new oases of life elsewhere?

The outcome depends on us. Wise choices will require the idealistic and effective efforts of natural scientists, environmentalists, social scientists and humanists – aided by the insights that twenty-first-century science will surely bring.

FURTHER READING

SIMON SCHAFFER

James Delbourgo, *A Most Amazing Scene of Wonders: Electricity and Enlightenment in Early America* (Cambridge, MA, Harvard University Press, 2006)

Patricia Fara, *An Entertainment for Angels: Electricity in the Enlightenment* (London, Icon Books, 2003)

Tal Golan, *Laws of Men and Laws of Nature: The History of Scientific Expert Testimony in England and America* (Cambridge, MA, Harvard University Press, 2004)

J.L. Heilbron, *Electricity in the 17th and 18th Centuries: A Study of Early Modern Physics* (Berkeley, University of California Press, 1979)

Bruno Latour, 'Why has critique run out of steam? From matters of fact to matters of concern', *Critical Inquiry*, 30 (2004), 25–48

Trent A. Mitchell, 'The politics of experiment in the eighteenth century: the pursuit of audience and the manipulation of consensus in the debate over lightning rods', *Eighteenth-Century Studies,* 31 (1998), 307–31

Sir Basil Schonland, *The Flight of Thunderbolts*, 2nd edition (Oxford, Clarendon Press, 1966).

RICHARD HOLMES

The author has drawn on the following sources for this article:

Thomas Baldwin, *Airopaedia, containing the Narrative of a Balloon Excursion from Chester,* with illustrations (London, 1786)

Joseph Banks, *The Scientific Correspondence of Joseph Banks 1765–1820*, edited by Neil Chambers, 6 vols (London, Pickering & Chatto Ltd, 2007)

Tiberius Cavallo FRS, *A Treatise on the History and Practice of Aerostation* (London, 1785)

Erasmus Darwin, *The Loves of the Plants* (London, J. Johnson, 1789)

Barthélemy Faujas de Saint-Fond, *Descriptions des Experiences des Machines Aerostatiques de MM Montgolfier* (Paris, 1783)

Nathan G. Goodman (ed.), *The Ingenious Dr Franklin: Selected Scientific Letters of Benjamin Franklin* (Philadelphia, University of Pennsylvania Press, 1931)

Richard Holmes, *The Age of Wonder* (London, HarperCollins, 2008)

Dr John Jeffries FRS, *A Narrative of Two Aerial Voyages with monsieur Blanchard, as presented to the Royal Society* (London, 1786)

Paul Keen, 'The Balloonomania: Science and Spectacle in 1780s England', *Eighteenth-Century Studies*, 39.4 (2006)

Mi Gyung Kim, 'Balloon Mania: News in the Air', *Endeavour*, 28, Issue 4 (December 2004)

Desmond King-Hele, *Erasmus Darwin: A Life of Unequalled Achievement* (London, Giles de La

Mare Publishers, 1999)

Vincenzo Lunardi, *My Aerial Voyages in England* (London, 1785)

Thomas Martyn, *Hints of Important Uses for Aerostatic Globes* (London, 1784)

John Southern, *A Treatise upon Aerostatic Machines* (London, 1786)

Encyclopaedia Britannica, 3rd edition, 1797. Major article on 'Aerostation'

Gentleman's Magazine (September 1784)

Le Tableau de Paris (1 decembre 1783)

Monthly Review, 69 (1783).

HENRY PETROSKI

Michael R. Bailey (ed.) *Robert Stephenson – The Eminent Engineer* (Aldershot, Hants., Ashgate, 2003)

Derrick Beckett, *Stephensons' Britain* (Newton Abbot, Devon, David & Charles, 1984)

Government Board of Engineers, *The Quebec Bridge Over the St Lawrence River Near the City of Quebec On the Line of the Canadian National Railways* (Ottawa, Department of Railways and Canals, 1918)

Francis E. Griggs Jr, 'Joseph B. Strauss, Charles A. Ellis and the Golden Gate Bridge: Justice at Last', *Journal of Professional Issues in Engineering Education and Practice,* in press

Peter R. Lewis, *Beautiful Railway Bridge of the Silvery Tay: Reinvestigating the Tay Bridge Disaster of 1879* (Stroud, Gloucestershire, Tempus, 2004)

A. Lucas, *John Lucas, Portrait Painter, 1828–1874: A Memoir of His Life Mainly Deduced from the Correspondence of His Sitters* (London, Methuen, 1910)

Sheila Mackay, *The Forth Bridge: A Picture History* (Edinburgh, Her Majesty's Stationery Office, 1993)

David McCullough, *The Great Bridge* (New York, Simon & Schuster, 1972)

Roland Paxton (ed.) 'Thomas Telford: 250 Years of Inspiration', *Proceedings of the Institution of Civil Engineers,* 160, special issue one (May 2007)

Henry Petroski, *Design Paradigms: Case Histories of Error and Judgment in Engineering* (New York, Cambridge University Press, 1994)

Henry Petroski, *Engineers of Dreams: Great Bridge Builders and the Spanning of America* (New York, Alfred A. Knopf, 1995)

Henry Petroski, *Pushing the Limits: New Adventures in Engineering* (New York, Alfred A. Knopf, 2004)

L.T.C. Rolt, *Isambard Kingdom Brunel* (Harmondsworth, Middlesex, Penguin Books, 1970)

Richard Scott, *In the Wake of Tacoma: Suspension Bridges and the Quest for Aerodynamic Stability* (Reston, Va., ASCE Press, 2001)

Mary J. Shapiro, *A Picture History of the Brooklyn Bridge* (New York, Dover Publications, 1983)

Frank L. Stahl, David E Mohn and Mary C. Currie, *The Golden Gate Bridge: Report of the Chief Engineer*, Volume II (San Francisco, Golden Gate Bridge, Highway and

Transportation District, 2007)

Denyan Sudjic *et al.*, *Blade of Light: The Story of London's Millennium Bridge* (London, Penguin Books, 2001)

John van der Zee, *The Gate: The True Story of the Design and Construction of the Golden Gate Bridge* (New York, Simon and Schuster, 1986)

W. Westhofen, 'The Forth Bridge', *Engineering*, 28 February 1890, pp. 213–83.

STEVE JONES

T. Andersen, J. Carstensen, E. Hernández-García & C.M. Duarte, 'Ecological thresholds and regime shifts: Approaches to identification', *Trends Ecol. Evol. Syst.*, 24 (2009), 49–57

E. Beninca, J. Huisman, R. Heerkloss, K.D. Johnk, P. Branco, E.H. Van Nes, M. Scheffner & S.P. Ellner, 'Chaos in a long-term experiment with a plankton community', *Nature*, 451 (2008), 822–6

J.C. Briggs, 'The marine East Indies: Diversity and speciation', *J. Biogeog.*, 32 (2005), 1517–22

S.J. Clark, 'Beyond neutral science', *Trends Res. Ecol. Evol.*, 24 (2009), 8–15

A. Dance, 'Soil ecology: What lies beneath?', *Nature*, 455 (2008), 724–5

D.F. Doak, J.A. Estes, B.S. Halpern, U. Jacob, D.R. Lindberg, J. Loworn, D.H. Monson, M.T. Tinker, T.M. Williams, J.T. Wotton, I. Carroll, M. Emmerson, F. Micheli & M. Novak, 'Understanding and predicting ecological dynamics: Are major surprises inevitable?', *Ecology*, 89 (2008), 952–61

R. Grenyer, 'Global distribution and conservation of rare and threatened vertebrates', *Nature*, 444 (2006), 93–6

L. Gross, 'Untapped bounty: Sampling the seas to survey microbial biodiversity', *PLoS Biol.*, 5 (2007): e85 doi:10.1371/journal.pbio.0050085

M.J. Heckenberger, C.J. Russell, C. Fausto, J.R. Toney, M.J. Schmidt, E. Pereira, B. Franchetto & A. Kuikuro, 'Pre-Columbian urbanism, anthropogenic landscapes, and the future of the Amazon', *Science*, 341 (2008), 1214–17

M. de Heer, V. Kapos & B.J.E. ten Brink, 'Biodiversity trends in Europe: Development and testing of a species trend indicator for evaluating progress towards the 2010 target', *Phil. Trans. R. Soc. B*, 360 (2005), 297–308

D. Jablonski, 'Extinction: Past and present', *Nature*, 427 (2004), 589 http://www.nature.com/nature/journal/v427/n6975/full/427589a.html

E.G. Leigh, 'Neutral theory: A historical perspective', *J. Theor. Biol.*, 20 (2007), 2075–91

M.W. McKnight, P.S. White, R.I. McDonald, J.F. Lamoreux, W. Sechrest, R.S. Ridgely & S.N. Stuart, 'Putting Beta-Diversity on the Map: Broad-Scale Congruence and Coincidence in the Extremes', *PLoS Biology*, 5(10) (2007): e272; doi:10.1371/journal.pbio.0050272

R. Muneepeerakul, 'Neutral metacommunity

models predict fish diversity patterns in Mississippi–Missouri basin', *Nature,* 453 (2008), 220–23

D. Nogues-Bravo, M.B. Araujo, T. Romdal & C. Rahbek, C., 'Scale effects and human impact on the elevational species richness gradients', *Nature,* 453 (2008), 216–20

W. Renema, D.R. Bellwood, J.C. Braga, K. Bromfield, R. Hall, K.G. Johnson, P. Lunt, C.P. Meyer, L.B. McMonagle, R.J. Morley, A. O'Dea, J.A. Todd, F.P. Wesselingh, M.E.J. Wilson & J.M. Pandolfi, 'Hopping hotspots: Global shifts in marine biodiversity', *Science,* 321 (2008), 654–7

M. Scheffer, S. Rinaldi, J. Huisman & F.J. Weissing, 'Why plankton communities have no equilibrium: Solutions to the paradox', *Hydrobiologia,* 491 (2003), 9–18

J. Schipper *et al.,* 'The status of the world's land and marine mammals: Diversity, threat, and knowledge', *Science,* 322 (2008), 225–30

F. Sergio, T. Caro, D. Brown, B. Clucas, J. Hunter, J. Ketchum, K. McHugh & F. Hiraldo, 'Top predators as conservation tools: Ecological rationale, assumptions, and efficacy', *Ann. Rev. Ecol. Evol. Syst.,* 39 (2008), 1–19

N.S. Sodhi, 'Tropical biodiversity loss and people: A brief review, *Basic and Applied Ecology,* 9 (2008), 93–99

S.Y. Strauss & R.E. Irwin, 'Ecological and evolutionary consequences of multispecies plant-animal interactions', *Ann. Rev. Ecol. Evol. Syst.,* 35 (2004), 435–66

J.E. Vermaat, J.A. Dunne & A.J. Gilbert, 'Major dimensions in food-web structure properties', *Ecology,* 90 (2009,) 278–82

E.O. Wilson & F.M. Peter, *Biodiversity* (Washington DC, National Acad. Press, 1988).

PHILIP BALL

Agricola, *De Re Metallica,* transl. H.C. Hoover & L.H. Hoover (New York, Dover, 1950)

F. Bacon, *Advancement of Learning and Novum Organum* (New York, Willey Book Co, 1944)

F. Bacon, *New Atlantis* (1627); available at http://www.gutenberg.org/etext/2434

R. Cotterill, *The Cambridge Guide to the Material World* (Cambridge, Cambridge University Press, 1985)

P. Medawar, *Pluto's Republic* (Oxford, Oxford University Press, 1984)

J. Meikle, *American Plastic: A Cultural History* (New Jersey, Rutgers University Press, 1995)

C.P. Snow, *The Two Cultures: And A Second Look* (Cambridge, Cambridge University Press, 1965)

L. Wolpert, *The Unnatural Nature of Science* (London, Faber & Faber, 1992).

OLIVER MORTON

David Oldroyd, *Earth Cycles: A Historical Perspective* (Westport, Greenwood Press, 2006)

Robert Poole, *Earthrise: How Man First Saw the Earth* (New Haven, Yale University Press, 2008).

LIST OF
ILLUSTRATIONS

HarperPress and the authors wish to thank the Royal Society for access to their tremendous Library and Archives collection at Carlton House Terrace for many of the images in this book.

The collections are of international importance in the history of science – an extraordinary and unrivalled record of the development of science that spans nearly 350 years. Resources include manuscripts, printed books and paintings, amassed to provide a record of scientific achievements.

HarperPress are also grateful to the following individuals and organisations for providing photographs and for permission to reproduce copyright material. While every effort has been made to trace and acknowledge copyright holders, the publishers would like to apologise for any omissions and will be pleased to incorporate missing acknowledgements in any future editions.

Images not listed below have been provided for use in this book by the Royal Society.

Abbreviations: t: top, b: bottom, l: left, r: right, c: centre.

40: © National Portrait Gallery, London; 43: Paramount Pictures; 49: © National Portrait Gallery, London; 50: © 2009 The British Library; 53: © 2009 The British Library; 63: Tony & Daphne Hallas/ Science Photo Library; 65: Time & Life Pictures/ Getty Images; 68: View of the chapel looking towards The Last Judgement, c.1305 (fresco), Giotto di Bondone (c.1266-1337)/ Scrovegni (Arena) Chapel, Padua, Italy / The Bridgeman Art Library; 69: View of the south wall depicting scenes from the Life of Joachim and Anna and the Life of Christ, c.1305 (fresco), Giotto di Bondone (c.1266-1337)/ Scrovegni (Arena) Chapel, Padua, Italy/ The Bridgeman Art Library; 70: Getty Images; 151: British Library MSS Add. 30094, fol. 220; 159(l): SSPL via Getty Images; 162: Science Museum Pictorial/ Science & Society Picture Library; 177: © National Portrait Gallery, London; 185: ©

The Natural History Museum, London; 187: © The Natural History Museum, London; 212: © The Natural History Museum, London; 217: © The Natural History Museum, London; 220: © The Natural History Museum, London; 225: Science Source/ Science Photo Library; 232: David Toase/ Getty Images; 236: Courtesy of the Institution of Civil Engineers; 239: Getty Images; 240(l+r): Courtesy of the Institution of Civil Engineers; 242(l): Jason Todd/ Getty Images; 242(r): From The Gate: The True Story of the Design and Construction of the Golden Gate Bridge by John van der Zee, iUniverse, Inc., 2000; 245: Topham Picturepoint; 248: Sami Sarkis/ Getty Images; 262: Courtesy of The Oxford Mail; 264(l): Courtesy of MRC Laboratory of Molecular Biology, University of Cambridge; 264(r): © BBC; 270: © Estate of Graham Sutherland; 276: © The Natural History Museum, London; 281: © The Natural History Museum, London; 282(tl): SSPL via Getty Images; 282(tr): AFP/ Getty Images; 282(bl): Daryl Benson/ Getty Images; 282(br): ZenShui/ Odilon Dimier/ Getty Images; 292: National Geographic/ Getty Images; 304: The Royal Society and courtesy of Bassano Portrait Studios; 306: Drawn by Philip Ball; 310: Science Museum Pictorial/ Science & Society Picture Library; 323: Time & Life Pictures/Getty Images; 327: Royal Astronomical Society/ Science Photo Library; 332: Drawn by Geoff Westby; 337: Courtesy NASA; 343(l+r): Courtesy NASA/JPL-Caltech; 345: Science Photo Library; 347: Royal Observatory, Edinburgh/ Science Photo Library; 358: Friedrich Saurer/ Science Photo Library; 366: Drawn by Geoff Westby; 370: Mehau Kulyk/ Science Photo Library; 376: Humanities and Social Sciences Library/ Rare Books Division/ New York Public Library/ Science Photo; 380: Drawn by Geoff Westby; 383: © R.V. Sole, reproduced by permission of the artist; 387: Courtesy NASA; 392: © Tim Knowles www.timknowles.com; 399: Getty Images; 402: Courtesy NASA; 407: Stocktrek Images/ Getty Images; 409: © HarperCollins Publishers, Glenn Beanland/Getty Images (Big Ben), Kaz Mori/Getty Images (wave); 413: Getty Images; 416: Harald Sund/ Getty Images; 422: The Last Man, 1849 (oil on canvas), Martin, John (1789-1854)/ © Walker Art Gallery, National Museums Liverpool / The Bridgeman Art Library; 439: Getty Images; 442: NI syndication; 447: Architect of the Capitol, http://www.aoc.gov/cc/photo-gallery/other_sculpt.cfm; 458: Stocktrek Images/ Getty Images; 471: Science Museum Library/ Science and Society Picture Library; 475: www.opte.org/; 477: Getty Images; 480: Wilfried Krecichwost/ Getty Images; 482: AFP/Getty Images; 483(l): Time & Life Pictures/Getty Images; 483(c): Science Photo Library.

The Obligation of the Fellows of the Royal Society

We who have hereunto subscribed, do hereby promise, that we will endeavour to promote, the good of the Royal Society of London for improving Natural Knowled. and, to pursue, the ends for which the same was founded; that we will carry out so far as we are able, those actions requested of us in the name of the Council; and, that we will observe the Statutes and Standing Orders of the said Society. Provided, that, whensoever any of us shall signify to the President under our hand that we desire to withdraw from the Society, we shall be free from this Obligation for the future.

2001

David Attwell
David Baulcombe
John Beddington
Robert J Bingeman
Michael Berol
Keith Burnett
Paul Callaghan
Graham Collingridge

Richard Dawkins
Roger Cox

R
Brian Eyre
Peter Gluckman
Charles Godfray
Brigid Hogan
John Hunt
Frances Kirwan
S. R. Kulkarni
Andrew Leslie

Michael Leitt
Robin Lovell-Badge
Paul Madden
M S Paterson
Bruce Ponder
Geoffrey Raisman

Dale Sanders
David W. Schindler
George Shedrick
Keith Sherlock
Tony Simpson

M.V. Srinivasan

Ian Stewart
Marc Feldmann

N K
William Unruh
B R Whittle

A J Wilkie
Alexei Abrikosov
James F. Crow
Alan B Ford
Greg J. Armstrong

Patrick Moore

Harry Kroto
Hugh Bostock
Robin Carrell
Mick Crawley
Stuart G Cull-Candy
John S. Dainty
Roger Davies
Anne Dell
David Dolphin
Harry Elderfield
David Fowler
Stephen Bryce Rutherford
Graham B Goodwin